Origins of Human Language: Continuities and Discontinuities with Nonhuman Primates

SPEECH PRODUCTION AND PERCEPTION

Edited by Susanne Fuchs and Pascal Perrier

VOLUME 4

Notes on the quality assurance and peer review of this publication:
Prior to publication, the quality of the work published in this series is double blind reviewed by external referees appointed by the editorship. The referee is not aware of the author's name when performing the review; the referees' names are not disclosed.

Louis-Jean Boë / Joël Fagot / Pascal Perrier /
Jean-Luc Schwartz (eds.)

Origins of Human Language: Continuities and Discontinuities with Nonhuman Primates

PETER LANG
EDITION

Bibliographic Information published by the Deutsche Nationalbibliothek
The Deutsche Nationalbibliothek lists this publication in the Deutsche Nationalbibliografie; detailed bibliographic data is available in the internet at http://dnb.d-nb.de.

Library of Congress Cataloging-in-Publication Data
A CIP catalog record for this book has been applied for at the Library of Congress

Printed by CPI books GmbH, Leck.

ISSN 2191-8651
ISBN 978-3-631-73726-2 (Print)
E-ISBN 978-3-631-73807-8 (E-PDF)
E-ISBN 978-3-631-73808-5 (EPUB)
E-ISBN 978-3-631-73809-2 (MOBI)
DOI 10.3726/b12405

Peter Lang Edition is an Imprint of Peter Lang GmbH.

Peter Lang – Frankfurt am Main · Bern · Bruxelles · New York · Oxford · Warszawa · Wien

This publication has been peer reviewed.

www.peterlang.com

Contents

Editors: Louis-Jean Boë, Joël Fagot,
Pascal Perrier, Jean-Luc Schwartz

Introduction to "Origins of Human Language: Continuities and Discontinuities with Nonhuman Primates"

There have been a number of contributions in the past years about language origins from various points of view. In this book, we intend to contribute to establish a state-of-the-art of the knowledge about the continuities and ruptures between communication in primates and language in humans. A major strength of the present book is to explore a diversity of perspectives on the origins of language, including the description of vocal communication in animals, mainly in monkeys and apes, but also in birds, the study of vocal tract anatomy and cortical control of the vocal productions in monkeys and apes, the description of combinatory structures and their social and communicative value, and the exploration of the cognitive environment in which language may have emerged from nonhuman primate vocal or gestural communication. Interestingly, this portrait emerges from a situation in which one long-standing hypothesis stating that a low larynx position was a prerequisite for the emergence of speech has been clearly discarded. Indeed, some contributors of this book have just participated to two papers showing that the monkey vocal tract was "speech ready" (Boë et al., 2017; Fitch et al, 2016). This renders the debates clearer, in that neurocognitive and social evolutions now unequivocally appear as the major potential sources of evolution towards language. The series of eleven chapters provides a rather complete portrait and elaboration on the facts, proposals, arguments and claims that pave the science way from animal communication to human language.

The book begins by a descriptive analysis of baboon calls by *Caralyn Kemp, Arnaud Rey, Thierry Legou, Louis-Jean Boë, Frédéric Berthommier, Yannick Becker and Joël Fagot*. In their study of the "*Vocal Repertoire of Captive Guinea Baboons (Papio papio)*", the authors provide ethograms and a prototypical description of twelve kinds of vocalizations emerging from the analysis of individual calls and call sequences in the vocal repertoire of a group of captive Guinea baboons. Typical sound examples of each type of vocalization are also provided in Supplementary Materials. This study will be of substantial value for students of primate vocalizations. More importantly in the context of the present book, it provides a concrete and significant example of the "phonetic" description of the vocal communication system in nonhuman primates, which contributes to the documentation of the precursors of human speech possibly enlightening the conditions of its emergence. Of importance here is the fact that exploitation of variations in various dimensions of the vocalizations appears as a possible way to increase the efficiency of communication without expanding the vocabulary of available units. Interestingly, the large co-variations between formants and fundamental frequency also suggest a non-independent mastery of vocal source and vocal tract configuration in baboons' vocalizations.

The next chapter is in continuity with the previous one, providing a zoom on one of the twelve baboon vocalizations. *Louis-Jean Boë, Thomas R. Sawallis, Jöel Fagot and Frédéric Berthommier* question "*What's up with Wahoo? Exploring Baboon Vocalizations with Speech Science Techniques*". Focusing on the "wahoo" vocalization, they analyze a corpus of 69 utterances of wahoo calls coming from the corpus of the previous chapter. Careful spectral analysis of these utterances provides major spectral peaks separately for the three proto-components {w}, {a} and {hoo}. These peaks are compared with those of a [wa.u] phonetic sequence uttered by a human speaker in various phonatory modes. In parallel, the authors propose an articulatory analysis of a film presenting a baboon uttering a wahoo vocalization. Altogether, they claim that these combined acoustic and articulatory analyses converge towards the assumption that baboon "wahoo" is rather similar to a human phonetic sequence that can be transcribed as [wa↓.u↑], with a first syllable chaining a back rounded semi-consonant /w/ and a front open /a/ produced in an ingressive way, and a back rounded /u/ produced in an egressive way.

The exploration of vocalizations in nonhuman primates continues with *Adriano R. Lameira* proposing a view on "*Origins of vowels and consonants: Articulatory continuities with nonhuman great apes*". From his study of the call repertoire of orangutans, the author introduces the idea that there could exist an articulatory homology between voiceless calls and human consonants on the one hand, and between voiced calls and human vowels on the other hand. Among the set of voiceless calls, Lameira focuses on whistles and shows clear learning abilities in captive orangutans, which relates to a number of reports of learning processes in other great apes. Concerning voiced calls, Lameira displays kinds of "babbling" vocalizations with rhythmic jaw movements similar to the ones displayed by infants, together with imitation games in which a captive orangutan is able to modify fundamental frequency in response to modulations of a human tutor. These plastic voiceless and voiced vocalizations could provide in the author's view "proto-consonants and proto-vowels" in a kind of language precursor in a human ancestor.

Importantly, vocalizations in nonhuman primates are constrained by the anatomy of the orofacial system. This is at the core of the contribution by *Frédéric Berthommier, Louis-Jean Boë, Adrien Meguerditchian, Thomas R. Sawallis and Guillaume Captier* dealing with "*Comparative Anatomy of the Baboon and Human Vocal Tracts: Renewal of Methods, Data, and Hypotheses*". This comparative anatomy aggregates a series of invaluable data enabling to qualitatively and quantitatively compare vocal tracts in baboons and humans. These data include (1) a dissection of two adult *Papio papio* heads, enabling detailed description of the vocal tract, the larynx and the tongue musculature, (2) fifty-six 3D MRI scans of *Papio anubis* baboons from 2 years to adulthood enabling authors to elaborate precise vocal tract biometry, (3) radiographic data for 127 human children from 3 to 7.5 years providing reference human biometry for comparison with the preceding set of Baboon data. This enables authors to claim that the hyoid bone would be placed one vertebra lower in human infants than in adult baboons – and also one additional vertebra lower in male human adults. The increase in the pharyngeal part of the vocal tract in humans would be accompanied by compensatory facial shortening, thus maintaining the vocal tract length similar in both species. On the basis of these data authors address the issue of how exaptation of articulatory patterns in feeding could have contributed to structure the articulation of speech sounds.

Vocalizations in nonhuman primates also depend of course a lot on the cortical and sub-cortical networks available for vocal and orofacial control. The question of cortical control is explored in the next two chapters. Firstly, *Veena Kumar and Kristina Simonyan* discuss in great detail the "*Evolu-*

tion of the laryngeal motor cortex for speech production". Their starting point is that, as already discussed in the first chapter by Kemp and coll., laryngeal control seems much more precise and stable in humans. Kumar and Simonyan analyze possible differences in laryngeal cortical control between humans and nonhuman primates. Firstly, they recapitulate several studies from their group leading to the conclusion that, while laryngeal motor control would be localized both in the primary motor cortex and in the premotor cortex in humans, localization would be reduced to the premotor cortex in apes and monkeys. Their hypothesis is that the premotor cortex would be sufficient for basic functions associated to e.g. breathing or physical effort, but the fine control in humans would require the additional involvement of the primary motor cortex. This evolution would be combined with the emergence in humans of direct cortical connections towards the brainstem, while they would be indirect in monkeys. Finally, the cortical network of connections between the laryngeal motor cortex and parietal and temporal regions necessary for learning and control would also be much more developed in humans.

William D. Hopkins then addresses the question of a potentially crucial cortical area for language, often considered as a potentially major piece in the emergence of language: Broca's area. His contribution, entitled "*Motor and Communicative Correlates of the Inferior Frontal Gyrus (Broca's Area) in Chimpanzees*", provides a rich synthesis of various types of comparative data about the Inferior Frontal Gyrus in monkeys, chimpanzees and humans. Firstly, he provides a detailed analysis of the literature on the morphology and cytoarchitectonics of Broca's area in primates and particularly in the species his group studied most, that is chimpanzees. While the Pars Opercularis (ParsO) and Pars Triangularis (ParsT) are difficult to define in the Inferior Frontal Lobe in monkeys, ParsO can be rather clearly defined in chimpanzees, but defining a ParsT homolog is less clear. Areas 44 and 45 present large volumetric expansions and more consistent leftward asymmetries in humans compared to chimpanzees, together with a larger amount of synaptic connections. The author then presents a number of results from his group displaying consistent correlations in chimpanzees between morphological properties of individual Inferior Frontal Gyrus and behavioral abilities associated with communicative actions (e.g. pointing manual gestures and attention-getting vocalizations) and tool-use. These correlations seem partly genetically heritable. He concludes by discussing the implications of these findings in theories of language emergence.

The next two chapters explore the way vocalizations could indeed constitute a real communication system likely to open the road towards human oral language. Firstly, *Camille Coye, Simon Townsend and Alban Lemasson* discuss the question of combination and compositionality, in their chapter entitled: "*From animal communication to linguistics and back: insight from combinatorial abilities in monkeys and birds*". From their analysis of the very wide literature on compositionality in bird songs and monkey calls, the authors attempt to carefully disentangle what could be a "phonological level" in which non-meaningful vocal units would be combined in various ways to provide meaningful sequences, and what could be a "morphosyntactic level", in which meaningful units would be combined for producing larger meaningful structures. They argue that most reports in the literature do not provide convincing examples of nonhuman compositionality in one of these two strict senses. Then they present some "promising examples" of proto-phonology in the composition of flight calls in chestnut-crowned babblers (Australian birds), and protomorphosyntax in the composition of meaningful calls both in southern pied babblers (South-African birds) and in Diana monkeys from forests in West Africa. Finally, they suggest some possible social pressures driving the use of compositionality, in relation with the complexity of the social organization, and the habitat constraints on communication pushing for complex vocal communication with low ambiguity and long-range facilities of use.

Klaus Züberbuhler then engages in a global evaluation of the ability of primate vocalizations to constitute "*The Primate Roots of Human Speech and Language*". For this aim, he reviews the continuities between nonhuman primate vocal communication and oral language, but also some major limitations that can be traced in these continuities. First, the vocal tract seems speech-ready but cortical control is not sufficient to master the vocal source and the vocal learning and combination processes required in human speech. Second, the communicative content of the calls seems likely to be interpreted and even modulated by monkeys and apes in relation to context and audience. However, vocal call exchanges appear to convey low levels of intentionality – in reference to Denett's scale – and nonhuman primates could lack the ability to share intentions and goals. The author also addresses the question of referential communication, a crucial component of human language. Monkeys and apes do display elements of referential communication, but rather focused on themselves. The lack of clear view on the nature of their "mental concepts" sets severe limitations on establishing strong links with human language.

At stage, where the focus all over this book has been mostly put on vocal communication, the contribution by *Katja Liebal* provides a timely and important comparative overview of "*What gestures of nonhuman primates can (and cannot) tell us about language evolution*". She begins by a review of arguments pros and cons for either vocal, gestural or orofacial communication as the possible unique precursor of human language, and she nicely shows that arguments in favor of one or the other are often partly incomplete or in some sense partial, and hence that no "unique precursor" theory is wholly convincing at this stage. Then, she focuses on what could be gained for a theory of language evolution by looking at gestural communication in monkeys and apes. Interestingly, this provides a number of echoes to the previous chapter by Zuberbühler, by discussing what aspects of gestural communication could display some continuity with human language. Intentionality is a basic component of communicative gestures, with clear evidence that both monkeys and apes monitor the attention of their partner and modulate communication accordingly. Flexibility – the ability to vary the context of use of a given stimulus – seems rather larger for gestures than for calls or orofacial productions. Gesture compositionality appears rather weak, with only one or two possible examples in the literature. Referentiality and iconicity are debated. Altogether, the author stays in a careful position in terms of the gestural vs vocal origin of human language.

The last two chapters open the angle of view even more widely by addressing the question of the cognitive environment required for the emergence of language. *Tecumseh Fitch* focuses on "*Dendrophilia and the Evolution of Syntax*". Syntax is classically considered as a highly specific property of human language, and Fitch continues his exploration of the specific cognitive abilities that make humans special and could trace a major discontinuity in the emergence of language. He introduces the assumption that this ultimate human cognitive ability consists in the capacity to manipulate "supra-regular" grammars, thanks to a structural working memory providing generalization and elaborating hierarchies. This is what he calls "dendrophilia" – a tendency to organize sensory flows into tree-like structures. The author reviews experimental data in which various animal species have been claimed to manipulate grammar-like structures. He raises objections to each of these studies, to argue that humans are the single species able to manipulate supra-regular grammars. He concludes on the likely implication of Broca's area in the neural implementation of this uniquely human process.

Finally, *Joël Fagot, Raphaëlle Malassis, Tiphaine Medam and Marie Montant* adopt the inverse perspective by "*Comparing human and nonhuman animal performance on domain-general functions: towards a multiple bottlenecks scenario of language evolution*". They propose an alternative to the search of a uniquely human capacity, and rather explore possible continuities and discontinuities in general cognitive abilities. They successively analyze integration in time and space, integration across sensory dimensions and sensory modalities, and various types of categorization processes. In each of these domains, they document resemblances between animals and humans, and aspects in which humans display a specific behavior. Humans appear better at processing and learning complex sequences, at extracting global aspects of visual scenes, at integrating sensory dimensions, at extrapolating perceptual properties in equivalence classes and elaborating qualitative rules and generalizing these rules across domains. This results in various types of "bottlenecks" that could have constrained the emergence of language. The authors conclude on the specific importance of attention and working memory in the bundle of factors that seem to have co-evolved in the route towards human language.

Although not exhaustive, we hope that the tour offered in this book will convey a clear sense of the progress that have been made in the field of language evolution, and also hope that this book will serve as resource for students and researchers in the field. We would like to thank all the contributors for their contributions.

Acknowledgements

Joël Fagot was supported by ANR-16-CONV-0002 (ILCB), ANR-11-LABX-0036 (BLRI) and ANR-11-IDEX-0001–02 (A*MIDEX). Jean-Luc Schwartz was supported by a grant from the European Research Council under the European Community's Seventh Framework Programme (FP7/2007–2013 Grant Agreement no. 339152, "Speech Unit(e)s").

References

Boë, L.J., Berthommier, F., Legou, T., Captier, G., Kemp, C., Sawallis, T.R., Becker, Y., Rey, A., and Fagot, J. (2017). Evidence of a Vocalic Proto-System in the Baboon (*Papio papio*) suggests Pre-Hominin Speech Precursors. *Plos One, 12(1),* e0169321. doi.org/10.1371/journal.pone.0169321.

Fitch, W.T., de Boer, B., Mathur, N., and Ghazanfar, A.A. (2016). Monkey vocal tracts are speech-ready. *Science Advances*, 2, 12, e1600723.

Caralyn Kemp[1,2], Arnaud Rey[2,3], Thierry Legou[2,4], Louis-Jean Boë[5], Frédéric Berthommier[5], Yannick Becker[2,3], Joël Fagot[2,3]

[1]*Taronga Conservation Society Australia, Sydney, Australia*

[2]*Brain and Language Research Institute, CNRS and Aix-Marseille University, France*

[3]*Laboratoire de Psychologie Cognitive, CNRS and Aix-Marseille University, Marseille, France*

[4]*Laboratoire Parole et Langage, CNRS and Aix-Marseille University, Aix-en-Provence, France*

[5]*GIPSA-Lab, CNRS and Grenoble Alpes University, Grenoble, France*

Vocal Repertoire of Captive Guinea Baboons (*Papio papio*)

Abstract: In order to study the evolution of language, it is useful to understand the communicative systems of nonhuman animals. To this end, descriptive ethograms of primate vocal repertoires are the ideal starting point. We examined the vocal repertoire of a group of captive Guinea baboons (*Papio papio*). Twelve vocalizations were readily distinguishable using individual call components and call sequences. Some of these vocalizations were sex and/or age specific (e.g., copulation grunts in females, moans in infants). We compared these vocalizations to those reported in wild Guinea baboons as well as the other baboon taxa. The Guinea baboons share the basic call units with the other baboon species. However, a large degree of variability occurs within call sequences (e.g. number of grunts within a bout, F0 and calling rate [number of grunts/second]). The baboons also showed vocal variability through the combination of different vocalizations (e.g. moans, screams and yaks in varying order and number within a bout) and the use of one vocalization (barks) in a new captive-specific context. The present study complements recent studies on the vocal productions of baboons, and opens several new perspectives on the evolution of language.

Keywords: vocal repertoire, baboons, primate vocalizations, language

1. Introduction

The evolution of speech from more simplistic primate communication may have been a pivotal transition for our species (Smith and Szathmary, 1995;

Snowdon, 2004). However, evidence for how this occurred is scarce, with only a handful of features which define language, such as rudimentary forms of syntax (Ouattara et al., 2009), found in the vocalizations of some primates. It is important to point out, though, that only a small proportion of primate species have had their vocal repertoire described and analyzed (Zuberbühler, 2012). Ethograms are the first step towards better understanding these vocal systems. They can provide the basis for comparative studies, and are especially useful for newcomers to the species and those who work closely with the animals (Fischer and Hammerschmidt, 2002). Careful analysis of vocal repertoires in nonhuman primates also provides the groundwork for systematically tracking the development of more complex vocal systems. Here we present the findings of a study on the vocal repertoire of a group of captive Guinea baboons (*Papio papio*), from which it was possible to determine their ability to produce vowel-like sounds (Boë et al., 2017).

The description of the vocal repertoire of baboons has a complex history due to the wealth of terminology used between and within taxa and researchers, and to indecision regarding species or sub-species status of this primate group. Regarding the latter, the so-called savannah species (Guinea: *P. papio*, Olive: *P. anubis*, Yellow: *P. cynocephalus*, Chacma: *P. ursinus*, Kinda: *P. kindae;* Hayes et al., 1990) have generally been considered to be relatively homologous subspecies with similar vocalizations while the hamadryas baboon (*P. hamadryas*) has been considered, and thus studied, separately as a full species with its own unique vocalizations (e.g., Estes, 1992). Recent genetic evidence suggests that the taxa should be considered as phylogenetic species or biological subspecies. Furthermore, this research has shown that hamadryas baboons have not greatly diverged from the other taxa and share genetic and physical characteristics with Guinea baboons (Newman et al., 2004). Vocalizations are particularly sensitive to the process of speciation (Lanyon, 1969) and their study may serve to provide additional information for baboon systematics. However, while there is a large degree of similarity in the vocalizations between baboon taxa (Maciej et al., 2013a), not all vocalizations seem to occur in all species (Estes, 1992).

The wild studies by Byrne (1981) and Maciej et al. (2013a) comprise the only published reports on the vocal repertoire of Guinea baboons, although notes and some analyzes on individual vocalizations have also been made by Anderson and McGrew (1984), Andrew (1962), Maciej et al. (2013b) and Maestripieri et al. (2005). From these studies, we can determine that vocalizations seem to be important for this species to maintain contact but also to warn off predators (Anderson and McGrew, 1984; Byrne, 1981). However, these studies have limitations. Byrne's ethogram did not include spectrograms or fine-detailed descriptions of all the vocalizations, Maciej et al. (2013a) did not include the vocalizations of juveniles and infants, and a variety of terminology has been used throughout the literature; this can make it difficult to compare the vocalizations or even determine how a particular call sounds.

Analyzes of the vocal repertoires of primates have been conducted by ear (e.g. Byrne, 1981), or using temporal and frequency measures of individual calls and bouts of calls from spectrograms (e.g. Bermejo & Omedes, 1999; Fischer and Hammerschmidt, 2002), and, more recently, cluster and principal component analysis (e.g. Gros-Louis et al., 2008; Maciej et al., 2013a). Using discriminate function analysis, Bezerra et al. (2010) showed that subjective differentiation of vocalizations – that is, by audible and visual inspection – is relatively reliable. Commonly considered structural parameters of vocalizations in these analyzes include duration, frequency range, modulation, harmonics, and noise.

The aim of our study is to identify the full range of vocalizations produced by captive Guinea baboons and provide descriptions of each. After a first presentation of the general principles of our methods, we report below an overview of the different vocalizations in three sections: 'Acoustic description' details the basic features of the vocalization, including variability; 'Context & usage' defines how and when the vocalization was used; the 'Terminology' section lists synonymous vocalizations and their terminology throughout the literature. We then present the results of the formant analyses which were conducted on several categories of vocalizations, including the grunts, barks, wa- (of wahoo), -hoo (of wahoo), yaks, and copulation calls. In our discussion, we will show that such detailed descriptions of the vocal repertoire of a nonhuman primate species open interesting perspectives on the evolution of language. Appendix 1 provides a glossary of the key terms used in this chapter.

2. Methods

This research adhered to the legal requirements of France and to the American Society of Primatologists Principles for the Ethical Treatment of Non-Human Primates.

2.1 Subjects and Housing

We recorded the vocal behavior of 31 Guinea baboons (12 males, 19 females, aged between 2 months and 27 years at the start of this study; Appendix 2) which are maintained within three groups at the Rousset-sur-Arc Centre National de la Recherche Scientifique (CNRS) primate center, France (see Fagot et al., 2014 for housing details). This center also houses olive baboons, which are within auditory but not visual range of our subjects.

2.2 Recording of Vocalizations

We recorded the vocalizations of the baboons from September 2012 to June 2013, with the behavior, social interactions and context noted. *Ad libitum* opportunistic sampling techniques of spontaneous vocalizations, which included social events and responses to stimuli occurring naturally within their environment (e.g., sheep [*Ovis aries*] passing next to the center), were used to record over 1000 vocalizations. The baboons were accustomed to humans standing and walking by the fence of their enclosures and the presence of the recorders and their equipment did not disturb the baboons from their natural daily activities.

Recording took place between 8:00 and 21:00. We particularly focused on the half hour prior to feeding (16:30–17:00) as the baboons were more vocal, and more consistently vocal, during this time. Recordings did not occur between 17:00 and 18:00 when the baboons were eating, so as to avoid potential distortion of the vocalizations due to chewing and full cheek pouches. Digital Zoom Handy Recorders (H4n) with a Me66 Sennheiser microphone was used to record the vocalizations. This is a super cardioid microphone with a high sensitivity (50mV/Pa ± 2.5dB) and a wide (40Hz – 23kHz) and flat ± 2.5dB frequency response. Recording was conducted at a distance from the baboons from 1m to 20m, with longer distances suitable only for the long-distance vocalizations.

2.3 Vocalization Analysis

After disregarding recordings where the caller could not be identified, or had poor signal to noise ratio, because of disruptive background noise or vocalizations overlapping each other, we created a database of over 1000 vocalizations. Male and female vocal productions were separated in the adult and sub-adult classifications, but were combined for juveniles and infants. This decision was based on the lack of body size differences between male and female juveniles and infants, and the similarity in F0. Vocalizations were then grouped using several methods: by ear, visual inspection using spectrograms, broad descriptive features, and detailed formant analysis. A minimum of 10 recordings per vocalization were analyzed for descriptive features. Our analyses focused on the fundamental frequency (F0), the number of individual call units per vocalization series, the duration of each call or phase, the duration of the interval between two calls in the same bout, the total duration of a calling bout, and formants (F1 and F2). The acoustical analyses of the vocalizations were performed using Praat 6.0.13 for spectrograms and high F0 vocalizations (bark, yak) and (wahoo). A problem using Praat for the measure of F0 is that it relies on the relative periodicity of the speech signal as computed based on short-term autocorrelation. This program is not adapted for inferring F0 for the grunts, barks and chattering, because these calls exhibit some irregularities, additive noise, or very low F0 values (< 60 Hz), i.e. long periods. In our study, F0 was inferred for these latter vocalizations with a home-made Matlab script, which computed F0 from a hand tagging of the periodicity of the acoustic signal (see Figure 1 for an illustration of this procedure).

Figure 1: Illustration of the method used to measure F0 for the grunts, barks and chatterings. The top panel, which shows the auditory signal for a chattering (see the definition of a chattering below), illustrates our hand tagging of the periodicity of the signal. The bottom panel shows the corresponding F0, which was calculated with our Matlab script from the interval durations (bottom panel).

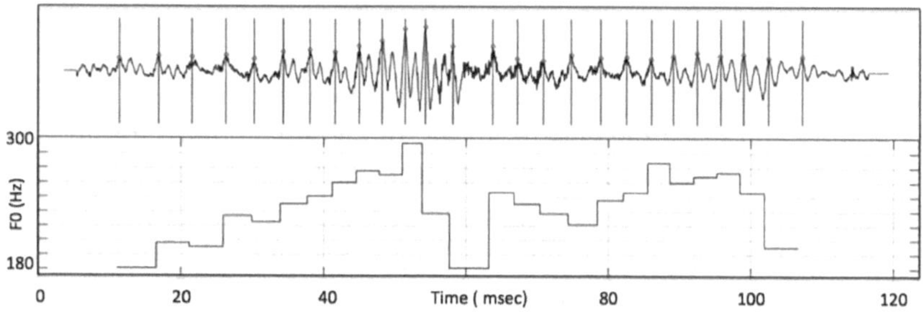

3. Results

3.1 Overview

Twelve distinct vocalizations were distinguishable in the captive Guinea baboons. Some vocalizations were age and sex specific. Table 1 provides the full list of vocalizations and their occurrence per age and sex. Illustrative audio files of each kind of vocalization can be found in the webpage (https://osf.io/nr2ye/) provided as supplementary material. While broad descriptive features of vocalizations were useful in grouping calls and creating distinct categories, formant analysis was only possible in 5 of the 12 vocalizations (grunts, wahoos, barks, yaks, and copulation calls, see below). We selected the clearest recordings for each vocalization per sex/age group for analysis of their broad descriptive features (a minimum of 10 separate recordings was possible).

Table 1: *Defining features of each vocalization within the repertoire of Guinea baboons. The calls found within each sex-age category are noted as the percentage recorded, with sample size taken into account. A '-' indicates that this characteristic was not applicable or measurable for that vocalization. ~ indicates that the vocalization was observed in this category but we did not record it. Note that no two vocalizations have the same characteristics.*

	Sex-age category							Phases		Number of calls in a vocalisation		Interval duration		F0		Formants			
	Adult males	Adult females	Sub-adult males	Sub-adult females	Juveniles	Infants	Total % recorded	1 phase	2 phases	Single call	Multiple calls	Even	Uneven	< 100	> 100	Present	Absent	Stable	Modulated
Rhythmic grunts	4.97%	2.93%	3.49%	0.52%	1.10%	2.09%	**15.10%**	*			*	*		*		*		*	
Barks	1.18%	5.70%	3.57%	2.09%	3.55%	4.88%	**20.97%**	*		*		–	–		*	*			*
Threat grunts	~	0.26%	0.70%	~	1.05%	0.35%	**2.36%**	*		*		–	–		*	*		*	
Yaks	0.13%	1.83%	0.09%	1.05%	0.64%	3.66%	**7.40%**	*			*	*			*	*			*
Scream	2.09%	2.56%	2.62%	3.14%	4.42%	8.72%	**23.55%**	*		*		–	–		*		*	–	–
Wahoo	6.80%	0.42%	3.57%				**10.79%**		*	*		–	–		*	*		*	
Roargrunts	0.65%						**0.65%**	*			*		*	*		*		*	

	Sex-age category							Phases		Number of calls in a vocalisation		Interval duration		F0		Formants			
	Adult males	Adult females	Sub-adult males	Sub-adult females	Juveniles	Infants	Total % recorded	1 phase	2 phases	Single call	Multiple calls	Even	Uneven	< 100	> 100	Present	Absent	Stable	Modulated
Male grunts	4.97%		0.26%				**5.23%**	*			*	*		–	–	*		*	
Two-phase grunts	0.39%						**0.39%**		*		*	*		*		*		*	
Copulation call		2.41%		5.23%	0.17%		**1.39%**	*			*		*		*	*		*	
Chattering					0.52%	0.87%	**7.81%**	*			*	*			*	*		*	
Moans						4.36%	**4.36%**	*		*		–	–		*	*		*	
	21.18%	**16.11%**	**14.3%**	**12.03%**	**11.45%**	**24.93%**													

3.2 Vocalizations produced by all or most age and sex categories

3.2.1 Rhythmic grunts

Acoustic description (see Figure 2): This tonal vocalization is characterized by the presence of multiple, single-phase calls of even temporal spacing, with clear formants (although sometimes only one formant could be detected, particularly when produced by adult males). F0 is low and sometimes changed within a bout but otherwise grunts were acoustically stable in their physical structure within the same bout. Grunt bouts did not vary much between contexts, although faster calling rates were found in contexts 3 and 4 (see below, 'Context & usage'). Calling rates were around 2.2 grunts/s in adults and sub-adults, 1.8 grunts/s in juveniles, and 1.01 grunts/s in infants. Infant grunts showed physical differences from adult and even juvenile grunts, with the loudness and F0 being much higher.

Context & usage: Rhythmic grunts were the most common affiliative vocalizations and were used by all age-sex groups in nine main contexts: 1) towards infants to elicit interaction, 2) towards mothers with infants, 3) after an infant scream, 4) by an individual, not the mother, usually an adult male, holding an infant to its chest, sometimes bouncing it, 5) between hugging adults, 6) by males eliciting a female to copulate, 7) by males after copulation, 8) from dominant animals (or males) to lower ranked females when approaching to groom or sit close by, and 9) by a non-moving group. Grunts were almost always produced as a series of calls (bout). Between 2 and 18 calls per bout were recorded; grunts were considered to belong to the same bout when they were less than 1.5s apart.

The grunts were soft and therefore used only as short distance contact call. The production of grunts by one individual typically did not elicit grunts from other individuals, although this did occur, specifically in contexts 4 and 9. The function of this grunting could not be determined. In this situation, several individuals were sitting within a meter from each other, looking in different directions; they were typically not physically interacting. One individual would begin grunting and the others would then join in. Infants only grunted in response to adult grunts. Grunts were produced with the mouth almost closed, the baboon's ears were twitched backwards and the eyebrows raised with each call.

Figure 2: Rhythmic grunt of an adult female. Audio signal (top panel), wide-band spectrogram (Praat) showing the first two formants as well as the characteristic vertical lines due to low F0 periodicity (middle panel), and F0 computed with our Matlab scripts (bottom panel).

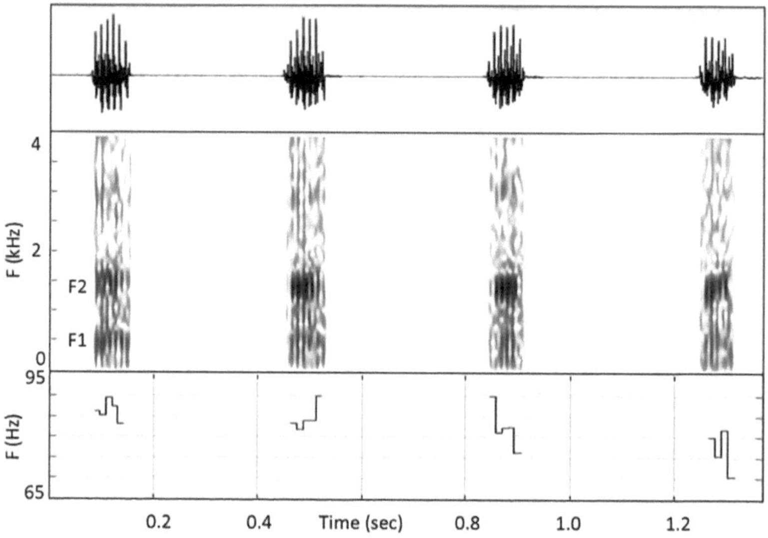

Terminology: This section lists the terminology used within the baboon taxa literature that, based on descriptions or spectrograms, appears to correspond to the vocalization described here. Grunts (*P. papio*: Byrne, 1981; *P. ursinus*: Cheney and Seyfarth, 2007; Rendall et al., 2005; *P. cynocephalus*: Hall and DeVore, 1965; *P. anubis*: Ey and Fischer, 2011), rapid grunt (*P. papio*: Byrne, 1981), rhythmic grunts (*P. hamadryas*: Ransom, 1981; Smuts, 1985), basic grunt (*P. anubis*: Ransom, 1981; Smuts, 1985), broken grunting (*P. anubis*: Ransom, 1981), low amplitude grunt (*P. ursinus*: Engh et al., 2006), soft grunts (*P. papio*: Anderson and McGrew, 1984).

3.2.2 Barks

Barks were recorded in two main contexts – prior to feeding in response to the presence of humans (contact barks) and in response to the presence of sheep (alarm barks). They were not distinguishable by ear, but the analysis did reveal differences in their acoustic structure. The difference between contact and alarm barks is described below. We observed barks to also occur in two other, albeit more rare, instances. One infant (Grimm), produced barks after his mother was removed from the troop due to illness. We deemed these barks also as a form of contact (see Cheney et al., 1996), but were not included in the analysis of the contact barks. The second instance was in response to the alarm wahoos of the nearby olive baboons. As it was not possible within the constraints of this study to determine if our subject group could distinguish between the contact and alarm wahoos of the olive baboons, we did not categorize their response barks as either contact or alarm. Only one adult male, Articho, produced barks (in both the contact and alarm contexts), but this was rare and this vocalization was more typically produced by females, juveniles and even one infant (Grimm). Barks were produced with a rounded 'O' shape mouth.

- *Contact barks*

Acoustic description (see Figure 3): This bark is sharp and clear, with a defined and modulated harmonic structure, and lower signal-to-noise ratio than observed in alarm barks. The F0 of contact barks typically followed a curved temporal pattern, rising in frequency (Hz) from the start of the call before returning to the starting frequency; this curved feature was less pronounced in adult male barks. The barks produced by Grimm after the removal of his mother were shorter (0.12 ± 0.01s) than those he produced prior to feeding (0.18 ± 0.01s), and the F0 was similar.

Figure 3: Contact barks. These five contact barks were produced by an adult male, sub-adult male, sub-adult female, juvenile and infant, respectively. Audio signal (top panel), wide-band spectrogram (Praat) showing the first two formants (middle panel), and F0 calculated with Praat (bottom panel).

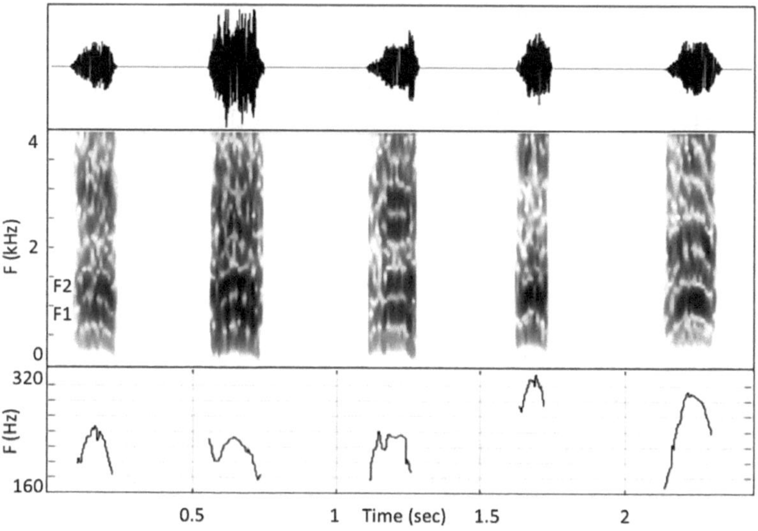

Context & usage: Contact barks were largely produced in the hour prior to feeding when the baboons observed humans leaving the office complex nearby and when the humans were preparing the food. The baboons visually fixated on staff when producing these barks. Barks by one individual could elicit barks in others to create a chorus.

Terminology: Clear bark (*Papio ursinus*: Ey and Fischer, 2011; Fischer et al., 2001b), dog-like bark (all savannah baboon species: Estes, 1992; *P. cynocephalus*: Hall and DeVore, 1965), contact bark (*Papio ursinus*: Cheney et al., 1996; Ey and Fischer, 2011; Fischer et al., 2001a), sharp bark (*P. papio*: Byrne, 1981).

- *Alarm barks*

Acoustic description (see Figure 4): With higher formants, alarm barks have quite the same general acoustical structure as the contact barks describe above; however, this bark type is noisier than the contact barks and less tonal.

Figure 4: Alarm barks. Audio signal (top panel), wide-band spectrogram (Praat) showing the first two formants and characteristic vertical lines due to low F0 periodicity (middle panel), and F0 calculated with Praat (bottom panel). These three alarm barks were produced by an adult female, a sub-adult male and a juvenile, respectively.

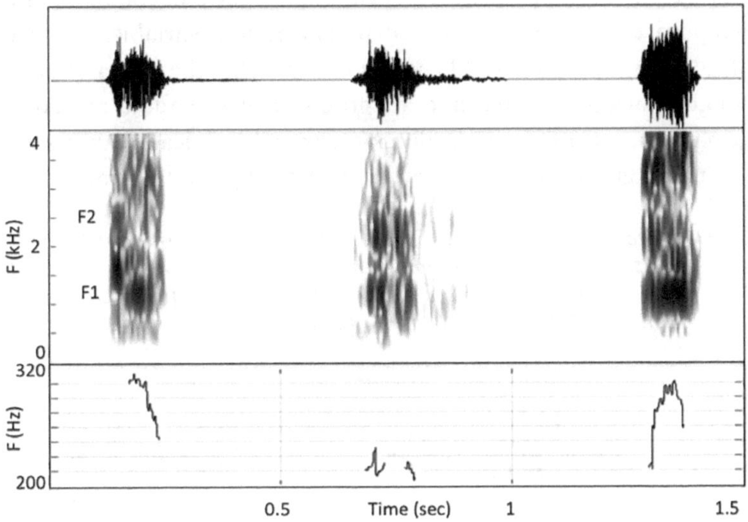

Context & usage: Alarm barks were produced when the sheep were heard approaching, grazing next to and passing by the primate center. Single barks were the norm, although barking bouts (up to 6, with less than 1.5 sec between calls) were recorded. The baboons visually fixated on the sheep, or in the direction from which the sheep could be heard approaching, when barking.

Terminology: Fear bark (*P. ursinus*: Cheney and Seyfarth, 2007), alarm bark (*P. ursinus*: Cheney and Seyfarth, 2007; *P. papio*: Byrne, 1981), cough-bark (*P. anubis*: Ransom, 1981), harsh bark (*P. ursinus*: Fischer et al., 2001a), shrill bark (all savannah baboon species: Estes, 1992; *P. cynocephalus*: Hall and DeVore, 1965; *P. anubis*: Ransom, 1981; Rowell, 1966; *P. ursinus*: Fischer et al., 2001a), sharp bark (*P. papio*: Byrne, 1981).

3.2.3 *Threat Grunts*

Acoustic description (see Figure 5): Threat grunts are a highly noisy call, with harsh but soft rolling egressive cough-like sounds. There were enough recordings of single call productions to suggest that a call should be considered as the vocalization; however, vocal bouts were still common, although the temporal connection between call units was quite variable. The F0 of this vocalization is low and unstable within each call, but the formants are stable. Although individual threat grunts are produced by sub-adults and adults as a single phase (i.e., continuous production), juveniles typically gave a seemingly double phase grunt, as if the sound hitched during production.

Figure 5: Threat grunts of an adult male. Audio signal (top panel), wide-band spectrogram (Praat) showing the first two formants and characteristic vertical lines due to low F0 periodicity (middle panel), and F0 computed with our Matlab script (bottom panel).

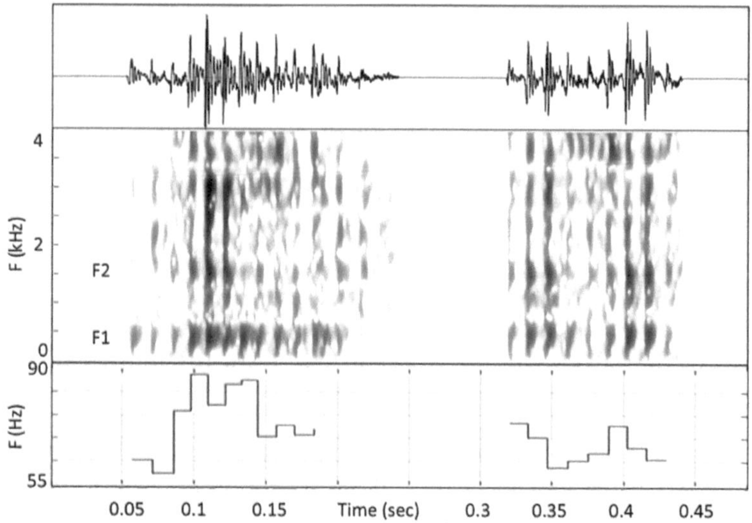

Context & usage: This vocalization was observed in two contexts. The first of these was in antagonistic situations between adult females, in which the aggressor produced the vocalization. The second context was in response to the sheeps; all sex-age groups produced this vocalization in this context, although it was rarer in adult males and infants. Threat grunts were often observed in conjunction with barks (juveniles, adult females and sub-adult males) and wahoos (adult males only). Two calls were often produced within 1.5 sec before a long pause until the next call.

Terminology: Threat grunts (*P. ursinus*: Cheney and Seyfarth, 2007; Engh et al., 2006).

3.2.4 Yaks

Acoustic description (see Figure 6): Yaks have an irregular harmonic structure. The F0 is also modulated, being highly variable within a single call, with a lower frequency at the beginning of the yak than at the end. This vocalization is typically produced as a series of high F0, single phase calls with even temporal distribution, although calling rate can increase with context intensity. Up to 50 yaks in a series were recorded, with calls being considered as part of the same series when produced less than 1.5 sec apart.

Context and usage: This vocalization was produced by individuals being threatened or in distress. The corresponding facial expression involved a strong retraction of the lips. It may be that the call is a form of appeasement, as suggested by Estes (1992). It did not appear to act as a recruitment vocalization. Infants produced yaks when they were rebuffed by their mother and were looking for comfort, often in the form of nursing. Yaks were produced as a long series of calls, but were also given in conjunction with screams and/or moans (infants only) in varying orders and numbers. Context suggested that yak-only series were given in lower intensity situations, especially in comparison to screams. Yaks were produced by adults with the teeth bared and the body often cowed and shoulders hunched, with the tail lowered and ears back. Yaks by infants were not given with the same body posture; instead, infants were usually running after their mothers.

Figure 6: Yaks of an adult female. Audio signal (top panel), wide-band spectrogram (Praat) visualizing the harmonics (middle panel), and F0 calculated with Praat (bottom panel).

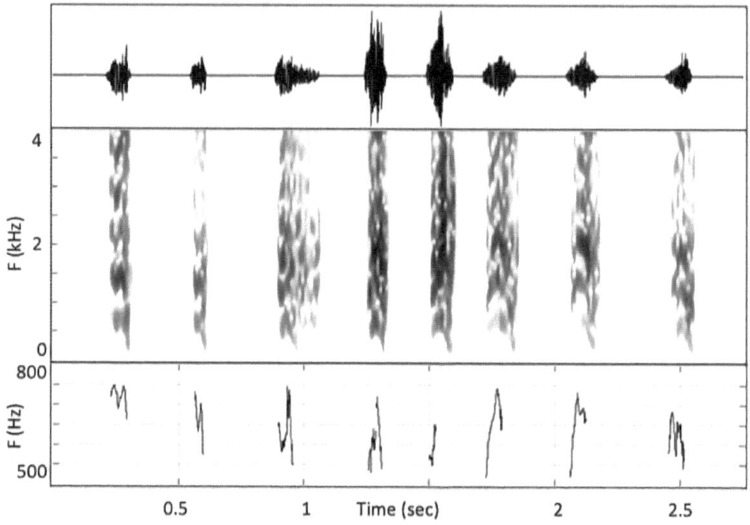

Terminology: Yak/yakking (*P. cynocephalus*: Hall and DeVore, 1965; all savannah baboon species: Estes, 1992), geck (infants only – *P. anubis*: Ransom, 1981; *P. papio*: Anderson and McGrew, 1984; *P. hamadryas*: Ransom, 1981; Smuts, 1985), chirplike clicking (infants only – *P. cynocephalus*: Hall and DeVore, 1965), ick (of the ick-ooer, infants only – all savannah baboon species: Estes, 1992), fear bark (*P. ursinus*: Cheney and Seyfarth, 2007), staccato coughing (*P. hamadryas*: Kummer, 1968), disjointed coughing (*P. hamadryas*: Ransom, 1981; Smuts, 1985), contact call (Rendall et al., 2009).

3.2.5 Screams

Acoustic description (see Figure 7): Screams were highly variable, probably the most variable vocalizations produced by the baboons. Calls could have either harmonics or no clear harmonics, with some recorded instances of calls having alternations of both characteristics. Durations were also variable, ranging from less than a second (quick yelps) to extended calls of over

2s. They could be produced as a single call or as multiple calls within a bout. The high F0 (~1kHz) did not allow for formants to be observed. The maximum frequency observed was very high (approaching 20kHz). Some screams (or scream sections) were noisy and harsh with no clear harmonic structure. Harmonic production could be either clear or mixed with some noise. Inspection of screams found that the baboons could change F0 quite rapidly and dramatically within a call. Screams were considered singular vocalizations that could be produced in bouts. Each call was analyzed separately.

Figure 7: Screams of a sub-adult male. Audio signal (top panel), wide-band spectrogram Praat) showing the harmonics (middle panel), and F0 (bottom panel) calculated with Praat. Note that F0 is too high in this example for visualizing the formants,

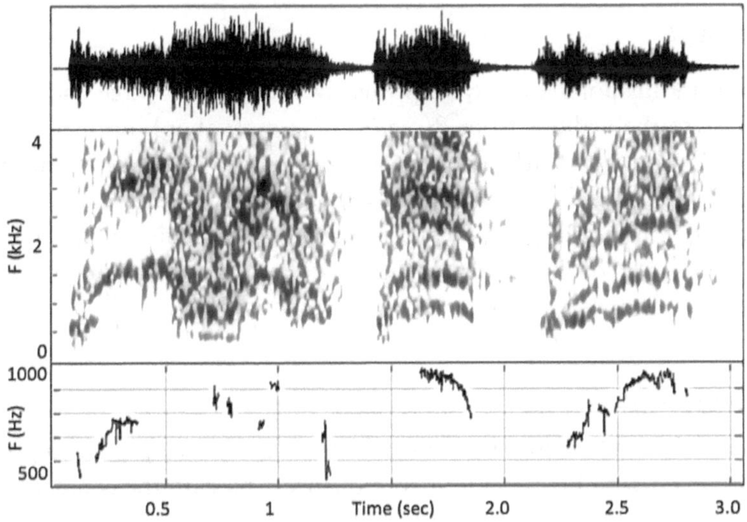

Context & usage: Screams were observed in three main contexts: surprise, fights and maternal rejections (i.e., produced by infants when their mother did not allow nursing or clinging). Screams produced when the baboon was surprised by an event, such as a sudden movement or shock, was more a 'yelp'-like sound. Regarding the second context, screams were occasionally produced by the aggressing individual, but it was far more typical for the scream to be produced by the individual being aggressed. Screaming from infants could produce reactions from adults and older juveniles, including grunting and physically comforting; screams due to maternal rebuffs rarely elicited a response from other baboons. These screams were strongly harmonic. Screams were often coupled with yaks and/or moans (infants only) in various combinations (e.g. yak-scream-scream-yak-yak-yak-yak-yak-yak-yak-scream-yak-yak-moan). A single yak often preceded a screaming bout. One sub-adult male baboon (Cloclo) and one juvenile (Feya) would produce a short scream after a single bark at feeding. Screams were produced with the teeth bared and the lips retracted.

Terminology: Scream (*P. anubis*: Ransom, 1981; *P. papio*: Byrne, 1981), screaming (Hall and DeVore, 1965), screeching (Estes, 1992).

3.3 Vocalizations produced by adults and sub-adults

3.3.1 Wahoo

Wahoos in our population were primarily produced in three contexts: in response to the wahoos from *P. anubis* (contest wahoo), prior to feeding in conjunction with barks (contact wahoo) and in response to the sheep (alarm wahoo). The contest wahoos were typically produced in low light, making identification of the vocalizing individual difficult. As we could not be sure in our recordings if any of these vocalizations came from our group of Guinea baboons or the nearby olive baboons, they are not included in our discussion here.

- *Contact wahoos*

Acoustic description (see Figure 8): Wahoos are a two-phase, single call vocalizations with high and low F0 sequences. As with the contact barks, these wahoos had a lower signal-to-noise ratio than those produced in the alarm context. The F0 varies from the 'wa' to the 'hoo', with the latter typically produced with a lower F0.

Figure 8: Contact wahoos of sub-adult males. Audio signal of a wahoo (top panel), wide-band spectrogram (Praat) showing the first two formants (middle panel), and F0 (bottom panel) calculated with Praat for the wa-, and with our Matlab program for the -hoo.

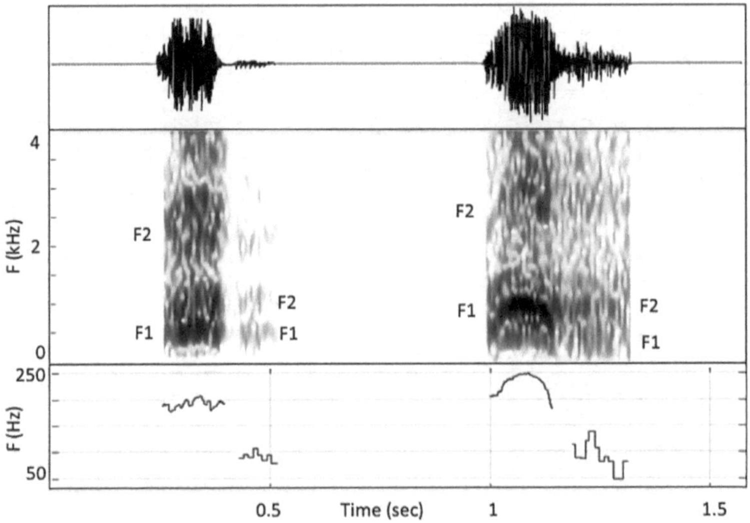

Context & usage: Contact wahoos were typically made by sub-adult males, although occasionally adult females also seemed to give a wahoo instead of a bark. However, it is important to note that while wahoos from adult females were often identified by ear, spectrogram analysis showed that these were more likely to be barks, with the 'hoo' sound being a faint continuation of the exhalation of breath. During production, the mouth was widely opened in an elongated vertical 'O' during the 'wa', before closing to a horizontal opening for the 'hoo'.

- *Alarm wahoos*

Acoustic description (see Figure 9): The 'wa' of alarm wahoos showed some similarities with the alarm barks, in that they were tonal with a large degree of noise. The 'hoo' production was distinct and of longer duration in this context, in comparison to the wahoos produced prior to feeding (Table 2).

Table 2: Results of the analyzes of the temporal features of each vocalization type for each age-sex category with a useable sample size of ≥10 separate vocalizations. That is, male grunts, two-phase grunts, and roar grunts were not included.

	Sex/age category	n_v	n_c	n_i	Fmx (kHz)	SE	C_{DUR} (sec)	SE	I_{DUR} (sec)	SE	T_{DUR} (sec)	SE	C/B	SE
Rhythmic grunts	AM	12	92	3	4.49	764.7	0.11	0.01	0.46	0.06	3.08	0.83	6.9	1.70
	AF	22	150	7	5.75	379.4	0.09	0.01	0.48	0.03	3.75	0.62	7.2	0.73
	SAM	10	22	1	7.47	1325.0	0.08	0.01	0.54	0.06	3.01	1.14	5.5	1.55
	J	10	60	4	7.20	548.3	0.09	0.004	0.59	0.05	3.45	0.70	6.0	0.9
	I	10	21	3	8.69	490.5	0.10	0.01	1.79	0.04	1.98	1.01	2.0	0.80
Barks -Contact	AF	11	n/a	5	19.15	751.2	n/a	n/a	n/a	n/a	0.16	0.01	n/a	n/a
	SAM	10	n/a	3	17.07	883.6	n/a	n/a	n/a	n/a	0.2	0.01	n/a	n/a
	J	12	n/a	6	15.53	886.6	n/a	n/a	n/a	n/a	0.13	0.01	n/a	n/a
	I	10	n/a	2	15.53	1569.8	n/a	n/a	n/a	n/a	0.18	0.01	n/a	n/a
-Alarm	AF	10	n/a	6	19.70	343.8	n/a	n/a	n/a	n/a	0.13	0.01	n/a	n/a
	SAM	10	n/a	3	20.90	425	n/a	n/a	n/a	n/a	0.13	0.00	n/a	n/a
	J	12	n/a	7	19.08	696.9	n/a	n/a	n/a	n/a	0.11	0.01	n/a	n/a
	I	10	n/a	2	20.66	168.3	n/a	n/a	n/a	n/a	0.11	0.01	n/a	n/a
Threat grunts	AF	19	n/a	4	2.93	107.2	n/a	n/a	n/p	n/p	0.08	0.03	n/a	n/a
	SAM	12	n/a	1	2.85	78.7	n/a	n/a	n/p	n/p	0.14	0.01	n/a	n/a
	J	27	n/a	4	18.40	546.1	n/a	n/a	n/p	n/p	0.10	0.01	n/a	n/a
Yaks	AF	10	200	8	19.55	725.5	0.11	0.01	0.37	0.02	14.49	1.06	23.6	9.1
	J	10	212	5	17.03	481.8	0.07	0.01	0.16	0.05	4.72	2.55	19.8	10.7
	I	10	250	4	16.22	564.2	0.04	0.02	0.19	0.03	12.84	3.27	22.9	13.2

	Sex/age category	n_v	n_c	n_i	Fmx (kHz)	SE	C_{DUR} (sec)	SE	I_{DUR} (sec)	SE	T_{DUR} (sec)	SE	C/B	SE
Screams	AM	10	n/a	2	18.01	167.4	n/a	n/a	n/a	n/a	0.77	0.11	n/p	n/p
	AF	18	n/a	7	19.00	404.5	n/a	n/a	n/a	n/a	0.49	0.15	n/p	n/p
	SAM	20	n/a	6	17.68	769.1	n/a	n/a	n/a	n/a	0.85	0.22	n/p	n/p
	J	28	n/a	5	19.22	398.2	n/a	n/a	n/a	n/a	0.64	0.09	n/p	n/p
	I	18	n/a	3	19.99	661.6	n/a	n/a	n/a	n/a	0.72	0.14	n/p	n/p
Wahoo -Contact	SAM -Wa	10	n/a	5	17.13	4020.0	0.16	0.01	n/a	n/a	n/a	n/a	n/a	n/a
	-Hoo	10	n/a	5	10.99	1589.4	0.15	0.01	n/a	n/a	0.31	0.01	n/a	n/a
-Alarm	AM -Wa	12	n/a	3	13.92	1318.8	0.18	0.01	n/a	n/a	n/a	n/a	n/a	n/a
	-Hoo	12	n/a	3	7.41	219.7	0.28	0.02	n/a	n/a	0.45	0.02	n/a	n/a
	SAM -Wa	12	n/a	5	17.30	2436.8	0.16	0.01	n/a	n/a	n/a	n/a	n/a	n/a
	-Hoo	12	n/a	5	8.03	1426.2	0.15	0.01	n/a	n/a	0.3	0.01	n/a	n/a
Copulation grunts	AF	20	177	7	12.49	544.0	0.07	0.003	0.18	0.01	2.04	0.26	8.9	1.09
	SAF	10	59	1	6.60	550.6	0.06	0.003	0.02	0.02	1.31	0.2	5.36	1.0
Chattering	J	10	53	5	3.03	678.2	0.04	0.02	0.37	0.09	4.66	0.54	9.9	2.3
Moan	I	23	n/a	3	6.15	593.1	0.91	0.05	1.78	0.51	14.23	1.85	n/a	n/a

n/a = not applicable; (n_v) = the number of vocalizations/series; (n_c) = the number of individual call units; (n_i) = the number of individual baboons whose calls were used in the analysis; (C_{DUR}) = the duration of each call or phase; (I_{DUR}) = the duration of the interval between two calls in the same bout; (T_{DUR}) = the total duration of a bout; (C/B) = the number of calls per bout; SE = standard error. Age/sex categories: AM – adult males, AF – adult females, SAM – sub-adult males, SAF – sub-adult female, J – juveniles, I – infants. SE – standard error.

Figure 9: Alarm wahoo of an adult male. Audio signal of a wahoo (top panel), wide-band spectrogram (Praat) for visualizing the first two formants (middle panel), and F0 (bottom panel) which was computed with our Matlab program.

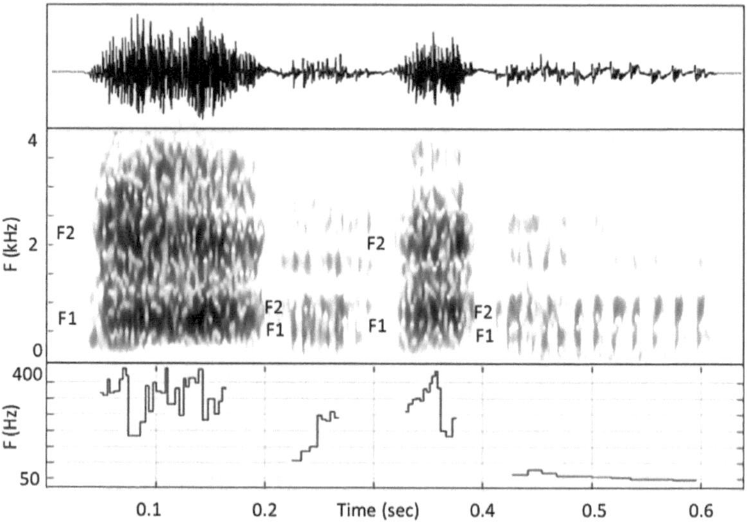

Context & usage: Like alarm barks, alarm wahoos were in response to the sound of and/or the presence of sheep. Although wahoos are typically considered as a single call vocalization, a series of three wahoos were observed on a few occasions and double wahoos – with the first wahoo shortened and immediately followed by the second wahoo – were also recorded.

Terminology: This terminology corresponds to both contact and alarm wahoos, as little to no differentiation in names have been noted within the literature. Wahoo bark (*P. papio*: Byrne, 1981), contact bark (*P. ursinus*: Cheney and Seyfarth, 2007; Ey and Fischer, 2011), wa-hoo (*P. anubis*: Ransom, 1981), wahoo (*P. ursinus*: Fischer et al., 2002 [note that the authors differentiate between 'contact', 'contest' and 'alarm' wahoos in terminology]; Kitchen et al., 2003), two phase/d bark (all savannah baboon species: Estes, 1992; *P. cynocephalus*: Hall and DeVore, 1965;

P. hamadryas: Ransom, 1981; Smuts, 1985), type 2 loud call (*P. papio*: Byrne, 1981), loud call (*P. ursinus*: Fischer et al., 2002; Kitchen et al., 2003) bahu bark (*P. hamadryas*: Kummer, 1968), oohu roar (*P. hamadryas*: Kummer, 1968).

3.3.2 Roargrunts

Acoustic description (see Figure 10): This is a series of loud grunts with a hum-like grunt typically preceding the call, a pause of 5–6 sec and then 4–6 grunts produced in close succession as a crescendo, with a final long grunt or roar, similar to a double wahoo. The call sequence is variable, with the hum difficult to discern or absent, and the concluding roar not always produced. We did not record enough of these vocalizations to determine why there was so much variability in the entire bout. However, the grunts that make up the majority of the vocalization were always present and produced consistently. Although we did not analyze this vocalization due to the small sample size, we did note that the grunts had a low F0 (~30Hz), with F1 typically around 440Hz and F2 at 1.1kHz. Each call within the bout was typically longer (0.34s) than those produced by the adult males during rhythmic grunting (0.11s), although the interval durations were similar (~0.45s). The notable features of this vocalization are the long grunts produced at a slow calling rate (~1.7grunt/s).

Context & usage: Adult males produced this call either prior to feeding or when the sheep were present, suggesting it is in response to high arousal level due to tension.

Terminology: Type 1 loud call (*P. papio*: Byrne, 1981), roargrunt (*P. papio*: Byrne, 1981; Maciej et al., 2013b; *P. anubis*: Ransom, 1981; *P. hamadryas*: Ransom, 1981; Smuts, 1985), hum-roargrunt (*P. anubis*; Ransom, 1971), roaring (all savannah baboon species; Estes, 1992), crescendo of two-phase grunts (all savannah baboon species; Estes, 1992), grating roar (*P. anubis*: Estes, 1992).

Figure 10: Roar grunt of an adult male. Audio signal (top panel), wide-band spectrogram (Praat) for visualizing the first two formants (middle panel), and F0 (bottom panel) computed with our Matlab program.

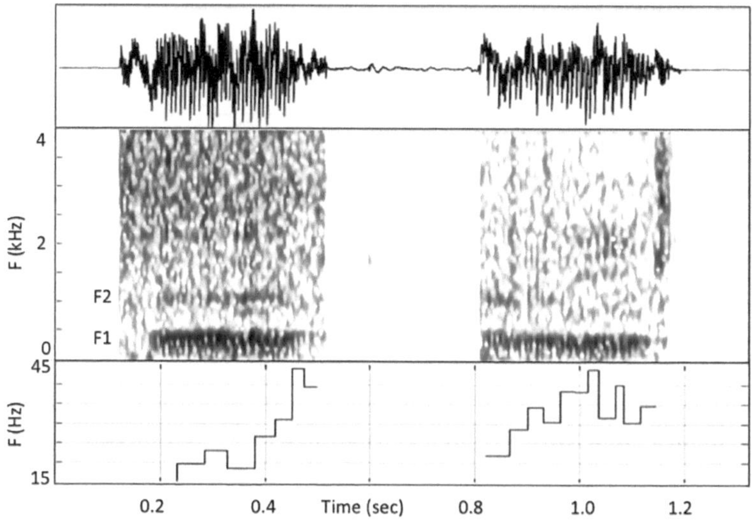

3.3.3 Male Grunts

Acoustic description (see Figure 11): This vocalization was particularly difficult to analyze as it was often produced with accompanying acoustic displays that interfered with the recorded signal (see below, 'Context & usage'). Therefore, we only provide spectrograms and audio files (see supplementary material) of prototypical examples of these calls, but no analysis was performed. The vocalization consists of a series of rapidly-produced, short (~0.05sec), breathy, strongly egressed grunts, which form a crescendo and sometimes end in a roar or double wahoo, similar to that heard at the end of some roargrunt sequences.

Figure 11: Male grunt of an adult male. Audio signal (top panel), wide-band spectrogram (Praat) showing the first two formants and the characteristic vertical lines due to low F0 periodicity (middle panel), and F0 computed with our Matlab program (bottom panel).

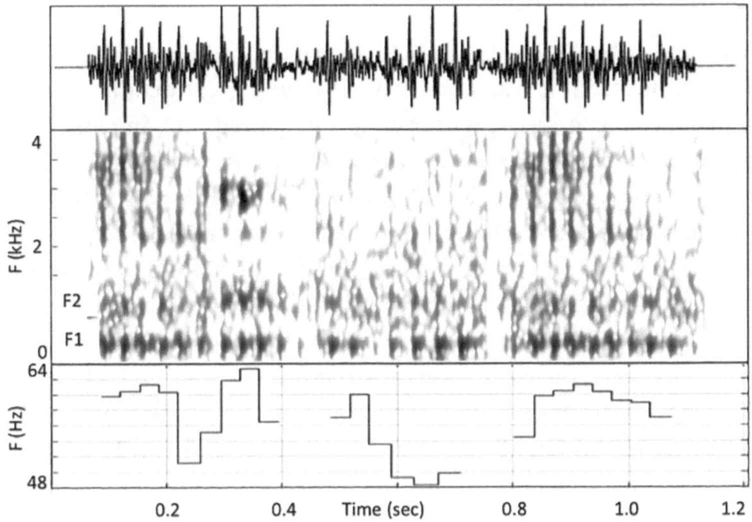

Context & usage: This is an adult and sub-adult male vocalization, albeit rarely in the latter ones, and typically performed together with power displays. These displays include fence shaking, jumping and rock throwing, and throwing the head back. Observations were made of both adult and sub-adult males as well as one infant (Grimm), one juvenile/sub-adult (Dan) and some adult females performing these displays without the vocalization or with only a subset of the full vocalization, suggesting that the coordination to do both required development and strength. This vocalization was produced after fights with other males, when the sheep were present for long periods of time, prior to feeding – especially if feeding was delayed – and when the computer systems (see Fagot and Palleressompoule, 2009), to which the baboons usually had access, were blocked. These contexts suggest that this vocalization is associated with high arousal levels and frustration, as well as indicators of male size and strength.

Terminology: Could not accurately determine corresponding vocalizations in other publications but may be the deep grunts described for *P. papio* by Byrne, (1981).

3.3.4 Two-Phase Grunts

Acoustic description (see Figure 12): As we only recorded a few instances of this vocalization, it was not analyzed in any great detail. These grunts were produced as a series, with two grunts paired, i.e., in close temporal proximity with short interval duration (~0.04s) then a longer interval duration (~0.15s) before the next pair. Duration of each pair was 0.4s, with the first grunt longer (~0.27s) than the second (~0.13s). It is therefore recommended that the grunt is analyzed like the wahoo, and considered, as the name suggests, as a two-part call. Due to the production of this vocalization (see below, 'Context & usage'), it is likely to be dismissed as panting. However, the clear formant structure (F1 = ~350Hz, F2 = ~2kHz) and controlled production indicates that this is a vocalization and not a consequence of running. Bouts were long (between 11 and 18 grunt pairs), with the first grunt being typically of a higher F0 (~60Hz) than the second grunt (~50Hz) within each pair.

Context & usage: Two-phase grunts are ingressive-egressive vocalizations, similar to panting. It was only produced by adult males, in contrast to the study by Byrne (1981), who found that all age-sex classes except infants produced this vocalization. The males were observed making this call while being chased by other adult males during fights.

Terminology: Two-phase grunts (*P. papio*: Byrne, 1981), pant-grunt (*P. anubis*: Ransom, 1971) uh-huh (all savannah baboon species: Estes, 1992; *P. cynocephalus*: Hall and DeVore, 1965), grunting (all savannah baboon species: Estes, 1992).

Figure 12: Two phase grunts of an adult male. Audio signal (top panel), wide-band spectrogram (Praat) showing the first two formants (middle panel), and F0 (bottom panel) computed with our Matlab program.

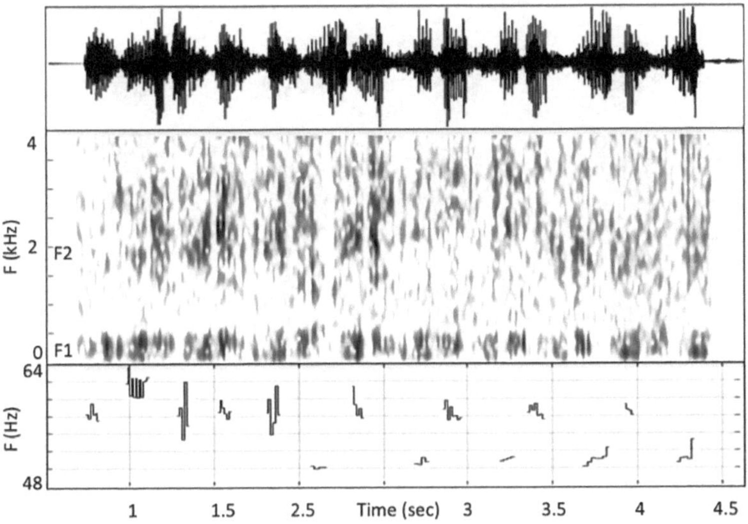

3.3.5 Copulation Calls

Acoustic description (see Figure 13): Copulation calls are defined by the production of a series of grunt calls (never singular), with fluctuating speed and F0. Egressed grunt-like breaths without formants were occasionally dispersed between the true grunts and/or at the end of the series. Copulation calls were typically tonal.

Figure 13: Copulation calls produced by adult females. Audio signal (top panel wide-band spectrogram (Praat) showing the first two formants (middle panel), and F0 (bottom panel) calculated with Praat.

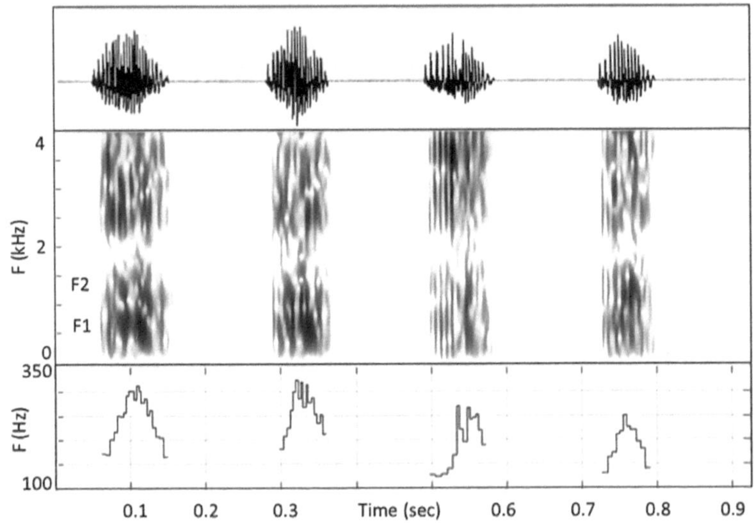

Context & usage: Adult, sub-adult and even juvenile (rare) females produced this vocalization toward the end of copulation, completing the call while running away from their male partner. It was also observed in one adult female (Mona) while being mounted by other females. The vocalization was preceded by a distinctive facial expression, in which the mouth was rounded into a 'O' shape, with the lips slightly pursed. Not all copulations were followed by copulation calls; however, the vocalization was produced more often than it was not and the facial expression was always present regardless of whether or not the vocalization was uttered.

Terminology: Muffled growl (all savannah baboon species: Estes, 1992; *P. cynocephalus*: Hall and DeVore, 1965), copulation grunts/call (*P. papio*: Byrne, 1981; *P. cynocephalus*: Hall and DeVore, 1965; Semple et al., 2002).

3.4 Vocalizations produced by infants and juveniles

3.4.1 Chatterings

Acoustic description (see Figure 14): Chattering is a series of unevenly spaced single phased calls, which have a chuffing-like sound, possibly ingressive-egressive due to the production method (see below, 'Context & usage'). The vocalization is quite soft in amplitude and not strongly harmonically structured (i.e., noisy). Formants and F0 were often difficult to discern, particularly in infants who produced much noisier calls than older juveniles.

Figure 14: Chatterings produced by a juvenile. Audio signal (top panel), wide-band spectrogram (Praat) showing the first two formants (middle panel), and F0 (bottom panel) computed with our Matlab program.

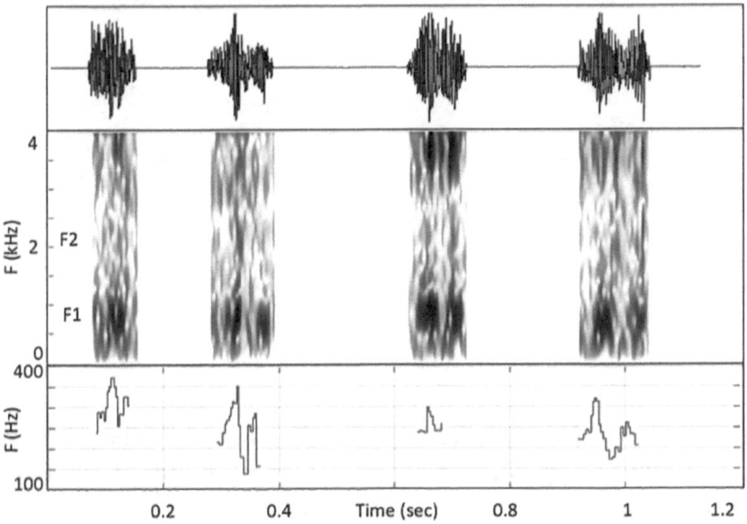

Context & usage: Chattering was used during play behavior, usually while running.

Terminology: Chattering (*P. ursinus*: Estes, 1992), panting (*P. anubis*: Ransom, 1981; *P. hamadryas*: Ransom, 1981; Smuts, 1985).

3.4.2 Moans

Acoustic description (see Figure 15): Moans are a single phase, single call vocalization. The call has a strong tonal structure with even formants and a gently arching, high F0. It sounds similar to a sheep vocalization.

Figure 15: Moans produced by an infant. Audio signal (top panel), wide-band spectrogram (Praat) showing the first two formants (bottom panel), and F0 (bottom panel) calculated with Praat.

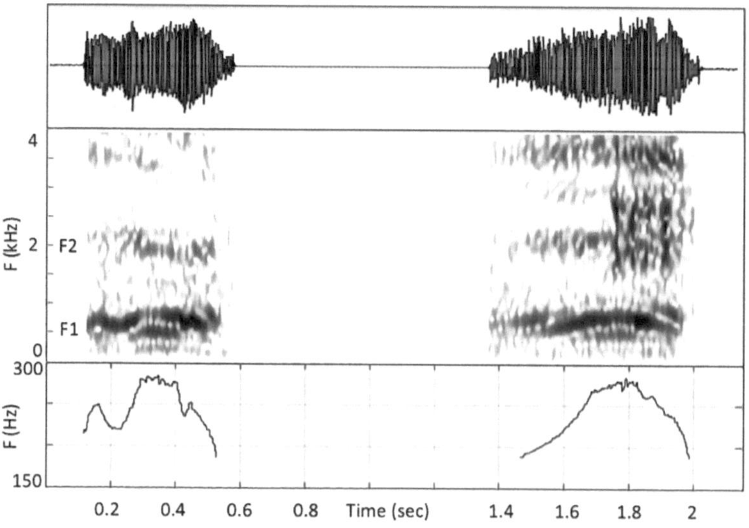

Context & usage: This vocalization was only observed to be produced by infants, usually in response to maternal rebuff or in distress situations. As with yaks, moans seemed to be produced when the mother was walking, not allowing her infant to be fed and hold on. Series of moans were observed; this seemed to be an extension of singular calls. Moans were often accompanied by yaks, to produce the 'ick-ooer' sound described by Hall and DeVore (1965). However, we noted that several yaks often preceded the moan. Occasionally, moans were accompanied by a scream after their production, but the two were not linked in the same way as the yak-moan sound.

Terminology: Ooer (of the ick-ooer – *P. cynocephalus*: Hall and DeVore, 1965), moan (*P. anubis*: Ransom, 1981; *P. hamadryas*: Ransom, 1981; Smuts, 1985), infant moan (*P. papio*: Byrne, 1981).

4. Formant analyses

4.1 Methods for formant analyses conducted on vocal categories

Formant analyses were performed for several classes of vocalization, including the grunts (two-phase grunts excluded), barks, wahoos, yaks and copulation calls. Vocalizations of infants and juveniles, and adult screams were not considered for these analyses because of their F0, sometimes approaching 1 kHz. The method used for formant analysis is explained in details in the supplementary material of Boë et al. (2017). For these analyses, the part of the vocalizations containing formants were grouped into one file per vocalization type (e.g., bark). The grunt file included the rhythmic, threat and roar grunts. These different types of grunts were grouped together because they were highly similar in their formants. The two-phase grunts were not included in the grunt file, because they differed slightly from the other grunts regarding their formants (see Figure 12). The bark files grouped the alarm and contact barks, and the wa- and -hoo files also grouped the alarm and contact wahoos. To limit the perturbation due to noise and to maximize the reliability of the LPC results and achieve the clearest possible characterization of the vocalizations, formant analyses were performed using frames from 0.5 to 2 seconds long, so that each frame encompassed several utterances. It was done with successive frames operating as a sliding window overlapping by 50%, and the results and subsequent processings were based on the frame outputs from this LPC processing. The frame database was then filtered to further control for detection errors, and all the frames missing F1 or with F1 or F2 values greater than 3 standard deviations from the means of their categories were eliminated from the dataset (see below). Also, F0 was measured in the same frames using autocorrelation and peak-picking. The detailed corpus characteristics and LCP settings are provided in Table 3. Interested readers are referred to Boë et al. (2017) for an in-depth discussion of our choice of variables, regarding for instance the number of poles.

Table 3: Corpus characteristics and LPC settings.

	Grunt ♂♀	Wa- ♂	-hoo ♂	Cop ♀	Bark ♀	Yak ♀
CORPUS						
N baboons	13	3	3	8	11	10
N vocalizations	522	69	69	124	116	504
Total duration (s)	65	11	15	10	29	69
Mean duration (ms)	125	159	219	81	250	137
LPC SETTINGS						
N poles	60	30	60	60	30	60
Frame duration (s)	1	1	1	0.5	1	2

4.2 Results of formant analyses conducted on vocal categories.

The acoustic results regarding F0 and the first two formants are reported in Table 4 for each class of vocalizations. Table 4 reveals that the baboons produced high- (i.e., ♂ wa-, ♀ bark, and ♀ yaks), and low vowel-like sounds (i.e., ♂ and ♀ grunts, ♂ -hoo, ♀ copulation calls), which are characterized by F1 formants in the high and low ranges, respectively. Table 4 also demonstrates the production of front and back vowel-like sounds, characterized by F2 formants in the high (♂ -wa, ♀ bark) and low ranges (♂ and ♀ grunts, ♂ -hoo, ♀ copulation call, and ♀ yak). We have no space here to present our analyses on formants in more details. However, note that this data set was analyzed in depth by Boe et al. (2017). Examining these vocalizations through modeling of their maximal acoustic space based on anatomical measures of the baboon's vocal tract, Boe et al. (2017) demonstrated that these vocalizations share the F1/F2 formant structure of the human [ɨ æ ɑ ɔ u] vowels. From these results, we can conclude that the baboons can produce several vocalic qualities differentiated by their formant structures, and that these structures are characteristic properties of vocalizations produced in distinct social contexts, or for different functions.

Table 4: Acoustic results obtained for vocalizations that could be analyzed for the first two formants and F0. The numbers in bracket indicate the standard deviations.

	Grunt ♀	**Grunt** ♂	**Wa-** ♂	**-hoo** ♂	**Cop** ♀	**Bark** ♀	**Yak** ♀
CORPUS STATISTICS							
Total nb of frames	39	76	18	26	36	50	19
ACOUSTIC RESULTS							
F1 (Hz)	476	392	948	552	583	1044	916
	(31)	(63)	(70)	(82)	(93)	(89)	(140)
F2 (Hz)	1440	1219	2165	1025	1211	2685	1500
	(129)	(137)	(112)	(66)	(119)	(121)	(116)
F0 (Hz)	64	61	417	121	133	431	—
	(20)	(20)	(105)	(37)	(56)	(45)	—

5. Discussion

5.1 On the Guinea baboon's vocal repertoire

We observed and recorded twelve vocalizations, easily distinguishable by both ear and production/acoustic characteristics, produced by our group of baboons. Two calls, the bark and the wahoo, showed slight differences in acoustic features when produced in different contexts (contact and alarm). Interestingly, more types of vocalizations were given by adult males in our study than any other sex/age category (Table 1) and constituted the second largest proportion of recorded vocalizations despite the small sample size. Eleven of the vocalizations in our repertoire had certainly been reported previously, either for Guinea baboons or other baboon taxa, but we could not find a clear analogy to the male grunt vocalization in any of the baboon vocal literature. It is a possibility that Byrne (1981) had referred to this vocalization as the 'deep grunt' but with only a description of "long, low pitched grunt, fluctuating in pitch and volume. Adult males only (?)" (p. 287) it is difficult to be sure.

Seven of the vocalizations we describe are short distance communications; that is, their production did not allow for long-distance detection. The baboons showed a large range of F0 production, from around 40 Hz for grunts to up to 1 kHz for screams. Feeding time and the occasional presence of sheep elicited the greatest variety of calls (barks, wahoos, threat grunts [in response to the sheep, only], roargrunts and male grunts) of any major contexts recorded. In regard to feeding, due to the captive environment, we are able to report the first known transfer of two vocalizations (barks and wahoos) to a new context in this species. It is known that baboons will use barks and wahoos to contact conspecifics when moving through dense vegetation (Cheney et al., 1996; Rendall et al., 2000) but this is the first time these calls have been reported to be used as a contact with caretakers.

We observed that some vocalizations could elicit vocal responses from conspecifics but found little evidence of communicative volleys between individuals. Some vocalizations (rhythmic grunts, screams, yaks, threat grunts, chattering and moans) could be directed towards specific baboons but they rarely elicited a vocal response. The bark or wahoo of one individual when observing (either visually or through auditory means) the approach of a human at feeding time or the sheep would often result in the production of these calls (usually barks) from other baboons. However, these calls were directed at external stimuli. Wahoos produced by adult males at night are known to create volleys whereby males from different groups produced wahoos back and forth (Anderson and McGrew, 1984; Byrne, 1981). We observed this occurring between our Guinea baboons and the olive baboons at night, but never just between the males within our group, and certainly the alarm wahoos in response to sheep never elicited a vocal response from the olive baboons. Although screams have been considered a recruitment call (e.g., in infants, Rendall et al., 2009), we found no particularly strong evidence to support this hypothesis; only some of the screams from infants and juveniles resulted in a vocal response (rhythmic grunts) or physical approach from adults (most screams were produced during conflicts and may better act as appeasement). However, it is important to note that rhythmic grunts directed towards individuals could elicit rhythmic grunts. For example, adult males grunting towards infants or juveniles would sometimes get grunts in return as the two animals approached each other. Hugging baboons would also often grunt. More research is required to determine the specific cues in the initial vocalization of one baboon that elicits the same vocal response, particularly when it is directed specifically to a conspecific rather than an external stimulus, in another baboon.

One vocalization that is produced by all age- and sex-groups is the yak. The term 'yak' has been typically used for the adult production of the infant/juvenile 'geck' vocalization. 'Geck' or 'gecker' is a common infant primate vocalization (see Jacobus and Loy, 1981; Patel and Owren, 2007) and is usually not produced by adults within these other species. Despite the alternative naming, it has long been suspected that the 'geck' and the 'yak' in baboons are equivalent. Certainly, we noted them in similar contexts, although infants have additional contexts (e.g., maternal rebuff). Our analysis suggests that the calls are the same, with acoustic structure differences due to the caller (i.e., age, size, development etc.). Meanwhile, after infancy the moan vocalization is no longer produced and chattering disappears at some point during the sub-adult stage.

In the literature, wahoos are typically differentiated between those made by adult females, juveniles and even sub-adult males from those produced by adult males, which are considered more stereotyped (e.g., Byrne, 1981; Fischer et al., 2002). These studies suggest that in adult female wahoos the 'hoo' is often missing or inaudible. We propose that these calls are more likely to be barks. Also, as the 'wa' of the wahoo is suspected to be ingressive (Gustison et al., 2012; also, personal observation – authors CK, TL and YB) but a bark is egressive, it is unlikely that these are the same vocalization and we therefore suggest that they should be more clearly differentiated in repertoires.

In a more general perspective on baboon's repertoire, detailing the vocal ethogram of Guinea baboons is a first step in better understanding the differences between the baboon taxa. It is important that full ethograms, including those from infants and juveniles, are reported for the other species so that we can better understand how the socio-ecological conditions, morpho-physiological and behavioral differences, as well as geographical variations, have affected vocal use for these closely related taxa.

5.2 On language evolution

The main strength of our study is the description of the acoustic parameters of the baboon's vocal productions, and the description of the ethological context in which these vocalizations were produced. In doing so, we followed a strategy which is not so different from language studies that try to map the acoustic features of the vocal production to meanings, as for example when phonology distinguishes the American words boat (/bot/) and bat (/bæt/) exclusively through the distinction between the /o/ and /æ/ vowel phonemes they contain. This approach suggests at least three lines of discussion regarding the evolution of language.

First, we note that the vocal repertoire of Guinea baboons is of a limited size (see McComb and Semple, 2005) for a species with a large social group size (Patzelt et al., 2011). The small repertoire of twelve vocalizations we report here is further constrained by the individual call types. That is, grunt-based vocalizations account for over half of the Guinea baboons' vocal repertoire. However, it appears that the baboons can increase their repertoire through the use of variability. Variability in vocal production occurred through changes of F0, tempo (calling rate), call duration, number of calls within a bout, and the combination of different vocalizations (e.g., the scream-yak-moan sequences of infants, bark-screams, double wahoos). More work is needed to identify whether the variability we observed in vocal production convey specific meanings. Addressing this question would require, for instance, comparing behavioral responses to long yak- or grunt series, in comparison to short series. The variations we observed in the baboon's vocalizations suggest that a first stage in language evolution might have been to introduce variations in the production and use of a limited set of vocal units, rather than expanding the number of different vocalizations. Considering context and variation within vocalizations may be essential to determine how nonhuman primates expand their limited repertoire to communicate with conspecifics, and to document the emergence of language.

Second, the analysis of the baboons' vocalizations has shown that several of them contain formants, and that these formants differ from one class of vocalization to the other (see Table 4). It has long been thought that nonhuman primates are incapable of producing sets of vowels-like sounds due to anatomical limitations (in particular, a too high larynx, Lieberman et al., 1969). The observation that baboons produce different vocal qualities, in different ethological contexts, shows that nonhuman primates can produce contrasting vowel qualities despite a high larynx (see Fitch et al., 2016 for converging results). This finding suggests homologies between baboons' vocalizations and human vowel systems, and more generally, that spoken languages could have evolved from an ancient vocal proto-system already present in our last common ancestor with baboons (Boë et al. 2017).

Third, Table 4 also reveals an interesting finding on language evolution. This table shows that F0 varied greatly both across (e.g., 64Hz for grunt 1 and 417 Hz for the wa (of wahoos)), and even within the vocalizations (417Hz for the wa- and 121Hz for the -hoo of the wahoos). In human languages, formants vary independently from laryngeal frequency, and the fundamental frequency of the baboons' vocal production was not as stable as found in speech. This finding suggests that the production of F0 and of formants could have been entangled during the early stages of language evolution. Clearer dissociations would have emerged later in the hominid lineage.

In summary, the data presented in this paper have two main functions. Firstly, this work aimed at serving as a reference guide for students of baboons' vocalizations and those interested in the communication systems of nonhuman animals. Furthermore, in documenting these aspects of baboons' vocal communication, this study also provides hypotheses on the emergence of speech. We believe that there is much to learn on these two aspects if this approach is replicated in other nonhuman primate species.

Supplementary material

Illustrative examples of the different vocalizations can be found at https://osf.io/nr2ye/

Acknowledgments

*Research supported by grants ANR-16-CONV-0002 (ILCB), ANR-11-LABX-0036 (BLRI) and ANR-11-IDEX-0001–02 (A*MIDEX).* Technical support from the staff of the Rousset-sur-Arc primate center is acknowledged.

References

Altmann, J., Altmann, S., and Hausfater, G. (1981). Physical maturation and age estimates of yellow baboons, *Papio cynocephalus*, in Amboseli National Park, Kenya. *American Journal of Primatology, 1*, 389–399.

Anderson, J.R., and McGrew, J.R. (1984). Guinea baboons (*Papio papio*) at a sleeping site. *American Journal of Primatology*, 6, 1–14.

Andrew, R.J. (1962). The situations that evoke vocalization in primates. *Annals of the New York Academy of Sciences*, *102*, 296–315.

Bermejo, M., and Omedes, A. (1999) Preliminary vocal repertoire and vocal communication of wild bonobos (Pan paniscus) at Lilungu (Democratic Republic of Congo). *Folia Primatologica, 70*, 328–57.

Bezerra, B.M., Souto, A.S., and Jones, G. (2010). Vocal repertoire of golden-backed uakaris (*Cacajao melanocephalus*): call structure and context. *International Journal of Primatology, 31*, 759–778.

Boë, L.J., Berthommier, F., Legou, T., Captier, G., Kemp, C., Sawallis, T.R., Becker, Y., Rey, A., and Fagot, J. (2017). Evidence of a Vocalic Proto-System in the Baboon (*Papio papio*) suggests Pre-Hominin Speech Precursors. *Plos One, 12(1)*, e0169321. doi.org/10.1371/journal.pone.0169321.

Byrne, R.W. (1981). Distance vocalisations of Guinea baboons (*Papio papio*) in Senegal: An analysis of function. *Behaviour*, *78*, 283–312.

Cheney, D.L., and Seyfarth, R.M. (2007). *Baboon metaphysics: the evolution of a social mind.* Chicago: University of Chicago Press.

Cheney, D.L., Seyfarth, R.M., and Palombit, R. (1996). The function and mechanisms underlying baboon 'contact' barks. *Animal Behaviour*, *52*, 1, 507–518.

Engh, A.L., Hoffmeier, R.R., Cheney, D.L., and Seyfarth, R.M. (2006). Who, me? Can baboons infer the target of vocalizations? *Animal Behavior*, *71*, 381–387.

Estes, R.D. (1992). *The behavior guide to African mammals. Including hooded mammals, carnivores, primates.* Berkeley: University of California Press.

Ey, E., and Fischer, J. (2011). Keeping in contact: flexibility in calls of olive baboons. In V. Sommer, and C. Ross (eds.), *Primates of Gashaka: socioecology and conservation in Nigeria's biodiversity hotspot.* (pp. 361–384), New York: Springer.

Fagot, J., Gullstrand, J., Kemp, C., Defilles, C., and Mekaouche, M. (2014). Effects of freely accessible computerized test systems on the spontaneous behaviors and stress level of social groups of Guinea baboons (*Papio papio*). *American Journal of Primatology, 76(1)*, 56–64. doi: 10.1002/ajp.22193.

Fagot, J., and Palleressompoule, D. (2009). Automatic testing of cognitive performance in baboons maintained in social groups. *Behavior Research Methods*, *41*, 396–404.

Fischer, J., and Hammerschmidt, K. (2002). An overview of the Barbary macaque, *Macaca sylvanus*, vocal repertoire. *Folia Primatologica*, *73*, 32–45.

Fischer, J., Hammerschmidt, K., Cheney, D.L., and Seyfarth, R.M. (2001a) Acoustic features of female chacma baboon barks. *Ethology, 107*, 33–54.

Fischer, J., Metz, M., Cheney, D.L., and Seyfarth, R.M. (2001b) Baboon responses to graded bark variants. *Animal Behaviour, 61*, 925–931.

Fischer, J, Hammerschmidt, K., Cheney, D.L., and Seyfarth, R.M. (2002). Acoustic features of male baboon loud calls: influences of context, age, and individuality. *Journal of the Acoustical Society of America, 111*, 1465–1474.

Fitch, W.T., de Boer, B., Mathur, N., and Ghazanfar, A.A. (2016). Monkey vocal tracts are speech-ready. *Science Advances*, 2, 12, e1600723.

Gros-Louis, J.J., Perry, S.E., Fichtel, C., Wikberg, E., Gilkenson, H., Wofsy, S., and Fuentes, A. (2008). Vocal repertoire of cebus capucinus: Acoustic structure, context, and usage. *International Journal of Primatology, 29*, 641–670.

Gustison, M.L., Le Roux, A., and Bergman, T.J. (2012). Derived vocalizations of geladas (*Theropithecus gelada*) and the evolution of vocal complexity in primates. *Philosophical Transactions of the Royal Society Biol. Sci., 367*, 1847–1859.

Hall, K.R.L., and DeVore, E. (1965). Baboon social behavior. In I. DeVore, (ed.), *Primate behavior*. (pp. 53–110). New York: Holt, Rinehart and Winston.

Hayes, V.J., Freedman, L., and Oxnard, C. E. (1990). The taxonomy of Savannah baboons: An odontomorphometric analysis. *American Journal of Primatology, 22*, 171–190.

Jacobus, S., and Loy, J. (1981). The grimace and gecker: a submissive display among patas monkeys. *Primates, 22*, 393–398.

Kitchen, D.M., Cheney, D.L., and Seyfarth, R.M. (2003). Female baboons' responses to male loud calls. *Ethology, 109*, 401–412.

Kummer, H. (1968). *Social organisation of hamadryas baboons*. Chicago: University of Chicago.

Lanyon, W.E. (1969). Vocal characters and avian systematics. In R.A. Hinde (ed.), *Bird vocalizations*. (pp. 291–310), Cambridge: Cambridge University Press.

Lieberman, P.H., Klatt, D.H., and Wilson, W.H. (1969). Vocal tract limitations on the vowel repertoires of rhesus monkey and other nonhuman primates. *Science*, *164*, 1185–1187.

Maciej, P., Ndao, I., Patzelt. A., Hammerschmidt, K., and Fischer, J. (2013a). Vocal communication in a complex multi-level society: constrained acoustic structure and flexible call usage in Guinea baboons. *Frontiers in Zoology*, *10*, 58–72.

Maciej., P, Patzelt, A., Ndao, I., Hammerschmidt, K., and Fischer, J. (2013b). Social monitoring in a multilevel society: a playback study with male Guinea baboons. *Behavioral Ecology and Sociobiology*, *67*, 61–68.

Maestripieri, D., Leoni, M., Raza, S.S., Hirsch, E.J., and Whitham, J.C. (2005). Female copulation calls in Guinea baboons: evidence for postcopulatory female choice? *International Journal of Primatology*, *26*, 737–758.

McComb, K., and Semple, S. (2005). Coevolution of vocal communication and sociality in primates. *Biology Letters*, *1*, 381–385.

Newman, T.K., Jolly, C.J., and Rogers, J. (2004). Mitochondrial phylogeny and systematics of baboons (*Papio*). *American Journal of Physical Anthropology*, *124*, 17–27.

Ouattara, K., Lemasson, A., and Zuberbühler, K. (2009). Campbell's monkeys concatenate vocalizations into context-specific call sequences. *Proceedings National Academy of Science*, *106*, 22026–22031.

Patel, E.R., and Owren, M.J. (2007). Acoustics and behavioral contexts if "gecker" vocalizations in young rhesus macaques (*Macaca mulatta*). *Journal of the Acoustical Society of America*, *121*, 575–585.

Patzelt, A., Zinner, D., Fickenscher, G., Diedhiou, S., Camara, B., Stahl, D., and Fischer, J. (2011). Group composition of Guinea baboons (*Papio papio*) at a water place suggests a fluid social organization. *International Journal of Primatology*, *32*, 652–668.

Ransom, T.W. (1971). Ecology and social behaviour of baboons (*Papio anubis*) at the Gombe National Park [dissertation]. Berkeley: University of California.

Ransom, T.W. (1981). *Beach troop of the Gombe*. East Brunswick: Associated University Press.

Rendall, D., Cheney, D.L., and Seyfarth, R.M. (2000). Proximate factors mediating "contact" calls in adult female baboons (*Papion cynocephalus ursinus*) and their infants. *Journal of Comparative Psychology, 114*, 36–46.

Rendall, D., Kollias, S., and Ney, C. (2005). Pitch (F_0) and formant profiles of human vowels and vowel-like baboon grunts: the role of vocalizer body size and voice-acoustic allometry. *Journal of the Acoustical Society of America, 117*, 944–955.

Rendall, D., Notman, H., and Owren, M.J. (2009). Asymmetries in the individual distinctiveness and maternal recognition of infant contact calls and distress screams in baboons. *Journal of the Acoustical Society of America, 125*, 1792–1805.

Rowell, T.E. (1966). Forest living baboons in Uganda. *Journal of Zoology, 149*, 344–364.

Semple, S., McComb, K., Alberts, S., and Altmann, J. (2002). Information content of female copulation calls in yellow baboons. *American Journal of Primatology, 56*, 43–56.

Sigg, H., Stolba, A., Abegglen, J-J., and Dasser, V. (1982). Life history of hamadryas baboons: physical development, infant mortality, reproductive parameters and family relationships. *Primates, 23*, 473–487.

Smith, J.M., and Szathmary, E. (1995). *The major transitions in evolution.* Oxford: WH Freeman.

Smuts, B.B. (1985). *Sex and friendship in baboons.* New York: Aldine de Gruyter.

Snowdon, C.T. (2004). Social processes in the evolution of complex cognition and communication. In D.K. Oller and U. Griebel (Eds.), *Evolution of communication systems: a comparative approach* (pp. 131–150). Cambridge: MIT Press.

Zuberbühler, K. (2012). Primate communication. *Nature Education Knowledge, 3*, 83.

Appendix 1. Glossary of terms used

Term	Meaning
Vocalization	Type of vocal production, with series of calls taken into account (i.e., copulation call). That is, a vocalization can either be comprised of a single call (e.g. wahoo) or a bout of calls which can be temporally connected (less than 1.5s apart) to be within the same vocalization (e.g. rhythmic grunts which are never produced as a single call)
Call	Individual unit within a vocalization (i.e., single grunt within a series).
Calling rate	Speed of call production within the bout (number of calls/s).
F0	Fundamental frequency. Measured in Hz.
Formant F1 F2	Acoustic resonances (first and second) of the vocal tract, affected by the position of the tongue, mouth cavity and lips. Measured in Hz.
Maximum frequency	The highest frequency (Hz) observable in our spectrograms.
Noise	Lacking harmonic structure
Harmonics	The simple periodic waves which make up the vocal signal, in which the F0 is the first harmonic and each subsequent harmonic repeats at the interval of the F0.

Appendix 2. List of the subjects involved in this study, their housing group, sex and age in months at the start of the study, as well as classification

The broad age classifications used (adult: 7+ years; sub-adult: 5–7 years; juvenile: 2–5 years; infant: < 2 years) were based on studies conducted on *P. cynocephalus* (Altmann et al., 1981) and *P. hamadryas* (Sigg et al., 1982). * indicates that these individuals moved up an age category during this study (age category given is that at the start of the study). ^ indicates that most of the vocalizations recorded for these individuals were after the move to the next age category. ° indicates that these individuals were selected for formant analysis. Any vocalizations of these individuals recorded around the time frame of their transition to the next category were carefully considered before analysis classification but we largely kept to the definition of category class.

Name	Group	Sex	Age (months)	Category
Pipo °	1	M	156	Adult
Vivien °	1	M	94	Adult
Bobo	1	M	73	Sub-adult
Dan	1	M	53	Juvenile*
Felipe	1	M	27	Juvenile
Filo	1	M	22	Infant*^
Grimm	1	M	12	Infant
Harlem	1	M	2	Infant
Kali °	1	F	204	Adult
Brigitte °	1	F	199	Adult
Michelle °	1	F	199	Adult
Mona °	1	F	186	Adult
Atmosphere	1	F	174	Adult
Petoulette °	1	F	162	Adult
Romy	1	F	149	Adult
Uranie °	1	F	104	Adult
Violette °	1	F	92	Adult
Angele	1	F	88	Adult
Arielle °	1	F	82	Sub-adult*
Dream	1	F	51	Juvenile*
Dora	1	F	49	Juvenile
Ewine	1	F	37	Juvenile
Fana	1	F	30	Juvenile
Feya	1	F	25	Juvenile
Flute	1	F	24	Juvenile
Hermine	1	F	6	Infant
Articho	2	M	82	Sub-adult*^
Barnabe	2	M	74	Sub-adult
Cloclo	2	M	66	Sub-adult
Cauet	2	M	65	Sub-adult
B06 °	3	F	332	Adult

Louis-Jean Boë[1], Thomas R. Sawallis[2],
Jöel Fagot[3], Frédéric Berthommier[1]
[1]*GIPSA-Lab, CNRS and Grenoble Alpes University,*
Saint-Martin-d'Hères, France
[2]*New College, The University of Alabama, Tuscaloosa,*
Alabama, United States of America
[3]*Cognitive Psychology Laboratory, CNRS and*
Aix-Marseille University, Marseille, France

What's up with Wahoo? Exploring Baboon Vocalizations with Speech Science Techniques

Abstract: The baboon repertoire includes around fourteen less vocalizations. One of them, the "wahoo", is referred to onomatopoetically, and is composed of three sounds in two syllables, which makes it interesting because it is the most complex baboon vocalization. We consider two hypotheses linking the baboon call and its human name: first, is there a demonstrable acoustical similarity with the human wahoo, and second, is there also further similarity related to articulation? We analyze both acoustic and articulatory information regarding why this vocalization is perceived phonetically as two separated syllables [wa.u]. This study corroborates the hypothesis of two equivalence levels: one acoustic-perceptual and the other related to the production mechanism in baboons and humans. This reveals an apparent similarity between a typical baboon vocalization and an utterance entirely typical of human languages, and thereby adds to the links between non-human primates vocalizations and human speech.

1. Introduction

Baboons are African and Arabian Old World monkeys which constitute the genus *Papio*, part of the subfamily *Cercopithecinae*. They can be classified into five species: *hamadryas*, *papio*, *anubis*, *cynocephalus*, and *ursinus*. They live in groups of 5 to 200 individuals and communicate with fourteen vocalizations (Hall and DeVore, 1965). It seems that except for *Papio hamadryas*, all baboons use comparable vocalizations. For half a century the vocalizations of baboons have been identified and associated with situations described ethologically (behavior and communication) (Hall and DeVore, 1965; Kemp et al., 2017; Zuberbühler, 2012). As with many animal sounds, baboon vocalizations are dependent on the sex, the status, and the age of the individual (e.g. copulation call for females, moans for infants). Vocalizations are named with reference to their production mode: barking, grunting, roaring, screeching, moaning, and yakking. One of them, wahoo, corresponds to a sequence of three sounds, perceived as [w], [a], and [u], while other vocalizations are simpler. This can be perceived by humans as the bisyllabic onomatopoeia "wahoo" (The same onomatopoeia occurs with different spellings in other languages.) Among the vocalizations of baboons, the grunt has been the most studied acoustically. It is a voiced call, in which the fundamental frequency (F0) can be accurately measured. As its spectrum displays the formant structure characteristic of vowels, it can be called a *vowel-like* vocalization (Owren et al., 1997). Wahoo has been less studied (Cheney et al., 1995; Cheney et al., 1996; Fischer et al., 2001; Fischer et al., 2002; Fischer et al., 2004; Kitchen et al., 2013; Maciej et al., 2013; Price et al., 2014).

We aim (i) to test the accuracy of the onomatopoeia of the call's name by acoustic and articulatory analysis (since the mimicry of other animal calls can differ greatly across languages), and (ii) to check the similarity of the production of the wahoo sequence as uttered by baboons and humans. More generally, this kind of study will help us better understand the specifics of communication by nonhuman primates. Observing such similarities and differences can also help in the search for any precursor elements of speech present in the vocalizations, and serve to infer the features of our common ancestral communication system (Fedurek and Slocombe, 2011; Fitch, 2002; Ghazanfar and Rendall, 2008; Zuberbühler, 2012). The comparison is especially interesting because of the phylogenic distance between baboons and humans, with a last common ancestor estimated at about 25 to 30 millions years ago. Better understanding of the analogy between human speech and a typical sequence of vocalizations in this distant phylogenetic cousin would add a valuable piece to the long history of studies comparing primate vocalizations to help illuminate the evolution of human oral language.

We conducted an analysis that led to formulation of two hypotheses: the first, based on acoustic and perceptual analysis, posits equivalence of acoustic and phonetic features for baboons and humans, and explains why we hear two separated syllables [wa.u]; the second posits, in addition, the equivalence of production processes between baboons and humans for this type of vocalization.

2. Data

Wahoos (also termed wahoo calls, double barks, loud two-syllable barks) are mainly produced by adult male baboons in various circumstances: danger situations (e.g. human or predator presence), intragroup aggression between adult males, attacks on females, or when splitting up a group into sub-groups out of visual contact (Byrne, 1981; Cheney et al., 1995; Rendall et al., 2000). The first part, {wa}, is similar to a bark, a typical alarm call produced by females (Fischer et al, 2001). The production of the first syllable {wa} has been described as pulmonic ingressive (Gustison et al., 2012), while the egressive nature of the bark, classically considered as similar, has been assumed but not investigated. The pulmonic ingressive mode of phonation is documented in many human languages (Eklund, 2004, 2008). However, it only occurs in paralinguistic contexts, and not phonemically. Furthermore, it has been little studied acoustically (Grau et al., 1995; Orlikoff et al., 1997). This ingressive production has also been noted in vocalizations of dogs, cows, horses, asses, and foxes, and also in purring in the domestic cat (*Felis catus*) and in the cheetah (*Acinonyx jubatus*) (Eklund et al., 2010; Peters, 2002). The second element, {hoo}, is more open than a grunt (i.e., a higher F1) and with a higher F0 as well (Boë et al, 2017).

We combined two types of data. The first set is a video, filmed in the wild and provided through YouTube (Larimer, 2012), of a chacma baboon P. ursinus, in which the wahoo is an indicator of male dominance (Fischer et al., 2002). The second set corresponds to 69 wahoos selected in the acoustic database of a previous study (Boë et al., 2017). Subjects were 3 adult males (6, 7, and 13 years) from a group of 24 guinea baboons (P. papio) living in semi-freedom, housed in a 750 m^2 outdoor enclosure at the CNRS primate center (Rousset-sur-Arc, France) (Fagot et al., 2014).

The acoustic database was crucial for providing F0 and spectral data in wahoo production by baboons compared with human speech (Section 3). The sample extracted from the video enabled articulatory analysis of great use in our articulatory analysis in Section 4.

3. Acoustical analysis

A short-window acoustic analysis was conducted to highlight the spectro-temporal structure of the representative example of wahoo extracted from the video. On the other hand a medium-window spectral analysis was conducted for the three segments {w}, {a} and {hoo} of the 69 wahoos in the database.

Figure 1: Acoustic analyses for wahoos from the video example (above, a – c) and from the database (below, d – f): (a) signal and F0 ({w} is initially unvoiced, then F0 for {wa} from Praat's autocorrelation routine and for {hoo}, from manually tagged periods); (b) spectrogram (5-ms window); (c) Welch spectrogram; (d – f) Welch spectrograms in grey (500-ms window) and their averages in bold for {w}, {a}, and {hoo}. P1, P2, & P3 correspond to spectral peaks. These are manually placed in (b – c) by analogy with (d – f).

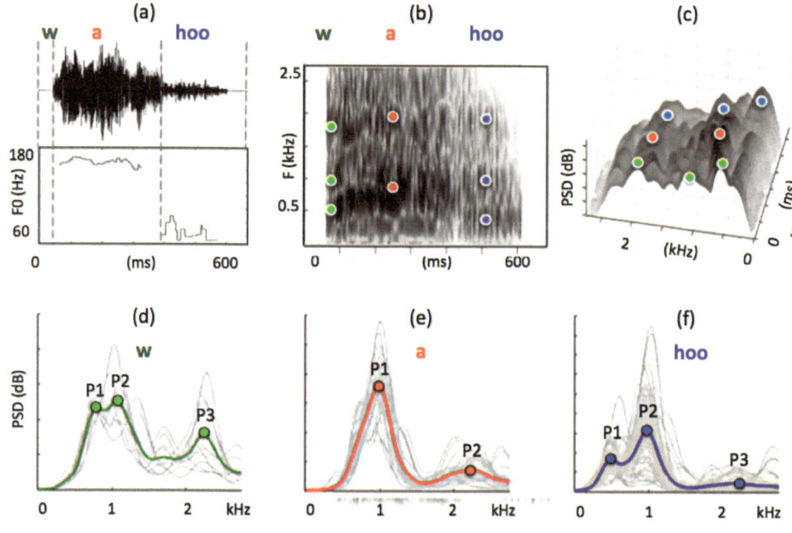

3.1 The video example

The analysis of the F0 was done by autocorrelation using the Praat software application (Boersma and Weenink, 2014) in the {wa} segment and by manually picking the fundamental periods for the {hoo} segment. The F0 plateau that makes up the majority of the {a} segment is centered around 170 Hz over a period of about 300 ms. The {hoo} part drops to around 60 Hz. The wahoo vocalization thus covers a range of about 1.5 octaves, a larger difference than typical wahoo produced by a human. A classic spectrogram obtained by the Praat software with a wideband 5-ms analysis window (Figure 1b) and a second Welch 3D spectrogram established for comparison between the example and the database analyses (Matlab®, Figure 1c), with a sliding window of 72 ms, allows identification of spectral peaks that may be either harmonics or formants. The spectral peaks shown in the 3D spectrogram have a similar complex spectro-temporal structure. They will be labeled by relying on the average spectra calculated from the database (Figure 1d–f).

3.2 Database

Each of the 69 wahoos of the database was divided into {wa} and {hoo} sections. Then the first 50 milliseconds of {wa} (corresponding to the initial F0 variation) were labeled {w} and the remaining was associated to the segment {a} (see an illustration in the sound from the video example in Figure 1a). These three segments {w}, {a}, and {hoo} were concatenated into separate files. For each of the three files, the Welch spectrogram (signal processing toolbox MatLab®) was generated after pre-emphasis (by differentiation) in long windows (500 ms) representing approximately two tokens of each segment. Finally overall average amplitude was determined in order to show the mean spectral peaks. The Welch spectrogram enables a better display of the global spectral structure. The spectra of {w}, {a}, {hoo} for each 500-ms segment vary widely (thin grey lines in Figure 1d–f), but the average of all segments (bold line) shows clear peaks, labeled P1, P2, & P3. These peaks were identified in the spectrograms (Figure 1b–c). Values of P1, P2 and P3 are given in Table 1, below, together with F0 values for each segment, measured with the method described above for the {w}, {a}, and {hoo} concatenated files.

Table 1: Data for wahoos from the database of 69 utterances by 3 male baboons. The duration of {w} is set arbitrarily to the first 50 ms of wahoo (thus, no variance in duration). For mean duration and mean F0, standard deviations are provided in parentheses.

label	{w}	{a}	{hoo}
total duration (s)	3.45	8.00	14.80
mean duration (ms)	50 (–)	116 (36)	214 (89)
P1 (kHz)	0.791	0.981	0.484
P2 (kHz)	1.054	2.275	0.972
P3 (kHz)	2.239	–	2.175
mean F0 (Hz)	333 (106)	308 (102)	109 (35)

4. Acoustico-phonetic question: Why do we hear [wa.u]?

We first wanted to understand why we perceive "wahoos" produced by baboons as the phonetic sequence [wa.u] (where the period represents a syllable break). Therefore, we compared the acoustic-phonetic characteristics of {wahoo} produced by baboons with a similar sound sequence produced by humans under several source conditions: (a) normal voice, (b) whispered voice, and (c) pulmonic ingressive voice on the first syllable, [wa] (and normal egressive voice on the [u]). This enables the comparison of F0 and spectral properties. As background, recall that (human) [w] is present in 74% of the representative sample of 451 human languages provided in the UPSID database (Maddieson, 1986). Phonetically, [w] is a glide, a voiced labio-velar approximant, which means that it is articulated with the back part of the tongue raised toward the soft palate, while rounding the lips. Acoustically (Calliope, 1989, pp. 118–119; Ladefoged, 2006; Potter et al., 1947, pp. 202–206), the first formant (F1) of [w] (around 0.3 kHz for a male speaker) is always more intense than both F2 (around 0.65 kHz) and F3 (around 2.5 kHz). F1 and F2 of [w] have a characteristic transition as an opening movement of the vocal tract (Figure 2). However, one can sometimes also perceive [wa] from a sequence [ua] in hiatus, i.e. when [u] is very short but without the rapid formant transition responsible for perception of [w] (Delattre, 1968).

Figure 2: Spectrogram of {wahoo} (Praat, 50 ms analysis window) (a) normal egressive speech, human male 1, (b) ingressive speech, human male 2, (c) whispered speech (i.e. with no laryngeal source, human male 2, (d) baboon, from the YouTube video. The four sets of peaks are manually placed in order to show their similarity below 1.5kHz.

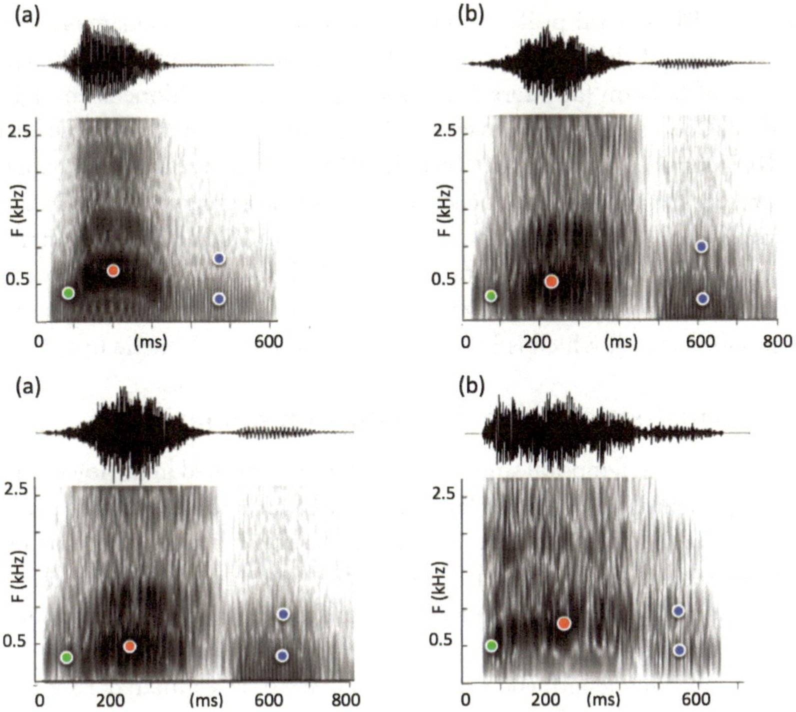

Hence, we analyzed [wa.u] sequences produced by a male speaker in normal modal voice or with ingressive [wa] (as in Figure 2 panels a & b). In our recorded human examples, the wahoo F0 contour covers a range of 0.75 octaves, much less than the 1.5 octaves in baboons. To simplify the spectral comparison between humans and baboons, we suppressed the signal above 1.5 kHz by a low pass filter, knowing that it would not significantly change the perception of the [a] (Delattre et al., 1952). We observe (with short-term windowing of 5 ms) that in the lower part of the spectrum the two spectro-temporal structures are quite similar. We hand-marked the spectral peaks that are common to the four utterances. The first peak of the speaker's [w] corresponds to the first formant F1. Along the first 50 ms, we observe changes in human speech that correspond to formant transitions proper, whereas for baboon vocalization, the trajectory involves also a spectro-temporal variation similar to a formant transition. We conjecture that human listeners perceptually infer a formant transition from this trajectory. The P1 spectral peak for baboon wahoo in {a} is also similar to the first formant of the speaker's [a], and in continuity with the P1 of {w}. In the case of baboon {a} filtered at 1.5 kHz, the P1 peak alone is enough to get the [a] percept, as is the case with F1 [a] for the human voice (Delattre, 1968). Then the continuity between the P2 of {w} and the P2 of {a} which is not present for baboon, is not necessary for the [a] perception. The P1 & P2 peaks for {hoo} uttered by baboons correspond to F1 and F2 for human [u] in all cases. This allows us to establish that all the acoustic features observed in the baboon {wahoo} vocalization can adequately explain the phonetic perception [wa.u], which is in line with our acoustic-phonetic hypothesis.

5. Articulatory question: Is production similar?

The spectral and temporal similarities that we established in the lower range of the spectrum (below 1.5 kHz), did not take into account the P2 peak of {a} around 2.3 kHz (Table 1). To explain the correspondence of this peak with the second formant in [a] we have to invoke an acoustic production model (Boë et al., 2013). This model lets us determine the maximal acoustic space within which the first two formants of all vowels occur for a vocal tract of the given length (Boë et al., 2017). We have estimated 13.4 cm as the length of the vocal tract of a male baboon. Figure 3 shows (in grey) the maximum acoustic space (MAS), which is the area enclosing all paired (F1, F2) values that could theoretically be produced by such a vocal tract, assuming total control of its shape and configurations (see Boë et al., 2017, for discussion). The spectral peaks P1 and P2 of {w}, {a}, and {hoo}, provided in Table 1 and interpreted as formants, are shown as they would be located inside this maximal acoustic space.

Figure 3: F1 and F2 for [w], [a], and [u], as shown in the maximum acoustic space of a 13.4 cm vocal tract, corresponding to that of an adult male baboon.

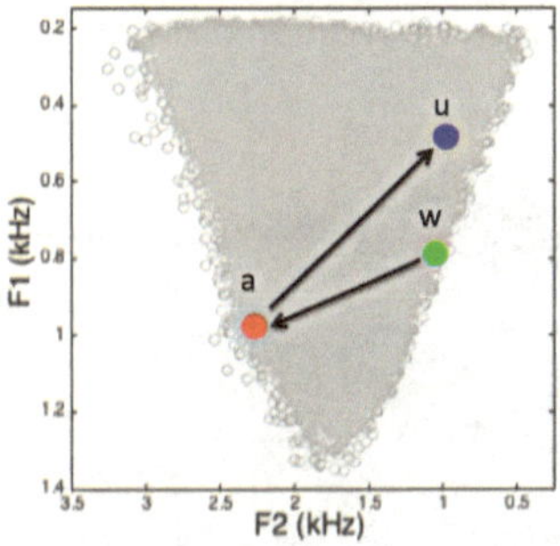

We observed that, in agreement with our production hypothesis, {a} uttered by baboons would correspond to an open front vowel. The {hoo} is located where we would expect the mid-high back rounded [o]. The fact that it does not seem to correspond really to the high back rounded [u] might be due to the fact that lips are not sufficiently closed and protruded. For {w}, the first peak, P1, is in continuity with P1 of {a} whereas the continuity between P2 of {w} and P2 of {a} is absent. Though this means we do not have an actual trajectory for [wa.u] in F1-F2 formant space, we have indicated an approximate trajectory schematically by arrows (Figure 3).

Figure 4: Distances a and b measure the displacement of the thoracic cage; distance c measures lip opening; distance d is from the center of the ear canal to the distal extremity of the upper lip and quantifies lip protrusion (images 6 and 15 of the video).

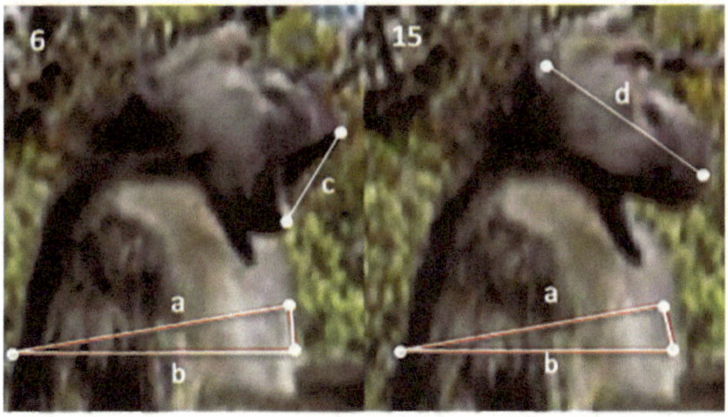

Figure 5: Articulatory analysis of the sample of "wahoo" extracted from the video. Synchronization of sound and image with spectrum (Praat, 50 ms analysis window), lip opening, rib cage variations, and lip protrusion (fitted with a 3rd degree polynomial curve).

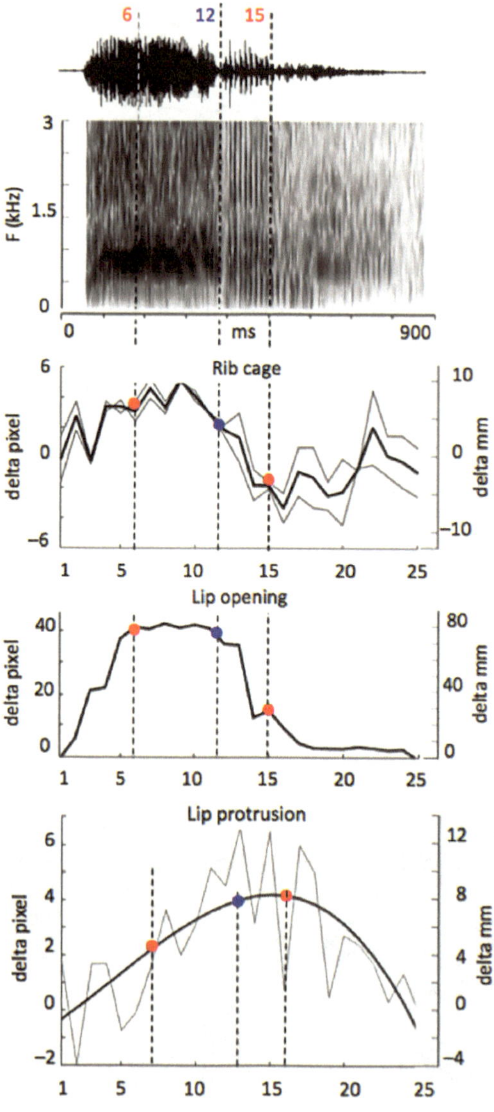

We also directly studied the wahoo production on the video clip. We manually estimated and reported three types of measurements from the video, image by image: (a) lip opening, (b) movement of the rib cage, and (c) lip protrusion (Figure 4). These measurements are synchronized with the sound, so as to allow analysis of the production sequence. From anatomical data (Boë et al., 2017), knowing that distance between the center of the ear canal and upper lip in the rest position is about 17 cm, we estimated the conversion factor at 5 pixels/cm. The lip opening gesture takes about 180 ms (5.5 images) and includes {w} and the beginning of {a}, with lips plateauing until its end (image 12), when the lips close to the beginning of the {hoo} (image 15). Ingressive airflow is shown by an increase of thoracic volume synchronized with {wa} and a plateau that ends at the beginning of {hoo} (image 12) together with the closing of the lips. The lip protrusion measurement is highly variable or noisy, with a trajectory fit (using a 3rd degree polynomial curve) which reaches its maximum when the lips close (image 15).

For the 3 types of measurement we can therefore estimate, relative to the rest position, a range of variation through 1 cm for lip protrusion, 8 cm for lip opening, and 1 cm for the rib cage movement (Figure 5). The {w} lip opening gesture is accompanied by a rise in F0 as well as a rise in the first formant. This effect is well-known for the [w] (Potter et al., 1947) and for stop consonants in general (Wang and Fillmore, 1961), and reflects the strong coupling between source and vocal tract. The fundamental frequency is correlated with the airflow at the glottis and the transition of the first formant corresponds to the opening of the vocal tract. The measurement of the protrusion supports the hypothesis of lip protrusion for {wa}. This protrusion also seems to involve a forward projection of all the tissue of the muzzle.

Overall, except for the difficulty of showing a clear F2 transition for {w}, our hypothesis of similar production mechanisms for baboon vocalizations and human speech is supported by our observations, with the caveat of course that pulmonic ingression is only paralinguistic for human speech.

6. Conclusions

There is an increasing interest from various research communities in animal vocalizations that are used for communication purposes. This is particularly true for researchers in primatology, due partly to increased interest in language evolution. We have here compared an understudied baboon vocalization, the "wahoo", and its onomatopoetic name from human speech. We have used standard methods, commonplace in speech research, to analyze and compare certain aspects of baboon productions to similar processes in human speech. In particular, we used acoustic analysis of F0 and of spectral characteristics of the baboon wahoo to understand how it is likely perceived by humans, and we showed that those acoustic traits indeed provide support to its onomatopoetic name. We also used a video of a baboon producing a wahoo to extract quantitative articulatory data allowing us to understand several interesting aspects of the baboon's production mechanism, and we showed that many of them are quite similar to human speech production mechanisms. We also verified that ingressive vocalization, which is found paralinguistically but is unusual in human speech, is common in the baboon wahoo. We thus transcribe baboon wahoo as [wa↓.u↑] with the ↓ down arrow indicating an ingressive initial syllable, and ↑ for the egressive second syllable.

We believe we have demonstrated that standard phonetic and acoustic methods developed for speech can be profitably used for the analysis of vocalizations in non-human animals, and we recommend further exploratory efforts in the same vein.

Acknowledgements

The authors are grateful to: Tom Larimer for the YouTube baboon video; Thierry Legou, Arnaud Rey, Caralyn Kemp, Yannick Becker for recording and labeling the vocalizations of *Papio papio* baboons; Guillaume Captier for the anatomical study; and Pierre Badin, Pascal Perrier, and Jean-Luc Schwartz for helpful discussions. This research was funded partly by ANR SkullSpeech grant ANR-13-TECS-0011 "e-SwallHome-Swallowing & Respiration: Modelling & e-Health at Home" in the "Technologies pour la Santé" program.

Research supported by grants ANR-16-CONV-0002 (ILCB), ANR-11-LABX-0036 (BLRI) and ANR-11-IDEX-0001–02 (A*MIDEX). Technical support from the staff of the Rousset-sur-Arc primate center is acknowledged.

7. References

Boë, L.J., Badin, P., Ménard, L., Captier, G., Davis, B., MacNeilage, P., Sawallis, T.R., and Schwartz, J.L. (2013). Anatomy and control of the developing human vocal tract: A response to Lieberman. *Journal of Phonetics*, *41*, 379–392.

Boë, L.J., Berthommier, F., Legou, T., Captier, G., Kemp, C., Sawallis, T.R., Becker, Y., Rey, A., and Fagot, J. (2017). Evidence of a vocalic proto-system in the baboon (Papio papio) suggests pre-hominin speech precursors. *PLoS One, 12*, e0169321.

Boersma, P., and Weenink, D. (2014). *Praat: doing phonetics by computer* [Computer program]. Version 5.3.63, retrieved 24 January 2014 from http://www.praat.org/

Byrne, R.W. (1981). Distance vocalisations of guinea baboons (Papio papio) in Senegal: an analysis of function. *Behaviour, 78*, 283–312.

Calliope (1989). *La parole et son traitement.* Paris: Masson.

Cheney, D.L., Seyfarth, R.M., and Palombit, R. (1996). The function and mechanisms underlying baboon 'contact' barks. *Animal Behavior*, *52*, 507–518.

Cheney, D.L., Seyfarth, R.M., and Silk, J.B. (1995). Responses of female baboons (Papio cynocephalus ursinus) to anomalous social interactions: evidence for causal reasoning? *Journal of Comparative Psychology*, *109*, 134–141.

Delattre, P., Liberman, A.M., Cooper, F.S., and Gerstman, L.J. (1952). An experimental study of the acoustic determinants of vowel color; observations on one- and two-formant vowels synthesized from spectrographic patterns. *Word*, *8*, 195–210.

Delattre, P. (1968). From acoustic cues to distinctive features. *Phonetica*, *18*, 198–230.

Eklund, R. (2004). Pulmonic ingressive speech: a neglected universal? *TMH-QPSR, KTH*, *50*, 21–24.

Eklund, R. (2008). Pulmonic ingressive phonation: diachronic and synchronic characteristics, distribution and function in animal and human sound production and in human speech. *Journal of the International Phonetic Association*, *38*, 235–324.

Eklund, R., Peters, G., Weise, F., and Munro, S. (2010). A comparative acoustic analysis of purring in four cheetahs. FONETIK 2012, Department of Philosophy, Linguistics and Theory of Science, University of Gothenburg, The XXVth Swedish Phonetics Conference, 41–44.

Fagot, J., Gullstrand, J., Kemp, C., Defilles, C., and Mekaouche, M. (2014). Effects of freely accessible computerized test systems on the spontaneous behaviors and stress level of guinea baboons (Papio papio). *American Journal of Primatology*, *76*, 56–64.

Fedurek, P., and Slocombe, K.E. (2011). Primate vocal communication: a useful tool for understanding human speech and language evolution? *Human Biology*, 83, 153–173.

Fischer, J., Hammerschmidt, K., Cheney, D.L., and Seyfarth, R.M. (2001). Acoustic features of female chacma baboon barks. *Ethology*, *107*, 33–54.

Fischer, J., Hammerschmidt, K., Cheney, D. L., and Seyfarth, R.M. (2002). Acoustic features of baboon loud calls: influences of context, age, and individuality. *Journal of the Acoustical Society of America*, *111*, 1465–1474.

Fischer, J., Kitchen, D., Seyfarth, R.M., and Cheney, D.L. (2004). Baboon loud calls advertise male quality: acoustic features and their relation to rank, age, and exhaustion. *Behavioral Ecology and Sociobiology*, *56*, 140–148.

Fischer, J., Metz, M., Cheney, D.L., and Seyfarth, R.M. (2001). Baboon responses to graded bark variants. *Animal Behaviour*, *61*, 925–931.

Fitch, W.T. (2002). Comparative vocal production and the evolution of speech: reinterpreting the descent of the larynx. In A. Wray (Ed.) *The transition to language* (pp. 21–45). Oxford: Oxford University Press.

Ghazanfar, A., and Rendall, D. (2008). Evolution of human communication. *Current Biology*, *18*, R457–R460.

Grau, S.M., Robb, M.P., and Cacace, A.T. (1995). Acoustic correlates of inspiratory phonation during infant cry. *Journal of Speech and Hearing Research*, *38*, 373–381.

Gustison, M.L., Le Roux, A., and Bergman, T.J. (2012). Derived vocalizations of geladas (Theropithecus gelada) and the evolution of vocal complexity in primates. *Philosophical Transactions of the Royal Society, B367*, 1847–1859.

Hall, K.R.L., and DeVore, I. (1965). Baboon social behavior. In I. DeVore (Ed.) *Primate Behavior: Field Studies of Monkeys and Apes* (pp. 53–110). New York: Holt, Rinehart and Winston.

Kemp, C., Rey, A., Legou, T., Boë, L.J., Berthommier, F., Becker, Y., and Fagot, J. (2017) Vocal repertoire of captive guinea baboons (*Papio papio*). In this volume.

Kitchen, D.M., Cheney, D.L., Engh, A.L., Fischer, J., Moscovice, L.R., and Seyfarth, R.M. (2013). Male baboon responses to experimental manipulations of loud 'wahoo calls': testing an honest signal of fighting ability. *Behavioral Ecology and Sociobiology, 67*, 1825–1835.

Ladefoged, P. (2006). *A course in phonetics.* Belmont: Thomson, Wadsworth.

Larimer, T.S. (2012). *The wa-hu shout.* https://www.youtube.com/watch?v=za839cpwUh0

Maciej, P., Ndao, I., Patzelt, A., Hammerschmidt K., and Fischer, J. (2013). Vocal communication in a complex multi-level society: constrained acoustic structure and flexible call usage in guinea baboons. *Frontiers in Zoology, 10*, 58–72.

Maddieson, I. (1986). *Patterns of sounds.* Cambridge: Cambridge University Press.

Orlikoff, R.F., Baken, R.J., and Kraus, D.H (1997). Acoustic and physiologic characteristics of inspiratory phonation. *Journal of the Acoustical Society of America, 102*, 1838–1845.

Owren, M.J., Seyfarth, R.M., and Cheney, D.L. (1997). The acoustic features of vowel-like grunt calls in chacma baboons (Papio cyncephalus ursinus): implications for production processes and functions. *Journal of the Acoustical Society of America, 101*, 2951–2963.

Peters, G. (2002). Purring and similar vocalizations in mammals. *Mammal Review, 32*, 245–271.

Potter, H.C., Kopp, R.K., and Green, G.A. (1947). *Visible speech.* New York: D. Van Nostrand Company, Inc.

Price, T., Ndiaye, O., Hammerschmidt, K., and Fischer, J. (2014). Limited geographic variation in the acoustic structure of and responses to adult male alarm barks of African monkeys. *Behavioral Ecology and Sociobiology*, *68*, 815–825.

Rendall, D., Cheney, D.L., and Seyfarth, R.M. (2000). Proximate factors mediating 'contact' calls in adult female baboons (papio cynocephalus ursinus) and their infants. *Journal of Comparative Psychology*, *114*, 36–46.

Wang, W.S., and Fillmore, C.J. (1961). Intrinsic cues and consonant perception. *Journal of Speech and Hearing Research*, *4*, 130–136.

Zuberbühler, K. (2012). Primate communication. *Nature Education Knowledge*, *3*, 83.

Adriano R. Lameira[1,2]
[1]*Department of Psychology and Neuroscience, St. Andrews University, St. Andrews, Scotland, UK*
[2]*Department of Anthropology, Durham University, Durham, UK*

Origins of Human Consonants and Vowels: Articulatory Continuities with Great Apes

Abstract: Science has a very pale idea of how proto-speech sounded like and how it was composed. The probability of an accurate reconstruction of speech and language evolution is, however, vanishingly small without such information. The original nature, form, and function of the building blocks of proto-speech directly determined which communicative operations and linguistic computations were possible. Knowledge about the sounds that composed the first syllables and words can offer, thus, insight into the chains of events that launched language evolution. Primate bioacoustics in our closest relatives – nonhuman great apes – represents a rich source of information on the probable composition of the ancestral great ape call repertoire that predated human speech. Here, I illustrate how the long-term inventory of the wild and captive call repertoire in orang-utans and African apes has unveiled a deep articulatory homology between great ape voiceless and voiced calls on the one hand (i.e. with and without vocal fold action as sound source, respectively) and human consonants and vowel on the other. This articulatory parallel offers a clearer view over the basic sounds that composed the "mother tongue" of all the world's spoken languages. The presence of proto-vowels and proto-consonants in the last common hominid ancestor spawns new questions regarding the steps that made up the process of speech and language evolution and their relative timing.

Keywords: proto-speech, orang-utans, African apes, call repertoire

1. Introduction

Given how fundamental language is to what defines us as humans, we know surprisingly little about its origin. Historically, the birth of language was something of remarkable. Since the first civilizations and across cultures over millennia, speech was a trait granted by divine forces and that separated humans from other animals (e.g. "In the beginning was the Word and the Word was with God and the Word was God", John 1:1). This theme is, for example, covered in manuscripts from the Indus Valley; the oldest philosophical essays that survive until present, dated to the ninth century BCE and predating Classical antiquity by hundreds of years (Favareau, 2010). An interesting reflexion of these concepts in folk culture can be found in fables and myths, where talking animals held a human mind and mental capacities (e.g. the Big Bad Wolf of Red Ridding Hood was Machiavellian, capable of deception, and disguise) and other uniquely human behaviours, such as standing upright and wearing clothing. On the other hand, human-like creatures that lacked speech preserved their monstrosity (e.g. Cyclops of Greek mythology).

In this chapter, I will lay out recent lines of evidence for the presence of evolutionary raw material for the emergence of human consonants and vowels among *nonhuman great apes* (hereafter *great apes*) – our closest living relatives – with a special focus on orang-utans (*Pongo* spp)[1]. Historically, these data represent a tipping-point in the theory of language origins; the way we view and think about human evolution, the emergence of our most diagnosing behaviour, and the origins of the most advanced communication system known in the natural world. This is an exciting period to be in in the field of language evolution, where both new and senior generations of researchers can overcome centuries-old mythical notions and pose, test, and advance new scientific hypothesis, and in this manner, make significant contributions to solving one of the oldest puzzles in human thought.

The plum tree and the plum seed

One of the dominant approaches to date to the study of language evolution has been through the identification of basic cognitive operations and computational mechanisms that underpin language use *today*. This approach has sought identifying what is truly and uniquely human and that makes language possible. The fundamental capacity for compositional

---

1 This chapter explores and focuses on possible parallels in articulation and acoustics, and remains agnostic on interpretations of call function and meaning (new efforts on this front have been done elsewhere (Schlenker et al., 2016)). A separate approach to these aspects of language evolution, i.e. articulation vs. meaning, has been defended in the Frame/Content Theory by MacNeilage (1998).

and combinatorial syntax has emerged as the key feature presumed to have transformed us into the talking ape (Hauser et al., 2002). From this position, scientists have searched for similarities within the primate order, but with little success (Fitch and Hauser, 2004) and some of its proponents have declared this approach to be an heuristic dead-end (Hauser et al., 2014).

Because human language *today* is so sophisticated, the weakness in this traditional approach can be understood with the aid of the following image. Think of a mature plum tree, or the highest sequoia. It is nearly impossible to conceive how such a tree could have arisen from its minute seed, particularly so without a priori knowledge about plant development. When we observe the tree, we see fruit, flowers, and photosynthesis. Yet, none of these features can be found back in the seed. Surely, then, the tree did *not* come from its seed!

A new approach to language evolution, resorting to comparative primate research, is seeking to study the development of the seed of language, starting from the seed's end and study its development forward in time, instead of starting from the fully grown tree and working backwards through its development. In this manner, we might be able to identify the conditions under which the seed germinates and takes the first roots, which structures capable of photosynthesis emerge first, and which shapes the first leaves take on. With this information, we might just be able to reconstruct how the language kernels that lay dormant within great apes may have become, over evolutionary time, the tall tree of human language.

What constitutes a probable precursor of speech?

Speech is fundamentally learned. Similar to cells that are substituted and renewed without the loss of their comprising tissues and organs, so too are languages. They are culturally renewed one generation after the other, through vocal learning by the young cohorts of their speakers. Each child, as a new member of a linguistic community, needs to receive acoustic input and feedback and will, through these means, learn to acquire the sounds that constitute her mother tongue.

Accordingly, it is helpful to assume that a particular primate call behaviour represents a conceivable precursor of a speech sound when, *at least*, that behaviour is the result of vocal learning. Learned calls contrast with other calls that are in essence genetically inherited, often termed *innate*. Although this dichotomy is not always clear-cut (Fitch, 2010), it is heuristically valuable as an entry point for screening primate call repertoires for potential language precursors. Innate calls emerge *inevitably* in normally developing individuals, without the need for relevant auditory input. Conversely, the process of acquisition of a new call into one's repertoire through vocal learning is underpinned and primarily driven by auditory input (Owren et al., 2011). *This input is mandatory*. Without it, no acquisition of new calls can occur and vocal learning can, then, manifest in two ways: via the capacity to acquire new calls (expanding in this way one's vocal repertoire) or via the ability to modify a call previously acquired (Pisanski et al., 2016).

The more we retreat in time along primate phylogenetic branches, the larger the uncertainty about how influential an ancient behaviour was in the process of language evolution. A higher level of attention and study should centre, thus, on vocal learning in great apes. Due to their close relatedness to our own species, their vocal behaviour and capacities will likely provide us with pertinent and plausible hypotheses for language evolution and a comparative platform to perhaps further explore older primate precursors. Among great apes, orang-utans have emerged as a particularly surprising species. Below, I review the most recent findings in this genus and how they relate with the latest studies involving African apes.

2. Are orang-utans a suitable model species for studying language evolution?

In the early 2000s, the study of the call repertoire of wild orang-utans started in its earnest in the swamps of the central province of Indonesian Borneo by a team of Dutch, Swiss, and Portuguese researchers (Hardus et al., 2009). Until then, all information that was available derived from pioneer work done back in the 60s and 70s (Mackinnon, 1974; Rijksen, 1978). These references sometimes lacked spectrograms and re-assessment had to be done strictly on the basis of written descriptions, which proved to be particularly challenging.

The reason why the first modern comprehensive description of the orang-utan call repertoire took decades longer to be published than, for example, the landing of the first man on the moon, was in part due to a known relationship between primate sociality and vocal complexity (McComb and Semple, 2005). This relationship postulates that the larger the typical group size of a primate species, the larger and richer its call repertoire tends to be. It followed from this correlation that orang-utans, being the only diurnal primate to exhibit solitary tendencies (Delgado and van Schaik, 2000), were predicted to produce a very small range of vocalizations and sounds. As it became apparent during the work in the swamplands of Borneo, this correlation fell short in explaining the high diversity and richness of the orang-utan call repertoire.

As the earliest diverging hominid genus, orang-utans were often argued by some to offer close to no insight to the question of language evolution. Some scholars contended this was because orang-utans show the least level of genetic similarity with humans. Interestingly enough, this is in fact inaccurate (!), specifically, with regards to the genetic mutations that presumably played a critical role in language evolution – mutations associated with the gene encoding the forkhead box protein 2 (*FOXP2*)(Enard et al., 2002).

African apes show a difference of two amino acid substitutions in this gene with humans, but a difference of three substitutions with orangutans. That is, orangutans do not share one amino acid difference that is common to all African apes. Thus, orangutans exhibit an extra amino acid substitution that is unique to *Pongo*. Because *FOXP2* gene is extremely conserved in mammals (Enard, 2011), the last great ape common ancestor exhibited in all likelihood an African-like *FOXP2* genetic profile, similar to that found in chimpanzees, for instance. This gene went on to undergo subsequent amino acid substitutions only in *Homo* and *Pongo* lineages – making orangutans *at least* as good model species as any African ape. If the number of mutations on FOXP2 can be assumed to reflect stronger selective pressures for vocal evolution, orang-utans stand then as *the* most promising model species among great apes for the study of language evolution.

One further reason why orang-utans make excellent model species for the study of language evolution derives directly from their vocal behaviour. It is in this domain that most progress has been made regarding our understanding of the sounds and structure of the human ancestral language.

3. Vocal differences between the Asian great ape and its African cousins (gorillas, bonobos, and chimpanzees)

3.1 Dialects

During the process of cataloguing the call repertoire of wild orang-utans, a remarkable feature of their vocal communication system transpired. Different populations produced different calls in the same context, whereas other populations used no call whatsoever in those same circumstances (Krützen et al., 2011; van Schaik et al., 2003; Wich et al., 2012) (Fig. 1). This pattern is identical to what classifies as dialects in humans (Lameira et al., 2010), where synonym words are used alternatively in different locations of the same language. For instance, the use of "trousers" and "pants" between British and American English respectively. In orang-utans, these are calls, for instance, that mother orang-utans produce to call their infants during travelling and calls produced during nest construction by adult individuals each day. As shown in Figure 1, while orang-utans produced raspberries (consisting of blowing air through pursing lips) during nest construction in Suaq and Sabangau – two far-distant research sites across the Karimata Strait between Sumatra and Borneo – nest smacks (seemingly produced in an similar way as a click consonant, with the tongue quickly shooting away from the palate) were produced instead up river from Sabangau, in Tuanan – a population of the same subspecies and likewise living in peat swamp. At the same time, orang-utan mothers produced local-specific throat scrapes towards their infants in Tuanan, and harmonic uhh's in Ketambe. Other populations in the same island of the same (sub)species produced no call in the same context.

Figure 1: First letter code refers to the kind of nesting call (R='Raspberry'; S='Nest Smack'; – = no call). Second letter code refers to the mother-infant call (U='Harmonic uuh'; T='Throat scrape'; – = no call). (Reproduced from Wich et al., 2012).

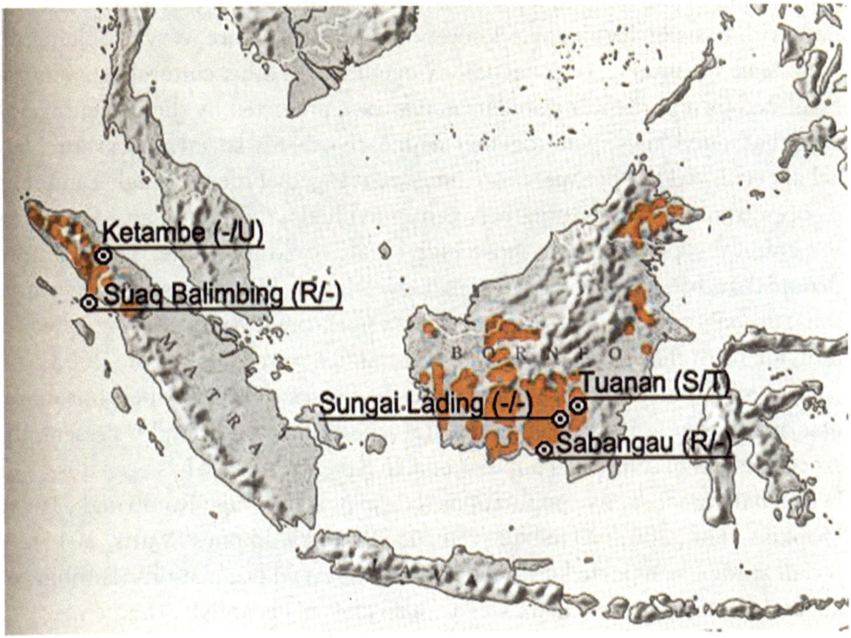

The geographic distribution of these "present1/present2/absent" patterns, or "synonym dialects" in orang-utans – where and which calls are present or absent – was not sufficiently explained by genetic or ecological divergence (Krützen et al., 2011). That is, levels of genetic and ecological dissimilarity between populations and sites did not correlate with the level of dissimilarity between the call repertoires of those populations (Wich et al., 2012).

In fact, differences of this nature would be nearly impossible to explain theoretically in terms of genetics or ecology. There is no knowledge of genetic mutations that turn on, or off, or replace specific calls strictly in one context and, simultaneously, leave the remaining of the call repertoire unmodified at the same time. On the other hand, orang-utan mother calls and orang-utan nest calls exhibit no ecological requirements. An orang-utan mother usually calls for her infant when the latter falls behind while travelling across the forest canopy, regardless of the species of tree they are crossing at that particular moment. Equally, orang-utans build nests across a huge range of different trees species *within a single site*. To explain the production of different mother calls, or nest calls, or no call at all *between sites* with basis on forest species composition is, therefore, very problematic.

Instead, the number of dialect calls (together with other cultural behaviours) exhibited by a particular population was best predicted by the percentage of time that individuals spent together with each other in social association (van Schaik et al., 2003). Because more time spent together meant a higher number of opportunities for learning between individuals, this led to an increase in the cultural repertoire that those individuals could assemble. These results denote that calls (when present) were indeed highly likely learned. They were, thus, probably maintained as local traditions through vocal learning between individuals of the same "linguistic" community (van Schaik et al., 2003).

Synonym dialects are still to be described among African apes, but many descriptions exist of a more fundamental type of dialect – "simple" present/absence patterns of call across populations. In *Pan*, this pattern has been observed between different captive populations (Hopkins and Savage-Rumbaugh, 1991; Hopkins et al., 2007; Marshall et al., 1999) and wild ones (Watts, 2015). A recent study has reported it in *Gorilla* between wild populations (Robbins et al., 2016), and some isolated cases are also known in captivity (Lameira et al., 2014; Perlman and Clark, 2015). Altogether, this evidence reflects the capacity of all great ape genera to generate and maintain local, population-specific vocal traditions (Robbins et al., 2016; van Schaik et al., 2003; Whiten et al., 1999). Accordingly, dialects seem to exhibit more complex patterns in orang-utans than in African apes and synonym dialects have only been described in wild orang-utans, thus far (Wich et al., 2012). However, a renewed interest and continued examination of the African repertoire will almost certainly reveal variation that has remained hitherto overlooked.

3.2 Accents

Data currently available suggest that chimpanzees and bonobos (gorilla data are much in need!) exhibit, instead, a richer variation in terms of what is characterized as *accents* in humans. Accents describe differences in the "pronunciation" of the same call, such as the difference between how

British English and American English speakers pronounce "tomato". This level of variation in African apes seems richer than what has been described so far in orang-utans, even though comparisons between call use and their possible function *between* species is highly challenging, also because the large majority of calls are not shared between species (Clay et al., 2011; Fedurek et al., 2015; Slocombe and Zuberbuhler, 2007).

In chimpanzees this level of variation has been shown to denote that vocal learning is involved in call production (Watson et al., 2015). This was exemplarily demonstrated in a unique study that managed to record chimpanzee food calls before and after a group housed in the Netherlands moved to the Zoo of Edinburgh (Watson et al., 2015). The food call variant directed for "apples" by the individuals at each group was, before the merge of the two groups, acoustically different. However, once the Dutch group moved to Scotland, their "apple" call gradually shifted over the period of some months to converge and become acoustically indistinguishable from the variant produced by the resident group (Watson et al., 2015). Characteristically, accents can thus manifest in the form of call convergence/divergence between groups and this has also been demonstrated to occur between individuals in wild chimpanzees (Mitani and Gros-Louis, 1998).

Accent differences tend to be more difficult to be detected by ear than dialects. While dialects are composed of distinct calls that facilitate their identification, differences in accent occur within one single call type and are therefore subtler. Experience (in the form of many hours of observation) helps a great deal in the audible detection of accent differences and acoustic analyses are typically required to verify the occurrence of accents (much as, similarly, hours of experience with a particularly novel language starts to allow us to understand where word and sentence cut-off points lay).

I recall a conference when a colleague, who is an expert in chimpanzee calls, gave a presentation where two different chimpanzee calls were played out to the audience. The presentation slides showed that the two calls differed in acoustics and in the way chimpanzees used either call, but I was baffled to how I was nearly incapable of detecting these differences in sound by ear. To someone who has made a scientific career by describing variation and differences between call types in orang-utans, this was admittedly embarrassing for me! This was a clear example of how there may possibly exist a quantifiable difference between the degree of graded-ness between orang-utans and their African cousins.

This called my attention to the fact that the vocal features exhibited by each species must be interpreted within the context of their natural call behaviour and the nature of their repertoire. Chimpanzees and bonobos, for example, exhibit a *graded* call repertoire where acoustic transitions between different calls are fine, sometimes elusive, but imbued with straightforward and powerfull differences in function. Receivers will react very differently to call variants that will sound much the same to an inexperienced human observer. Conversely, orang-utans exhibit a more *categorical* call repertoire. This means that functional differences between calls predominantly involve the use of clear-cut acoustic differences that unambiguously demark two different call types.

Under this light, a richer dialect or accent variation does not necessarily mean that one species is "better" model, or more advanced, than the other. If a species like orang-utans produces a repertoire typically exhibiting categorical differences between calls, then it can only be expected that orang-utans exhibit a rich variation in dialects because dialects involve differences *between* calls. If a species like chimpanzees or bonobos produces a repertoire typically exhibiting graded differences between calls, then it should be expected that chimpanzees or bonobos exhibit a rich variation in accents because accents involve differences within calls. Future research will be needed to accurately quantify these seeming differences between graded and categorical repertoires in great apes (Wadewitz et al., 2015). Nevertheless, the data coming in, thus far, suggest that interpreting any feature or trait yet to be found will need to be done with caution. Common capacities and skills (e.g. vocal learning) may in fact manifest differently across species, depending on the features of the repertoire of those species.

For now, great apes prove to display remarkable features in the interception of local traditions and vocal behaviour, which is exactly where potential language precursors are to be found. To understand how vocal evolution in great apes took shape and how language evolution ensued, we need an ever more comprehensive, inclusive, and integrated framework that will hopefully include a growing amount of audio recordings and behaviour data from *all* living great apes. This endeavour is bound to bring new insights into the evolution of language in our lineage.

4. Discovery of two major call categories in orang-utans

One of the most challenging tasks during the inventory of the orang-utan call repertoire (besides waking up at 4am and spending the day knees-deep in a swamp for 10 months!) was the fact that we were commonly faced with calls that exhibited features not described in the primate literature. In an age when we stand on the shoulders of giants, the lack of previous data dwarfed us.

Besides voiced calls, typically termed "vocalizations" and characteristically produced by all primates and terrestrial mammals (Taylor and Reby, 2010), the repertoire of orang-utans was proving to be particularly rich in noisy smacks, clicks, kiss sounds, and raspberries. These calls did not involve individuals' voice (which is the result of the regular oscillation of the vocal folds). Instead, call production resulted from the action of the supralaryngeal articulators – the lips, tongue, and jaw – and airflow generated by their own manoeuvres or by the action of abdominal musculature (e.g. diaphragm). Laid out in a spectrogram (i.e. a graphic means to visualize sound) these unvoiced calls typically exhibit distinctive traits from their voiced counterparts. Their distinct articulatory nature inescapably generates distinct acoustics.

At the time, we were unsure the extent to which these two different means of how orang-utans engage in vocal production – via voiced and voiceless call production – could be meaningful.

4.1 Speech building blocks

This binary aspect of the orang-utan vocal system might become clear when we take a closer look into the world's languages. Each and every human language is inherently, and by definition, composed by vowels and consonants – *without exception*. Acoustic and articulatorily, these two building blocks of speech are not equivalent. Vowels are characteristically voiced, using the activation of the vocal folds and their regular oscillation, whilst voiceless utterances in humans are characteristically consonants.

In fact, voiceless consonants dominate over other types of consonants in a large sample of human languages – 64% of the plosives, 72% of the fricatives and 74% of the affricates are unvoiced (Vallée et al., 2002). Moreover, voiceless consonants are found universally across the world's languages, while other types of consonant may or may not be present in particular languages (Lameira et al., 2014). In addition, evidence indicates that the original language before the exodus out of Africa, between 140,000 and 60,000 years ago, was particularly rich in voiceless consonants that characterize several African languages today (Atkinson, 2011; Knight et al., 2003). Even though many *modern day* consonants engage the voice, and voiced consonants exhibit a rich variation across the world languages (see for example, www.internationalphoneticassociation.org), the wide, rich, and time-deep presence of voiceless consonants in humans supports the view that great ape voiceless calls could be used as a desirable model to study the production and use of consonant-like calls in ancestral hominids during the first stages of language evolution.

There is, thus, an apparent parallel between the composition of the orang-utan vocal system and human language with regards to the articulatory commands and acoustic output underlying the two elementary particles of both systems. However, establishing an evolutionary link between the two requires establishing that great ape consonant-like and vowel-like calls are the result of vocal learning, as it occurs in humans.

4.2 First pillar: Consonant precursors

In orang-utans, voiceless calls preponderate orang-utan synonym dialects (Wich et al., 2012). In other words, when orang-utans produce population-specific calls as part of synonym dialects, these tend to be voiceless calls. This observation hints that voiceless calls in great apes are the result of vocal learning, and therefore, that they could allow establishing a link with human consonants. Empirical tests in captivity, that present the benefit of controlled settings, proved to be essential in testing this possibility.

The scientific discovery of the first whistling orang-utan provided the ideal conditions for a deeper investigation (Wich et al., 2009). Bonnie, a captive orang-utan, was known for many years among her caretakers to know how to whistle like a human. We were, however, perplexed when we caught wind of the news since this directly suggested that vocal learning was operating in captivity, allowing individuals to enlarge their repertoire with new calls as we had observed in the wild. Bonnie protruded her lips and, with gentle blows of air through the space in between, produced whistles.

The likelihood that Bonnie had acquired whistling through vocal learning was *very high* because whistles are *very* particular calls. Whistles are melodic and tonal as voiced calls commonly are, but they do not involve the voice (as anyone who can whistle knows) nor do they have formants (frequency resonance bands) that are characteristic features in voiced calls. They are the result of the airstream's periodic vortex shedding at the lips opening and generally exploit a very narrow frequency band where most of the acoustic energy is concentrated. They qualify as a rather unique type of voiceless call. No call with these features is known to exist in the primate order, with the obvious exception of human whistles, of course. Accordingly, Bonnie had very likely learned this call from humans.

The premise for our sound tests with Bonnie formulated the following: if Bonnie had indeed learned how to whistle through vocal learning, *then* she should be able to exert sufficient control over whistle production to alter some of its main acoustic parameters. In order to non-invasively prompt Bonnie to produce whistles, we presented her with a "do-as-I-do" paradigm. In this test setting, a human demonstrator produces model calls, implicitly requesting the subject (in this case, Bonnie) to produce back the same type of call. Great apes do particularly well at this imitation game and promptly understand what is wished from them. Through these means, Bonnie produced single whistles in response to single human whistles, double whistles in response to double whistles, short to short, and long to long. These results proved that Bonnie aptly controlled whistle production with enough accuracy to match simple human whistles.

After Bonnie, we have come to know the existence of, at least, ten captive orang-utans who have learned how to whistle, including some who learned from other whistling orang-utans (Lameira et al., 2013) (see video clip here: https://youtu.be/rMuLKoKILBw). Audio and video recordings of these individuals has shown that each individual uses its own lip style to produce whistles, producing for instance whistles with in- and out-airflow. Subsequent tests with some of these orang-utans have shown they too can match in-out and triple whistles produced by a human under the "do-as-I-do" settings. Altogether, these data confirm that vocal learning is a widespread phenomenon in orang-utans, and that individuals can acquire a very fine level of motor control over their lips' positioning and movement, as well as over the muscles involved in creating airflow through the vocal tract, including the diaphragm and other abdominal musculature.

With this new insight in our minds we started to prospect the repertoire of the other great apes in search of voiceless calls. If orang-utans were learning voiceless calls at much higher rates than ever believed to be possible in captivity, and if these calls had indeed an evolutionary link with human consonants, then African great apes ought to be expected to exhibit some of the same flexibility too. Some studies at the time had already confirmed the production of voiceless calls by captive chimpanzees (Hopkins et al., 2007). New examples continue now to emerge in chimpanzees in the wild (Watts, 2015), as well as in wild gorillas (Robbins et al., 2016) and some cases reported in captive gorillas (Lameira et al., 2014; Perlman and Clark, 2015). All these studies maintained that some level of vocal learning was necessary to explain the production and use of voiceless calls by some individual(s)/population(s) but that were otherwise absent elsewhere.

In fact, chimpanzee research has brought a new line of evidence supporting the view that great ape voiceless calls are vocally learned. Exciting new advances in the field of neuroscience have confirmed that chimpanzees who have learned to produce voiceless calls (Taglialatela et al., 2012) exhibit different neural networks in their brain from those individuals who have not learned how to produce voiceless calls (Bianchi et al., 2016). Critically, vocal learning individuals exhibit *increased* grey matter in the ventrolateral prefrontal and dorsal premotor cortices – constituent regions of the equivalent area to Broca's in the human brain. These regions with observed reorganization are responsible for orofacial motor control, demonstrating that these individuals required practice and development of enhanced control to produce voiceless calls.

Given the possible articulatory and acoustic homology with human voiceless consonants and accumulated evidence showing that great ape voiceless calls are learned, these calls can therefore be sensibly advanced putative precursors of human speech sounds. Notably, I suggest that great ape voiceless calls can be advanced as putative precursors of human consonants.

4.3 Second pillar: Vowel precursors

Once the evolutionary link between great ape voiceless calls and human consonants is suggested, the second link stands out conspicuously: primate voiced calls have probably given raise to human vowels. The articulatory and acoustic parallel between the two is not new (Owren et al., 1997). Indeed, voiced calls are characteristic of virtually all mammals. In primates, however, they seem pervasively innate. Establishing this second evolutionary link between primate voiced calls and human vowels requires, thus, some evidence for an active role of vocal learning. Vowels precursors should be the result of vocal learning and maintained across peers and generations through cultural mechanisms. Could it be instead, however, that motor control over vocal fold action is so difficult that learned voiced calls in great apes are simply absent?

We did not need to search far to confirm that great apes can learn voiced calls! It was during our efforts of registering all known whistling orang-utans that we knew Tilda. She is a wild-born orang-utan now well into her forties. There are no known records of her arrival in Europe. As any great ape smuggled into Europe in the pet trade, she was probably brought in as a baby and the first information we found only related to her adolescence onwards, when she was acquired by a private collector. As far back as we could verify, Tilda was known to whistle. There we were then, in the Cologne Zoo, Germany, where she lives now (painting and selling canvas to raise funding for her family members still living in the forests of South-east Asia!). Audio recorders ready, cameras rolling, we were all set. We confirmed her whistling capacity but then we were shown something that surpassed anything we had seen.

...

We were left speechless.

Tilda babbled to us (see video clip here: https://youtu.be/ab59zcsV35k)!

As in the case of Bonnie and all other whistling orangs, her caretakers knew for years that she could do this and that this was part of how she gathered human attention to request for food (Lameira et al., 2015). In a display that we had never observed in the wild or captivity, Tilda moved her lips and jaw at a very fast pace, producing vocalizations and sounds that could have been easily attributed to a Disney movie character if we were standing in front of the TV instead of an amazing great ape facility. Based on the video recordings collected, we analysed the pace at which Tilda moved her mouth (Lameira et al., 2015). Results showed that both calls that she produced in this manner showed a rhythm of about five calls per sec, with one call being voiced and the other voiceless. This fast pace is equivalent to the same at which you and I produce vowels and consonants during normal speech, which translates into five vowels and five consonants per second.

These findings are theoretically important because they verify that the acoustic and articulatory parallel proposed between voiceless and voiced calls on the one hand, and vowels and consonants on the other, makes empirical sense. Notably, these data provided a new level of similarity between these elements, notably, regarding their articulatory rhythm. The two great ape call categories can be produced at the same delivery rhythm as the two human speech elementary blocks. In comparison with the known orang-utan repertoire, the observed rhythm in Tilda's calls was seven-fold faster than any call described before (Lameira et al., 2015). Hence, the most likely and evident explanation is that Tilda acquired these calls through vocal learning from humans. Besides a new rhythmic parallel with human utterances, here we had the first indication in captive orang-utans that vocal learning of voiced calls – requiring activation of the vocal folds – was in reach. This mounted on preliminary data from the wild, where one of the existing synonym dialects involved a voiced call (Wich et al., 2012).

As in the case of orang-utan whistling, this evidence required experimental confirmation and we wanted to double-check our premise. If orang-utans in captivity were learning new voiced vowel-like calls, *then* they ought to be able to control vocal fold action to an extent where they could alter key voice parameters of the calls and match human demonstrations in real-time. However, we wanted to respect Tilda's life history, which very likely involved dubious relations and experiences with humans and so we pre-empted ourselves from running tests with Tilda for ethical reasons.

However, we knew Rocky! Rocky is an orang-utan teenager who, by the age of four had already been in photo-shoots with the famous pop music band Black Eyed Peas and made Levi's TV commercials. Rocky and his mom (who is a known whistler) retired from the entertainment business and were received at the former Great Ape Trust of Iowa. Today they live in one of the most advanced great ape facilities in the world at the Indianapolis Zoo. We had known Rocky since his arrival at the Trust. By then, he already produced a very distinctive call that we also had never heard in the wild or captivity. We coined these calls "wookies" because they repeatedly recalled us of the orang-utan-looking Star Wars character, Chewbacca. Wookies provided a perfect opportunity to address the question that was confronting us. Rocky was young, active, and eager to engage with humans and we knew he would take full advantage of any opportunity to show off his talents.

The first step in our research plan was to confidently confirm that wookies were indeed unique to Rocky. We wanted to make sure that wookies were not an otherwise common, but misclassified orang-utan call. We scanned our database for the closest known call produced by orang-utans in the wild. We then acoustically compared this call with wookies. Results indicated that their acoustics and underlying articulation was distinct and that they constituted two different call types (Lameira et al., 2016). Wookies were, thus, unique to Rocky and they represented a new call hitherto not described in orang-utans.

The second step in our test agenda was to present Rocky with the "do-as-I-do" imitation game, as we had done with whistling orang-utans. In this context, a human demonstrator presented Rocky with high and low pitch approximations of wookies as an implicit request for him to reply with wookies of similar acoustics (Fig. 2). To produce high and low frequency (Hz) wookies, Rocky was required to exert real-time and fine control over his voice. Specifically, Rocky had to be able to control the tension of the laryngeal muscles associated with his vocal folds and adjust their oscillation frequency to that of the human demonstrations. Results confirmed our suspicions – Rocky had vocally learned to produce these new calls. Whenever the human demonstrator would raise or lower the pitch of her voice, so did Rocky within centiseconds (see video clip here: https://youtu.be/Lg50_1RSc0E). The voice frequency of the human demonstrator and of Rocky significantly correlated positively with each other, and the high and low pitch wookies that Rocky produced were dramatically different from each other, as well as from wookies that Rocky produced spontaneously (that is, not in reply to human demonstrations). Indeed, we can see on Figure 2 that the correlation is highly significant, with a Spearman R above

0.5, meaning that the human voiced explained the majority of the variation in Rocky's responses and that no other factor could have exhibited a higher explanatory power. This meant that the human demonstrations were in fact guiding Rocky's voiced, namely, away from his typical voice range used during wooky production, and not the other way around.

Fig 2: Maximum frequency of human wookie demonstrations against maximum frequency of Rocky's match wookies (linear trend line with intercept suppressed). (Reproduced from Lameira et al., 2016).

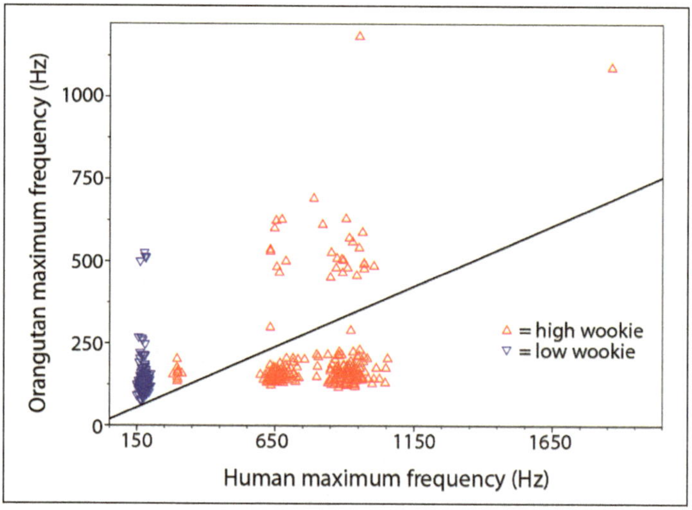

Even if one wonders that Rocky could have pulled off a "clever Hans" trick (i.e. anticipating by means of slight cues what the human demonstrator was about to do), this would still not affect our fundamental conclusion that Rocky controls voluntarily his voice. This is because, in such supposed scenario, Rocky would be responding to other cues or signal than the human voice, but one would still be left with explaining why Rocky's high and low calls were significantly different from spontaneous ones that were not produced in response to human demonstrations. There would be further issues with such hypothetical interpretation. One would need to explain why, instead of a direct voice-voice match, Rocky's was employing a multimodal match, using a cue or signal in a sensorial channel but, nevertheless, responding acoustically.

With this evidence, we can now cement the cornerstone of the second pillar of the new theoretical edifice of language evolution. Together with the first pillar, sufficient is now known about the call repertoire of great apes for new evolutionary links and research venues to be drawn. Human vowels and unvoiced consonants are to a certain extent homologous in terms of articulation, acoustics, and rhythm to voiced and voiceless calls in great apes. Claims that no evolutionary seeds for human spoken language can be found within our closest phylogenetic branch are precipitate and uninformed (and typically made by scholars who have never studied great apes!). Such past claims were based on absence of evidence, not evidence of absence. As we have seen so far, a new generation of great ape studies is starting to flow in and fill what was more of a deep gap in our knowledge about great ape vocal behaviour, than an evolutionary gap between human spoken language and great ape vocal systems. Now that this new evidence is emerging and amassing, we cannot afford to continue ignoring great ape vocal capacities and turn our backs to the news frontiers they unlock if we are to crack the evolution puzzle of language evolution.

5. Evolutionary trajectories

As we have seen, as a plum seed to a full-grown plum tree, it is important to retain the notion that language proto-stages do not need to have exhibited in the past the same features of full blown speech as today. Consonants exhibit today varied forms beyond those that are voiceless and many consonants engage vocal fold action – as in the voiced plosive /b/ or the nasal plosive /m/ –as vowels characteristically do. For this reason, linguists rather delineate the difference between *modern day* consonants vs. vowels as a measure of the constriction of the vocal tract required for production, with consonants being produced with relatively closed vocal tracts, and vowels with relatively more open configurations. Linguists do not rely, therefore, on a definition based on voiced-ness, as great ape data suggest may have been the case regarding proto-consonants vs. proto-vowels.

Today, consonants and vowels are also produced at a fast pace in intricate and swift alternation to compose words and sentences. Their transition is fluid. This rhythmic aspect of speech has also been suggested as a evolutionary forerunner of speech, in the milestone theory known as "Frame-

Content" (MacNeilage, 1998). Fundamentally, this theory suggests that the continuous mouth open-close alternation characteristic of speech derives from ancient mammal behaviour, such as suckling and chewing, which then took on communicative functions. Due to the predominant role that articulation plays in this theory, it has been one of the few that has so far proposed equal and parallel importance to consonants and vowels in the process of language evolution (Lameira, 2014; Lameira et al., 2014, 2016). Namely, it has recognized that consonants and vowels are respectively associated with the closed and opened phase of the mouth cycle. Linguists proponent of this theory commonly do not rely, therefore, on a definition of consonants and vowels where they occur separated and unchained.

Linguistic evidence and comparative data on great ape vocal behaviour do not need to be incoherent with one another, however, nor do great ape data challenge the Frame-Content Theory. Let us see why this is the case.

Importantly, one must appreciate the point in the timeline of language evolution at which linguists work and that at which great ape vocal researchers do. Linguistics, through the reconstruction of language-trees can recede up to 50 thousand (50,000) years ago (Gray and Atkinson, 2003). Great ape researchers, using our closest relatives as living models of ancient hominids, work within a frame up to 10 million (10,000,000) years ago – the time when our last great ape common ancestor lived. These timeframes differ by several orders of magnitude. Converted to seconds, linguistics work down to 14 *hours* ago from the present. Great ape researchers work at a point 4 *months* ago from the present. It is futile and heuristically unproductive to argue that these two points in language evolution are not connected simply because they are not *fully* aligned. Indeed, no one would expect 9,950,000 years of evolution go by without change or advance. In other words, seeds should not be expected to be able of photosynthesis! This is exactly what the reconstruction of language evolution is all about and what great ape researchers and linguistics are expected to do: better understand the path that bridges our last great ape common ancestor and us.

A possible parsimonious hypothesis that integrates these two lines of work could be as follows. Initially, proto-consonants and proto-vowels were produced and used either separately (as observed in all great apes today) or in relatively simple syllable-like call combinations (as observed in wild orang-utans today). The overall acoustic range of proto-consonants and proto-vowels was most probably much more limited than present day forms. With increasing selective drive for effective and efficient (social) communication, the use of these two proto-elements recruited an ancient mammal behaviour – fast paced mouth oscillations as argued in the Frame-Content Theory (MacNeilage, 1998) – to increase their production rate (Ghazanfar et al., 2012, 2013). Orang-utan vocal data suggest this recruitment may have involved a seven-fold increase in the delivery rate of successive consonants and vowels (Lameira et al., 2015). Once stringed together by fast open-close mouth alternations, the acoustic features of (voiceless) consonants and (voiced) vowels started to fuse. Voiced-ness lost, then, its signature role in dividing consonants and vowels. As a consequence, what was a stark division in terms of voiced-ness, became a graded one based on degrees of vocal tract openness.

While tentative for now, some support for this possible scenario is found in everyday discourse, notably, paralinguistic elements of human vocal communication. Examples of these sounds are, for instance, "Shhh" to demand silence at the start of a movie, "Mmmmm" to approve mother's cooking, "Ahhhhh" when we finally get the solution for a difficult quiz. Articulatorily and acoustically, they correspond to proto-consonant and proto-vowel sounds, and unlike typical consonants and vowels, they are not stringed together to form of a word or a sentence. Paralinguistic sounds seem to represent, therefore, relicts of former stages in language evolution when proto-consonants and proto-vowels were used separately, as observed in great apes. One must still explain the occurrence of voiced consonants as paralinguistic sounds, such as "Mmmmm." According to the timeline mentioned above, these sounds should have only emerged once proto-consonants and proto-vowels were stringed together. How could voiced consonants, then, be used separately today? The key word here is *today*. Modern humans have developed a tremendous degree of vocal control (Gokhman et al., 2017), demonstrated excellently by any opera singer or beat-boxer and the cultural variation in speech sounds across the world's languages. We can, today, deploy many different types of sounds communicatively in many different forms. No hominid ancestor is expected to have exerted such degree of vocal control.

This example illustrates that the endeavour of reconstructing language evolution will be one of "likely" scenarios and parsimony, not one of absolutism. Researchers and everyone interested in this fascinating topic will have to remain opened to the new data coming in, both from linguistics and comparative great ape vocal research, regularly check its axioms and predictions, and foremost, exercise time-perspective taking across the 9,950,000 years gap with our hominid great (ape) grand parents.

6. Concluding remarks

This chapter proposes a new heuristic framework for language evolution. It explains how consonant-like and vowel-like calls are present in great apes and proposes that these calls can serve as models to study ancient proto-consonants and proto-vowels that existed in a human ancestor. With this proposal in place, we can start investigating for the first time questions that have remained, thus far, unformulated. For instance, why did proto-consonants and proto-vowels came together to compose the first syllable or word (Lameira et al., 2017)? When did this occur? Where there particularly ecological conditions that made these combinations more prone to occur? The latest evidence reviewed in this chapter stands as proof that a renewed interest on great ape behaviour will yield important clues as we progress in reconstructing language evolution.

References

Atkinson, Q. D. (2011). Phonemic diversity supports a serial founder effect model of language expansion from Africa. *Science*, *332*, 346–349.

Bianchi, S., Reyes, L. D., Hopkins, W. D., Taglialatela, J. P., and Sherwood, C. C. (2016). Neocortical grey matter distribution underlying voluntary, flexible vocalizations in chimpanzees. *Scientific Reports*, 6, 34733.

Clay, Z., Pika, S., Gruber, T., and Zuberbuhler, K. (2011). Female bonobos use copulation calls as social signals. *Biology Letters*, *7*, 513–516.

Delgado, R. A., and van Schaik, C. P. (2000). The behavioral ecology and conservation of the orangutan (Pongo pygmaeus): A tale of two islands. *Evolutionary Anthropology: Issues, News, and Reviews*, *9*, 201–218.

Enard, W. (2011). FOXP2 and the role of cortico-basal ganglia circuits in speech and language evolution. *Current Opinion in Neurobiology*, *21*, 415–424.

Enard, W., Przeworski, M., Fisher, S. E., Lai, C. S. L., Wiebe, V., Kitano, T., et al. (2002). Molecular evolution of FOXP2, a gene involved in speech and language. *Nature, 418*, 869–872.

Favareau, D. (2010). Introduction: An Evolutionary History of Biosemiotics. In M. Barbieri (Ed.), *Essential Readings in Biosemiotics: Anthology and Commentary* (pp. 1–68). Dordrecht, the Netherlands: Springer.

Fedurek, P., Slocombe, K. E., and Zuberbuhler, K. (2015). Chimpanzees communicate to two different audiences during aggressive interactions. *Animal Behaviour, 110*, 21–28.

Fitch, W. T. (2010). *The evolution of language*. Cambridge: Cambridge University Press.

Fitch, W. T., and Hauser, M. D. (2004). Computational constraints on syntactic processing in a nonhuman primate. *Science, 303*, 377–380.

Ghazanfar, A. A., Morrill, R. J., and Kayser, C. (2013). Monkeys are perceptually tuned to facial expressions that exhibit a theta-like speech rhythm. *Proceedings of the National Academy of Sciences, 110*, 1959–1963.

Ghazanfar, A. A., Takahashi, D. Y., Mathur, N., and Fitch, W. T. (2012). Cineradiography of monkey lip-smacking reveals putative precursors of speech dynamics. *Current Biology, 22*, 1176–1182.

Gokhman, D., Agranat-Tamir, L., Housman, G., Nissim-Rafinia, M., Nieves-Colón, M., Gu, H., et al. (2017). Recent regulatory changes shaped human facial and vocal anatomy. *bioRxiv*, 106955. http://doi.org/10.1101/106955

Gray, R. D., and Atkinson, Q. D. (2003). Language-tree divergence times support the Anatolian theory of Indo-European origin. *Nature, 426*, 435–439.

Hardus, M. E., Lameira, A. R., Singleton, I., Morrough-Bernard, H. C., Knott, C. D., Ancrenaz, M., et al. (2009). A description of the orangutan's vocal and sound repertoire, with a focus on geographic variation. In S. Wich, T. Mitra Setia, S.S. Utami, and C. P. Schaik (Eds.), *Orangutans* (pp. 49–60). New York: Oxford University Press.

Hauser, M. D., Chomsky, N., and Fitch, W. T. (2002). The faculty of language: What is it, who has it, and how did it evolve? *Science, 298*, 1569–1579.

Hauser, M. D., Yang, C., Berwick, R. C., Tattersall, I., Ryan, M. J., Watumull, J., et al. (2014). The mystery of language evolution. *Frontiers in Psychology*, *5*, 401.

Hopkins, W. D., and Savage-Rumbaugh, E. S. (1991). Vocal communication as a function of differential rearing experiences inPan paniscus: A preliminary report. *International Journal of Primatology*, *12*, 559–583.

Hopkins, W. D., Taglialatela, J. P., and Leavens, D. A. (2007). Chimpanzees differentially produce novel vocalizations to capture the attention of a human. *Animal Behaviour*, *73*, 281–286.

Knight, A., Underhill, P. A., Mortensen, H. M., Zhivotovsky, L. A., Lin, A. A., Henn, B. M., et al. (2003). African Y chromosome and mtDNA divergence provides insight into the history of click languages. *Current Biology*, *13*, 464–473.

Krützen, M., Willems, E. P., and van Schaik, C. P. (2011). Culture and geographic variation in orangutan behavior. *Current Biology*, *21*, 1808–1812.

Lameira, A. R. (2014). The forgotten role of consonant-like calls in theories of speech evolution. *Behavioral and Brain Sciences*, *37*, 559–560.

Lameira, A. R., Delgado, R., and Wich, S. (2010). Review of geographic variation in terrestrial mammalian acoustic signals: Human speech variation in a comparative perspective. *Journal of Evolutionary Psychology*, *8*, 309–332.

Lameira, A. R., Hardus, M. E., Bartlett, A. M., Shumaker, R. W., Wich, S. A., and Menken, S. B. J. (2015). Speech-like rhythm in a voiced and voiceless orangutan call. *PLoS One, 10*, e116136.

Lameira, A. R., Hardus, M. E., Kowalsky, B., de Vries, H., Spruijt, B. M., Sterck, E. H. M., et al. (2013). Orangutan (Pongo spp.) whistling and implications for the emergence of an open-ended call repertoire: A replication and extension. *Journal of the Acoustical Society of America*, *134*, 1–11.

Lameira, A. R., Hardus, M. E., Mielke, A., Wich, S. A., and Shumaker, R. W. (2016). Vocal fold control beyond the species-specific repertoire in an orang-utan. *Scientific Reports*, *6*, 30315.

Lameira, A. R., Maddieson, I., and Zuberbuhler, K. (2014). Primate feedstock for the evolution of consonants. *Trends in Cognitive Sciences*, *18*, 60–62.

Lameira, A. R., Vicente, R., Alexandre, A., Campbell-Smith, G., Knott, C., Wich, S., and Hardus, M. E. (2017). Proto-consonants were information-dense via identical bioacoustic tags to proto-vowels. *Nature Human Behaviour*, *1*, 0044.

Mackinnon, J. (1974). The behaviour and ecology of wild orang-utans (Pongo pygmaeus). *Animal Behaviour*, *22*, 3–74.

MacNeilage, P. F. (1998). The frame/content theory of evolution of speech production. *Behavioral and Brain Sciences*, *21*, 499–511.

Marshall, A. J., Wrangham, R. W., and Arcadi, A. C. (1999). Does learning affect the structure of vocalizations in chimpanzees? *Animal Behaviour*, *58*, 825–830.

McComb, K., and Semple, S. (2005). Coevolution of vocal communication and sociality in primates. *Biology Letters*, *1*, 381–385.

Mitani, J. C., and Gros-Louis, J. (1998). Chorusing and call convergence in chimpanzees: tests of three hypotheses. *Behaviour*, *135*, 1041–1064.

Owren, M. J., Amoss, R. T., and Rendall, D. (2011). Two organizing principles of vocal production: Implications for nonhuman and human primates. *American Journal of Primatology*, *73*, 530–544.

Owren, M. J., Seyfarth, R. M., and Cheney, D. L. (1997). The acoustic features of vowel-like grunt calls in chacma baboons (Papio cyncephalus ursinus): Implications for production processes and functions. *Journal of the Acoustical Society of America*, *101*, 2951–2963.

Perlman, M., and Clark, N. (2015). Learned vocal and breathing behavior in an enculturated gorilla. *Animal Cognition*, *18*, 1165–1179.

Pisanski, K., Cartei, V., McGettigan, C., Raine, J., and Reby, D. (2016). Voice modulation: a window into the origins of human vocal control? *Trends in Cognitive Sciences*, *20*, 304–318.

Rijksen, H. D. (1978). *A field study on sumatran orangutans (Pongo pygmaeus abelii, Lesson 1827): Ecology, behavior, and conservation.* Wageningen, the Netherlands: H. Veenman and Zonen.

Robbins, M. M., Ando, C., Fawcett, K. A., Grueter, C. C., Hedwig, D., Iwata, Y., et al. (2016). Behavioral variation in gorillas: evidence of potential cultural traits. *PLoS One*, *11*, e0160483.

Schlenker, P., Chemla, E., and Zuberbuhler, K. (2016). What do monkey calls mean? *Trends in Cognitive Sciences*, *20*, 894–904.

Slocombe, K. E., and Zuberbuhler, K. (2007). Chimpanzees modify recruitment screams as a function of audience composition. *Proceedings of the National Academy of Sciences of the United States of America, 104*, 17228–17233.

Taglialatela, J. P., Reamer, L., Schapiro, S. J., and Hopkins, W. D. (2012). Social learning of a communicative signal in captive chimpanzees. *Biology Letters, 8*, 498–501.

Taylor, A. M., and Reby, D. (2010). The contribution of source–filter theory to mammal vocal communication research. *Journal of Zoology, 280*, 221–236.

Vallée, N., Boë, L. J., Schwartz, J. L., and Badin, P. (2002). The weight of phonetic substance in the structure of sound inventories. *ZAS Papers in Linguistics, 28*, 145–168.

van Schaik, C. P., Ancrenaz, M., Borgen, G., Galdikas, B., Knott, C. D., Singleton, I., et al. (2003). Orangutan cultures and the evolution of material culture. *Science, 299*, 102–105.

Wadewitz, P., Hammerschmidt, K., Battaglia, D., Witt, A., Wolf, F., and Fischer, J. (2015). Characterizing vocal repertoires--hard vs. soft classification approaches. *PLoS One, 10*, e0125785.

Watson, S. K., Townsend, S. W., Schel, A. M., Wilke, C., Wallace, E. K., Cheng, L., et al. (2015). Vocal learning in the functionally referential food grunts of chimpanzees. *Current Biology, 25*, 495–499.

Watts, D. P. (2015). Production of grooming-associated sounds by chimpanzees (Pan troglodytes) at Ngogo: variation, social learning, and possible functions. *Primates, 57*, 61–72.

Whiten, A., Goodall, J., McGrew, W. C., Nishida, T., Reynolds, V., Sugiyama, Y., et al. (1999). Cultures in chimpanzees. *Nature, 399*, 682–685.

Wich, S. A., Krützen, M., Lameira, A. R., Nater, A., Arora, N., Bastian, M. L. et al. (2012). Call cultures in orang-utans? *PLoS One, 7*, e36180.

Wich, S.A., Swartz, K.B., Hardus, M.E., Lameira, A.R., Stromberg, E., and Shumaker, R.W. (2009). A case of spontaneous acquisition of a human sound by an orangutan. *Primates, 50*, 56–64.

Frédéric Berthommier[1], Louis-Jean Boë[1],
Adrien Meguerditchian[2], Thomas R. Sawallis[3],
Guillaume Captier[4]
[1]GIPSA-Lab, CNRS and Grenoble Alpes University, Saint-Martin-d'Hères, France
[2]Cognitive Psychology Laboratory, CNRS and Aix-Marseille University, Marseille, France
[3]New College, The University of Alabama, Tuscaloosa, Alabama, United States of America
[4]Anatomy Laboratory, Montpellier University, Montpellier, France

Comparative Anatomy of the Baboon and Human Vocal Tracts: Renewal of Methods, Data, and Hypotheses

Abstract: This chapter focuses on the emergence of speech during human evolution, revisiting exaptation hypotheses (Fitch, 2010; MacNeilage, 1998) with new data from comparison with baboons. Speech necessarily evolved to be compatible with aero-digestive anatomy, reusing its functions of suction, chewing and swallowing. The tongue is involved with every feeding gesture, and also has a central position for speech. We analyze the evolution of the tongue position taking into account the distinction between the morphogenetic fields of HOX and non-HOX genes involved in the development of the pharyngeal arches and the cephalic structures, anatomical and neurological components, and functional support for breathing and swallowing. The hyoid bone is the locus of insertion of the tongue muscles as well as a precise marker of the glottis position. It is not fixed because it partly depends on the development of the facial area controlled by non-HOX genes. In contrast, the vertebral column has stable dimensions because it is controlled by HOX genes. After a detailed presentation of a baboon head dissection, we present a new method for mapping hyoid bone position relative to the vertebral axis, applied to MRI images. This is compared to a set of radiographs of 3–7.5 year human children. We observe that the hyoid bone is 1 vertebra lower in human infants than in adult baboons. The normalized oral cavity length is shorter, in agreement with prognathism reduction as controlled by non-HOX genes. Using the cervical vertebrae and their axis as a reference allows the conclusion that there is indeed laryngeal descent from baboons to humans and that it is accompanied by compensatory facial shortening.

This preserves the vocal tract length as well as the relationship between the tongue and the oropharyngeal cavity, which is important for swallowing and other feeding gestures.

Keywords: exaptation, baboons, vocal tract anatomy, HOX genes, laryngeal descent

1. Introduction

1.1 Why link speech emergence and primate vocalizations?

The existence of speech as a characteristic of the human species raises a series of questions that, for the most part, have remained open and unanswered for several centuries. What are the anatomical and cognitive prerequisites for vocal communication? When, where, and how did this type of communication arise? By what steps has this evolution taken place? Did gestural communication originate earlier? Or did gestures and vocalizations arise simultaneously?

Researchers have at their disposal human fossils which, though rarely complete, do allow us, to some extent, to trace the anatomical evolution of the head and neck, and thus the architecture of the vocal tract. Obviously, there are no recordings of their sound productions.

Already by the second third of the 19th century, Youatt (1835) had trouble understanding why the chimpanzees lacked the power to speak while they were able to shout loudly. We can understand why the anatomy of the vocal organs of chimpanzees has since then aroused great interest (Vrolik, 1841), but what explains that with very similar organs, these primates cannot use them in the same way humans do? More generally, for insights into the evolution of the cerebral cortex and cognition in human ancestors, researchers have long studied the comparative anatomy of the chimpanzee brain (Clark et al., 1936; Falk, 2014; Walker and Fulton, 1936).

Since we share common ancestors with both apes and monkeys we hypothesize that the current vocalizations of these primates provide us with an underexploited window for exploring the nature of speech, and can inform us about the stages of its emergence. Indeed, we assume that the system of speech communication was gradually established over the course of the millions of years of evolution that separate us from our common ancestors. Animal communication has evolved on several levels:

anatomical, cognitive, ethological, all under the constraining influence of the environment.

On the other hand, the other descendants of these common ancestors would not have followed the same evolution. We can therefore assume that their vocalizations have changed little. The vocalizations of present-day monkeys would thus be *relics* (Pisanski et al., 2016) of earlier vocal tract abilities and, we could say metaphorically, *fossil traces* of the communication of our common ancestors.

Monkey and ape vocalizations depend on the sex, status, and age of the analyzed individual. Among primates, baboons produce a repertoire of fourteen vocalizations identified and associated with situations ethologically described (including behavior and communication) (Hall and DeVore, 1965; Zuberbühler, 2012). There are several acoustic analyzes of baboon vocalizations (e.g. Andrew, 1976; Fischer et al., 2002; Owren et al., 1997; Rendall et al., 2005) and more recently it has been shown that they can produce five differentiated vocalizations corresponding to five different ethological situations (Boë et al., 2017).

1.2 The exaptation hypothesis

This chapter does not focus on the acoustic analysis of baboons vocalizations but rather on the anatomical aspects that enable and condition the production of these vocalizations, that is to say on the anatomy of the larynx and on the vocal tract and its position with respect to the larynx and to the cervical vertebrae.

Indeed, the vocalizations of mammals, and thus of human and non-human primates, are all produced by the same process. The sound generated by the vibration of the vocal folds (the source) is acoustically filtered by the resonance characteristics of the vocal tract (the filter), that extends from the glottis (the gap between the vocal folds) to the lips which radiate the filtered signal: this is the source-filter theory (Fant, 1960). By controlling the action of the vocal folds, by modifying the vocal tract shape through control of the articulators (tongue, mandible, lips), or by engaging the nasal passages (through lowering of the velum) it is thus possible for humans to articulate sufficiently differentiated vowels and consonants and to combine them in syllables and syllable sequences.

The question, then, is whether anatomical reasons explain why primates would not be able to produce differentiated vocalizations. The question arises all the more so since for almost 50 years a widespread and longstanding theory (Lieberman et al., 1969; Lieberman, 1975, 1984, 1998, 2007, 2015) has claimed that nonhuman primates, including premodern hominids, were incapable of producing systems of vowel-like sounds involving control of their vocal tract, due to their high larynx position and the resulting articulatory anatomy.

The comparative study of the anatomy of the upper aero-digestive tract of *Papio papio* and *Papio anubis* baboons and humans reveals similarities and differences. The first difference is the transition to the upright posture, which caused the centering of the foramen magnum, and triggered reductions in prognathism and the weight of the face. There is therefore a modification of the aerodigestive crossroads at the level of the epiglottis which ends up in a lower position and which is no longer in contact with the soft palate in humans. The second difference is the less flexed skull base, which has the biomechanical consequence of modifying the position of the hyoid bone (Reidenberg and Laitman, 1991). (Note that skull base flexion is measured as the angle of the orbital plane with that of the foramen magnum; the increased flexion in humans indicates a ventral displacement of the foramen magnum to accommodate upright posture.)

Hence, the observed human-baboon similarities are constrained by the primary functions, but the critical point is to consider these *similarities* as the *true prerequisites* to the production of speech. In contrast, many authors assumed that the differences are the markers of limits on the speech production ability. In this context, we will describe the evolution of the laryngeal elements (vocal folds, cartilages, thyro-hyoid membrane), the hyoid bone and the oral elements (tongue, palate). We will use a common reference frame for vertebrates based on the axis of the cervical vertebral column, with the apex of the odontoid as its origin. This landmark is under the control of the HOX development genes, as are the hyo-branchial apparatus and the larynx (Figure 1).

Figure 1: HOX and non-HOX zones delimited by anatomical landmarks on the skeleton. Insets: top, Drosophila, which is an important model for understanding body generation, and below, mouse and human embryos, which present HOX genes.

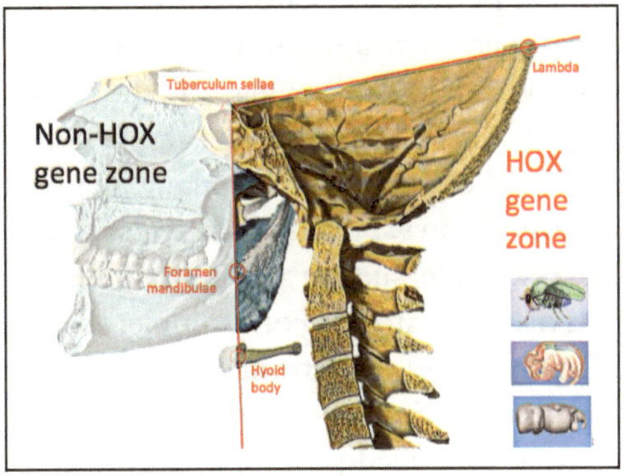

After the first 15 to 20 days following fertilization in vertebrates (Couly et al., 1993; Couly and Bennaceur, 1998; Couly et al., 2002), the HOX genes are responsible for embryo development and determine its anterior-posterior and dorsoventral organization, and thus the placement of the base of the skull, the head, and the body. Consequently, these genes are involved in the growth of relevant bones, which form the framework in which the vocal tract is situated. This system is highly conserved in vertebrates (McGinnis et al., 1984) and it can be assumed that this regulation maintains a suitable morphology for swallowing and protecting the airways.

Conversely, located in non-HOX areas, the oral part of the vocal tract is considered variable. This oral part, derived from the first pharyngeal arch, is under the control of a variety of genes and must be negative HOX for normal development (Chai and Maxson, 2006; Kuratani et al., 1997).

Speech would have evolved through possibilities and constraints external to speech: "speech from nonspeech" (MacNeilage and Davis, 2000a, 2000b), hence the interest in finely analyzing the anatomical structures of the vocal apparatus of non-hominin primates, because they are likely to enlighten us regarding the path followed during the emergence of speech. Thus, gestures of the tongue, the mandible, and the lips were compared across feeding and speech production (Green and Wang, 2003; Hiiemae, 2000; Hiiemae et al., 2002; Hiiemae and Palmer, 2003; Serrurier et al., 2012). Part of the control might also have been exapted (for discussion, see Ballard et al., 2003; Bunton, 2008; Folkins et al., 1995; Martin, 1991; Ziegler, 2003).

The vocal tract's original and still primary function is digestive. It is divided into two main parts that evolved with their own constraints and their own regulatory genes. The anterior part is dedicated to feeding, with suction and chewing as well as swallowing, and the posterior part is mainly related to swallowing. This chapter revisits the hypothesis of exaptation (Gould and Vrba, 1982) of speech from tongue anatomy as well as from these functions in several ways. First, speech gestures may be derived from feeding gestures. For example, suction and lip rounding are related. Second, they can reuse the existing anatomy. For example, the ability of the tongue for swallowing, which guides food from the anterior to the posterior, is related to its musculature. For speech, this permits constrictions inside the vocal tract at well-controlled positions. Third, the skill at chewing a variety of foods has an impact on the agility of tongue, as well as on the development of oral somatosensory perception and feedback necessary for speech. We now continue with an anatomical description of these anterior and posterior components of the vocal tract, followed by a quantitative analysis of their evolution from baboon to human.

1.3 The central position of the hyoid bone

The functional requirements of vocalization involve mobilization of the air source, used for breathing, and of the vocal folds, which protect the airway during swallowing. Unlike chimpanzee or other mammalian larynges (Harrison, 1995; Kelemen, 1969), and contrary to general anatomy (Swindler and Wood, 1973), the baboon hyoid bone and larynx seem, with a few minor exceptions, to have been relatively little studied (Nishimura, 2003ab, 2005, 2006).

The functional imperatives of swallowing and breathing require a close anatomical relation between the oral cavity, the base of the tongue, the pharynx, and the larynx. Presumably, "spatial constraints related to deglutition impose greater restrictions on the rate and degree of hyo-laryngeal descent than do adaptations for vocalization" (Lieberman et al., 2001). The oral and pharyngeal phases in swallowing allow passage of a food bolus from the oral cavity to the esophagus while protecting the airways. The position of these anatomical structures is determined by their insertions on the skeleton, especially on the base of the skull, the mandible, and the hyoid bone. The positions will be identified relative to the cervical spine (mainly C2, C3, and C4), which has been shown to be highly similar between baboons and humans (Tominaga et al., 1995).

Comparing baboons with humans reveals major morphological differences, in the less flexed base of the skull and in the face, that are associated with the arrangement of the muscle insertions and in the position of the hyoid bone in the baboon. It has been established that prognathism involves differences of the insertions of the muscles of the tongue and supra-hyoid muscles.

Figure 2: Functional anatomy of the components involved in swallowing, vocalization, and articulation for speech. The two main composite structures, the vocal tract (left) and the larynx (right), are at the top, with their specific anatomical components immediately below, with explicit ties to their functions at the bottom. This shows the great overlap between structures for speech and swallowing. Note that the hyoid bone is a key to this overlap because it is the insertion of the tongue root at the same time as its anatomical relationship with the epiglottis is important for swallowing.

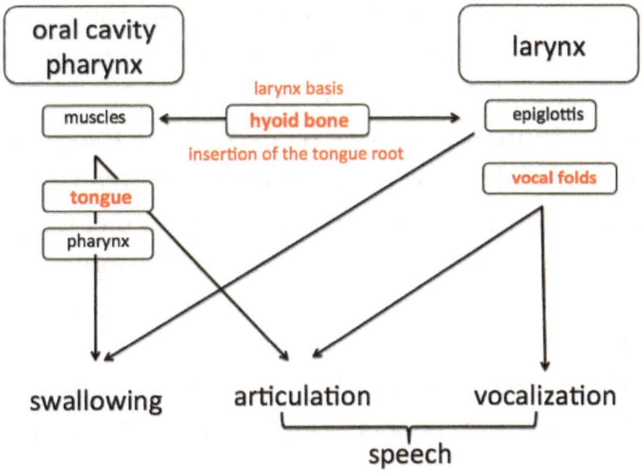

The hyoid bone plays an important role in the functional anatomy of swallowing, vocalization, and speech production, since it supports the base of the tongue (Figure 2). It is an isolated and fragile bone, of which only a few fossilized specimens have been found: a complete Neanderthal hyoid (Kebara2, 60 kya), a partial Neanderthal hyoid (Asturias, Spain, 43 kya), two hyoids assigned to *Homo heidelbergensis* (Sierra de Atapuerca, Spain, 530 kya), and a complete "chimpanzee-like" hyoid assigned to *Australopithecus afarensis* (Dikika, Ethiopia, 3.3 mya). These discoveries have renewed the interest in having such fossils for the debate on the origin of speech (Alemseged et al., 2006; Arensburg et al., 1989, 1990; D'Anastasio et al., 2013; Martinez et al., 2008; Rodriguez et al., 2003).

Indeed, the position of the hyoid bone relative to the cervical vertebrae has varied during phylogeny, and it also varies during ontogeny (Lieberman et al., 2001), both in humans and in non-human primates (Nishimura, 2006). The position of the hyoid is an important indicator to consider, perhaps more than that of the larynx itself, because it anchors the tongue root. Its position was considered as a marker of the speech production ability, according to the laryngeal descent hypothesis (Fitch, 2010, p. 312).

2. Descriptive anatomy of baboon vocal tract

An accurate comparison of the anatomy of the larynx and of the tongue musculature of baboons and humans is crucial for the discussion of the origin and evolution of speech, considering the crucial role played by these organs in speech. However, such comparative anatomy has been insufficiently described, notably less than for humans vs. chimpanzees (Hofer et al., 1990; Swindler and Wood, 1973; Takemoto, 2008).

The present description was based on two adult *Papio papio* heads, from one male and one female who died naturally in the UPS CNRS Primatology Station, Rousset, France, where various monkeys, including baboons, are kept. The two baboon heads were scanned at the Montpellier CHU in bone fenestration (General Electrics, cut 0.5 mm) when fresh, then sectioned in the strict median sagittal plane when frozen at the anatomy laboratory in Montpellier. Thawing was done in 10% formalin to perform the dissection that was conducted with binocular loupes in both *Papio papio* specimens.

2.1 General description of the vocal tract and larynx

The anatomical relations of the pharynx and the larynx are shown on a 3D reconstruction of a male *Papio papio* skull (Figure 3) incorporating a 3D reconstruction of the hyoid bone and the larynx (Figure 4). All 3D reconstructions are performed with Myrian® software. Additionally, a 3D reconstruction of the airways of the female *Papio papio* is superimposed on a sagittal section (Figure 5), and a zoomed portion of the larynx is compared to the median sagittal section (Figure 6). Details are provided in the figure captions.

The anatomy of the oral part of the baboon vocal tract is observed using a medial sagittal section of the female head (Figure 7). It is essentially equivalent to that of the human in its basic elements, but not in its proportions, as the tongue is longer and lower, the hard palate is flatter and the velum (soft palate) more horizontal. In humans, the genioglossal muscle (GG) has three groups of fibers (GGa, GGm, GGp) that form a fan-shaped structure with an anterior GGa vertical (Testut, 1897). In the baboon, the GG has the same structure with GGa composed of vertically oriented fibers and few or no fibers towards the tip of the tongue. The anterior portion of the tongue is proportionally larger, in concordance with the prognathism. The insertion on the superior mental spine is done by means of an aponeurosis on which the muscular fibers are fixed. The geniohyoid muscle, which is not a tongue muscle *per se*, is highly developed and accounts for more than half the tongue height. The styloglossus and digastric muscles were found at dissection to have a more horizontal orientation than those of humans. The hyoglossus presents, as in humans, two components, one inserting on the body of the hyoid bone and the other along the full length of the great horn.

Figure 3: 3D reconstruction (Myrian® software) of a male Papio papio skull. Left, the skeletal profile shows the entire mandible with a strong ramus but a minimally pronounced coronoid process and mandibular notch. Right, the posterior part of the mandible has been digitally removed after segmentation to show the hyoid bone, the larynx and the trachea. The lowest projection of the hyoid bone occurs above the lower edge of the mandible. The body of the hyoid bone is located forward and above the thyroid cartilage with a portion that descends forward. The hyoid's greater horns follow the upper border of the thyroid cartilage towards the thyroid's upper horns.

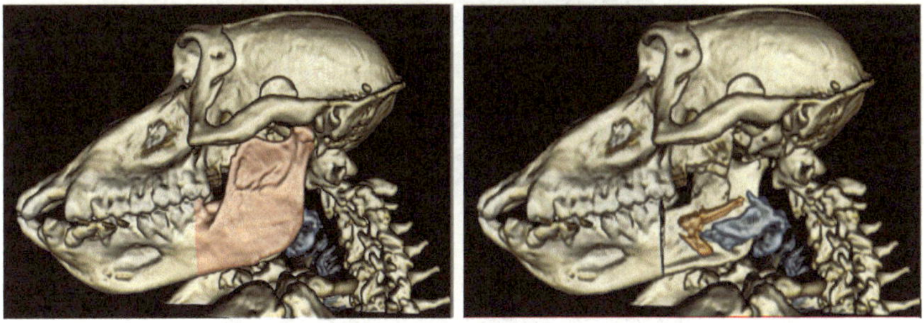

Figure 4: Comparing structure across species. Left panel, 3D reconstruction (Myrian® software) of the hyoid bone and the larynx of the male Papio papio. In this ¾ view, we find that morphology of the hyoid bone is grossly similar to that of the human with a median body, two greater horns and two lesser horns. The body of the hyoid bone has an inferior enlargement situated in front of the thyroid cartilage. The view of the larynx reveals a thyroid cartilage that is longer than it is high. It articulates with the cricoid cartilage and one can infer the position of the arytenoid cartilages. Right panel, in comparison to humans (right), the baboon larynx (left) shows that the thyroid cartilage is inset into a relatively larger hyoid bone, with the body enlarged inferiorly and total loss of the thyrohyoid membrane. This disposition is not found in great apes such as chimpanzee and gibbon, but it is found in the stump-tailed macaque, Macaca arctoides, and the white fronted capuchin, Cebus albifrons (Nishimura, 2003ab, 2005, 2006), shown in the inset (respectively left and right).

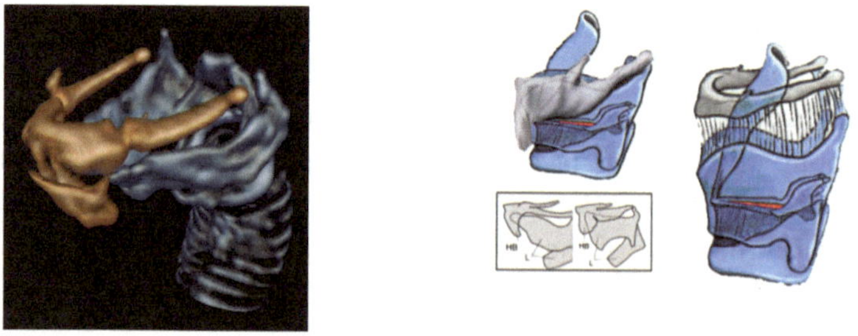

Figure 5: Sagittal section and 3D reconstruction (Myrian® software) of the airways of the female Papio papio. The airways were detected semi-automatically. The outlining of the upper airways and of the air in the oral cavity allows close study of their morphology. From behind the hyoid body and in front of the thyroid cartilage, the air sac and its conduit (circled) communicate with the larynx through a passage between the two cartilaginous plates of the thyroid cartilage. We see that the skull base is flattened: the angle shown is greater for all monkeys and apes than it is for humans. The axes of the hard palate and of the cervical vertebrae, both used in this study, are also displayed.

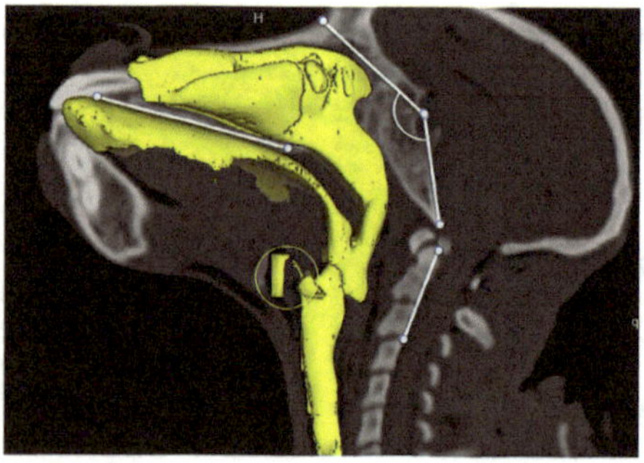

Figure 6: Laryngeal details of Papio papio. Left: The anatomical medial sagittal section of the larynx for a female Papio papio shows the hyoid bone (Hy) with the body anterior to the thyroid cartilage (Th). The air sac (AS) is between the two. The vocal folds (VF) are positioned at the middle of the thyroid cartilage's vertical dimension. The baboons' vocal folds measure 16.5 mm for the male and 11 mm for the female, in the same range as those of adult humans (Roers et al., 2009). The cricoid cartilage (Cr) is also found with its posterior part relatively high, as is the case for the epiglottis (E). The supraglottal portion of the larynx is very short. Right: Enlargement of the 3D reconstruction (Myrian® software) of the airway of the female Papio papio showing the connection from the air sac (AS) to the larynx, with the larynx divided into supraglottic (SG) and infraglottic (IG) portions. Connection is at the level of the glottis, and presents small laryngeal sacs (LS) laterally. The impression of the tongue (To) determines the oro-pharyngeal space that communicates with the piriform recess (PR).

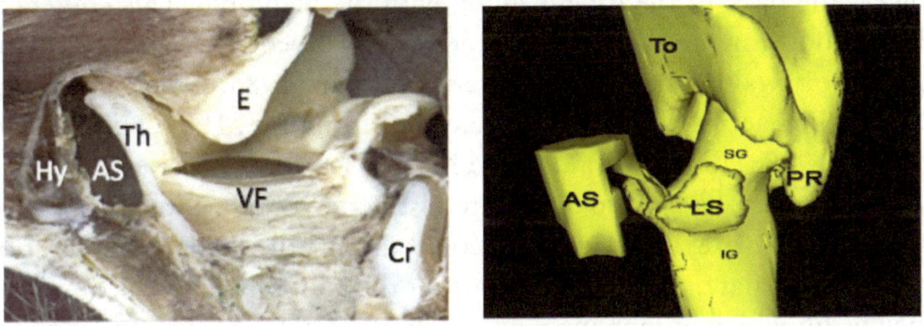

Figure 7: Anatomical medial sagittal section of a female Papio papio. The soft palate or velum (V) is at rest and disengaged from the pharyngeal wall (Ph), with the epiglottis (E) in contact with the uvula at the level of the atlas (C1). The anterior (GGa), middle (GGm) and posterior (GGp) parts of the genioglossus (GG) muscle of the tongue (To) are clearly discernible. The geniohyoid muscle (GH) is inserted from the symphysis (Sy) to the hyoid bone. (N.B. the left panel of Figure 6 is enlarged from this figure.)

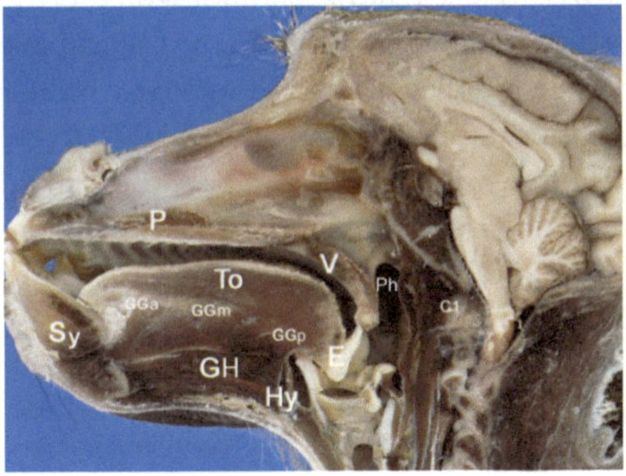

2.2 Tongue musculature and consequences for vocalization

The tongue musculature in baboons was examined by a dissection protocol similar to that used in humans. It appears that tongue musculature is structurally similar in humans and baboons, with the styloglossus and the three parts of the genioglossus, although the external shapes differ: the baboon tongue is flatter while the human tongue is rounded. The muscular hydrostat theory of the tongue shape suggests that, as in chimpanzees, the primary actions available to the baboon tongue are protrusion and retraction (Takemoto, 2008). In addition, the extrinsic muscles raise the back of the tongue through the action of styloglossus, while jaw opening lowers the back of the tongue along with the mandible. This confers to the baboon tongue the necessary degrees of freedom of movement required for swallowing (Crompton and German, 1984; Green and Wang, 2003; Hiiemae, 1967, 2000; Hiiemae and Crompton, 1985; Hiiemae and Palmer, 1999; Hiiemae et al., 1995; Martin 1991; Serrurier et al., 2012), which can then be used to articulate distinctive vocalizations combining two axes (Figure 8). Taking into account the length and flat configuration of the hard palate, it is not clear whether the baboon is capable of sounds such as /i/, which in humans require a long apical constriction along the alveopalatal area.

Overall, these considerations are nonetheless compatible with exaptation. Boë et al. (2017) discussed how "[t]he baboon's muscle fiber orientation allows tongue motion along two main axes." Antagonistic activation of the GGam, and SG tongue muscles produces changes in the vocal tract allowing both a front/back contrast homologous to the human [æ] ⇔ [u] along the first axis and the posterior constriction needed for [u]. The GGp and HG tongue muscles produce homologs to the human [ɑ] ⇔ [i] contrast through vertical tongue displacement along the second axis. These two axes do have different orientations in baboons and humans, due to the differences in the inclination of the styloglossus and the shape of the tongue (Badin and Serrurier, 2006; Buchaillard et al., 2010; Denny and McGowan, 2012; Honda, 1996).

The laryngeal descent hypothesis asserted that human-like speech production was not possible because of the lack of a posterior cavity allowing a second independent axis:

> This lower, pharyngeal portion of the vocal tract provides a whole new dimension: by moving the tongue backwards and forwards this lower tube can be *independently* modified (this arrangement is thus dubbed a "two-tube" tract) ... Thus, the descent of the tongue root and larynx provides an additional degree of freedom, a new dimension of control, compared to the capabilities inferred for a normal mammalian tract. (Fitch, 2010, p. 312, italics in the original).

We disagree, and have concluded that the second axis is not related to laryngeal descent or to the increase in the posterior cavity, but to the tongue muscle structure itself.

Figure 8: Human (left) and baboon (right) muscles and axes producing vowel contrasts. Note that the main axis of the styloglossus muscle (SG) is nearly horizontal in the baboon, with an action resulting in tongue retraction rather than elevation.

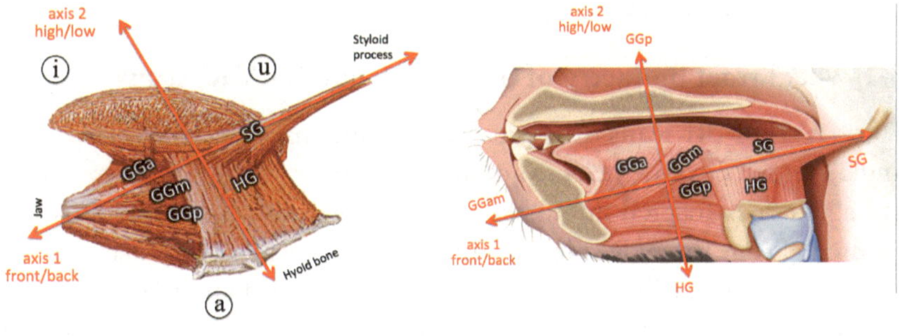

3. Quantitative comparative anatomy of baboons and humans

We now proceed to quantitative analysis of anatomical landmarks associated with the hyoid bone, glottis, prosthion, palatal plane, oral and pharyngeal cavities, and the C2, C3, and C4 vertebrae. We are looking for an adapted anatomical description by: (1) measuring the length of the vocal tract; (2) decomposing it into an oral part (oral cavity length, OCL) and a pharyngeal part (larynx height, LH); (3) locating the glottis and the hyoid relative to the cervical vertebrae; and (4) recapitulating these observations by age and sex for comparison of baboons with human females, males, and children.

Vocal tract length, measured from the glottis to the lips, depends on the bony structure, soft parts, and their interactions during growth (Scammon, 1930). The vocal tract is composed of an oral part, from the lips to the pharyngeal point (defined below), and a pharyngeal part, from the pharyngeal point to the glottis. In humans, the growth of these parts is heterochronic and there is sexual dimorphism: during puberty, the pharyngeal part develops more than the oral part, especially in males (Goldstein, 1980).

Importantly, in modeling, vocal tract length happens to be a key parameter for the estimation of potential capacities for production of formant resonances in humans through constrictions and cavities (Boë et al., 1989; Bonder, 1983; Liljencrants and Lindblom, 1972) and also for characterizing the acoustic structure of vocalizations in nonhuman primates (Boë et al., 2017; Riede et al., 2005).

3.1 Vocal tract biometry of baboons from 3D MRI scans

Fifty-six 3D MRI scans of *Papio anubis* baboons, consisting of 23 males and 33 females from 2 years to adulthood (at least 6 years), were performed at La Timone Hospital in Marseille. For the MRI scans the baboons were sedated. From the MRI files, a multiplanar analysis with OsiriX® Lite software (a DICOM viewer) served to locate the median sagittal plane. To do so, we first located median structures in an axial plane: we display in Figure 9, in front, the superior mental spine (EM; note that *spine* is *epine* in French), and in back, the middle of the vertebrae on the axial section. The median sagittal section passes through the mandibular symphysis at the level of the insertion of the genioglossus on the superior mental spine.

Figure 9: Reference sections of the baboon head. The median sagittal plane must be determined using three points across which there is general bilateral symmetry. The plane thus determined invariably sections the hyoid. The final settings and sections are shown here.

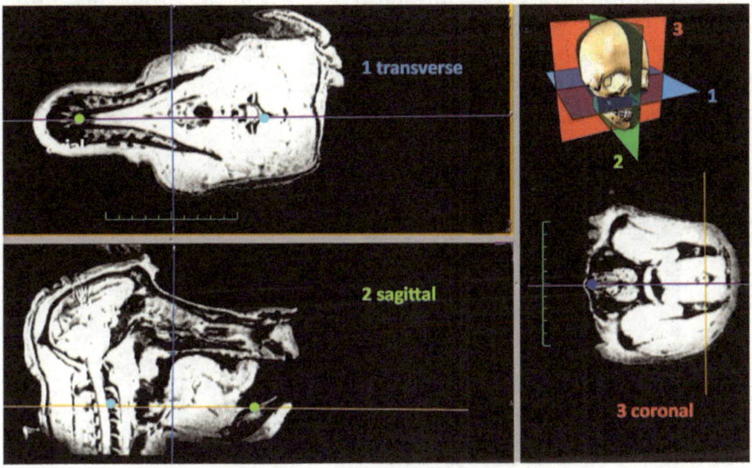

In the median sagittal plane, we located the following landmarks, using a program written in Matlab® (see Figure 10): the apex of the odontoid (Od), the posteroinferior edge of the vertebral bodies of C2, C3, and C4, the anterior nasal spine (ENa), the posterior nasal spine (ENp), the prosthion (Pr), the upper mental spine (EM), and the top of the upper edge of the hyoid bone (Hy). The following other references are obtained by construction (semi-landmarks): the occlusal plane (which is parallel to the palatal plane ENa-ENp) is evaluated and the pharyngeal point (PPh) is defined as the intersection of the occlusal plane with the pharyngeal wall in the sagittal plane. (Note that the occlusal and palatal planes are presumed perpendicular to the sagittal plane.) The position of the larynx in relation to the cervical spine changes with flexion and extension of the neck (Reidenberg and Laitman, 1991; Westhorpe, 1987).

Figure 10: Hyoid positioning under head flexion. As explained in the text, a circle was struck at the hyoid position, Hy, with Od at the center. A correction based on the angle of Od-C4 and the palatal plane was then applied to correct for head inclination and find the corrected hyoid position, Hyc. The projected hyoid position, Hyp, was at the orthogonal projection of Hyc on the Od-C4 line. For adult humans Hyp is consistently found at the level of C4, regardless of the angle of head inclination. In the right-hand panel, we note that the angle is close to 90 degrees, so only a minimal correction is necessary from Hy to Hyc.

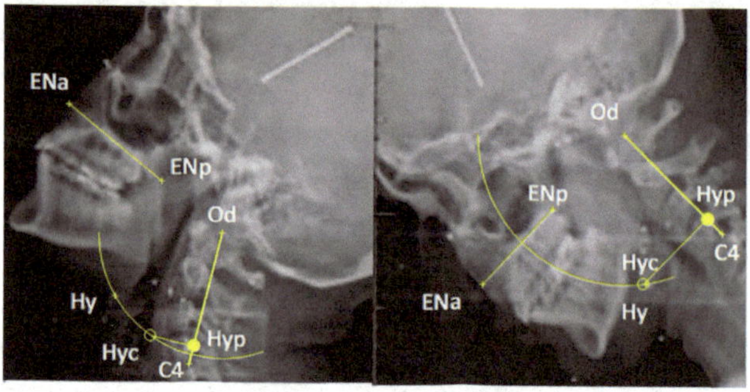

In previous primate studies the inclination of their head relative to the cervical vertebrae has not always been well controlled, which has allowed vertical displacement of the hyoid bone according to the inclination of the head. To correct such differences, we adopted a new procedure and validated it first for humans. Working from a profile view, it consisted in locating the palatal plane (using the anterior and posterior nasal spines ENa and ENp) and then in drawing a circle centered at the apex of the odontoid (the estimated center of rotation) and passing through the top edge of the body of the hyoid bone. The angle between the palatal plane and the vertebral column (defined as Od-C4), which is compared to a theoretical angle of 90° (to make comparisons with humans), is then measured. The difference is the correction angle applied to the hyoid bone and after rotation (Hy => Hyc), this point Hyc is then projected orthogonally onto the vertebral column to get the final landmark Hyp (Figure 10). The procedure was tested on different radiographs and partly on MRIs of vowels /i a u/ pronounced by a French speaker. The position of Hyp along the Od-C4 axis was found constant. We do not apply this procedure for finding the position of the glottis, which descends slightly with the inclination of the head, but the larynx is generally much more stable than the hyoid bone.

In order to compare vocal tract and other anatomical dimensions across species, ages, and subjects, we have normalized all distances based on the distance between the apex of the odontoid and C3 inf (dOC3). Thus, we analyze the variations in vocal tract dimensions with respect to cervical vertebrae and their axis taken as a reference. The normalized measures will be expressed as ratios without units, and the position of C3 inf is always taken as 1. This choice is motivated by the consideration that the vertebral axis should be a stable reference in evolutionary terms because of its location in the HOX zone (Figure 1), which is known to have undergone few mutations over a very long time, including the period of mammalian emergence. Conversely the oral part of the vocal tract, located in the non-HOX areas, is considered variable, as well as its pharyngeal and laryngeal parts which are influenced by the hyoid bone position. The normalization operation makes it possible to quantify these variations, under the assumption that the contribution linked to isometric growth of the vocal tract cavities and the cervical vertebrae is thus suppressed, keeping only the relative variations with respect to the vertebral axis. The size differences between subjects are of course eliminated at the same time. This differentiates our study from those of the Japanese chimpanzee and macaque (Nishimura 2003a, 2003b, 2005, 2006; Nishimura et al., 2003, 2008). This approach allows us to evaluate laryngeal descent relative to the odontoid along the vertebral axis, and also indirectly relative to the palate.

Figure 11: Landmarks used to characterize baboon vocal tract geometry (15 year old adult female): Prosthion Pr; Pharyngeal Point PPh (located at the intersection of the occlusal plane and the pharyngeal wall); Glottis Gl; Hyoid Hy, corrected Hyoid Hyc, and projected Hyoid Hyp; Odontoid Od; inferior edges of the cervical vertebrae C2, C3, C4; upper mental spine EM; and anterior and posterior nasal spines ENa and ENp.

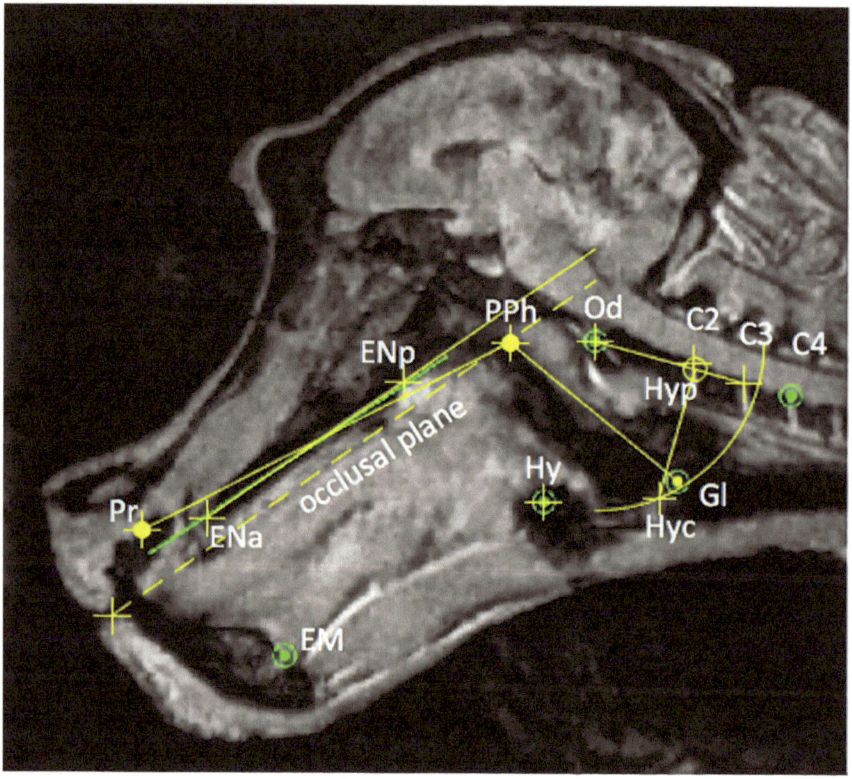

Figure 12: Biometric results for baboons. (See text for discussion of normalized measures in panels 1–4, where, on the abscissa, 0 corresponds to the odontoid and 1 to C3 inf.) (1) C2 inf position. (mean = 0.61); (2) Hyoid bone vertebral projection, Hyp, apparently bimodal; (3) Glottis position, which appears centered on C3 inf; (4) Gaussian modeling of data in panels 1, 2, & 3, with hyoid and glottis data split by age (<= and > 6 years); Note that the bimodal appearance of the panel 2 hyoid data is partially related to age, and that age also affects glottis position; (5) and (6) Single-sigmoid fits of projected Hyoid, Hyp, and Glottis, Gl, by age (red dots for females, blue for males); (7) Oral Cavity Length (OCL, the distance between Prosthion, Pr, and Pharyngeal Point, PPh); (8) OCL across age, with a variation represented by two levels, one up to 6 years and the other beyond; (9) LHI (ratio between Larynx Height, LH, defined as the distance between the pharyngeal point and the glottis, and the OCL) across age, which is approximately constant, with an average value of 0.43.

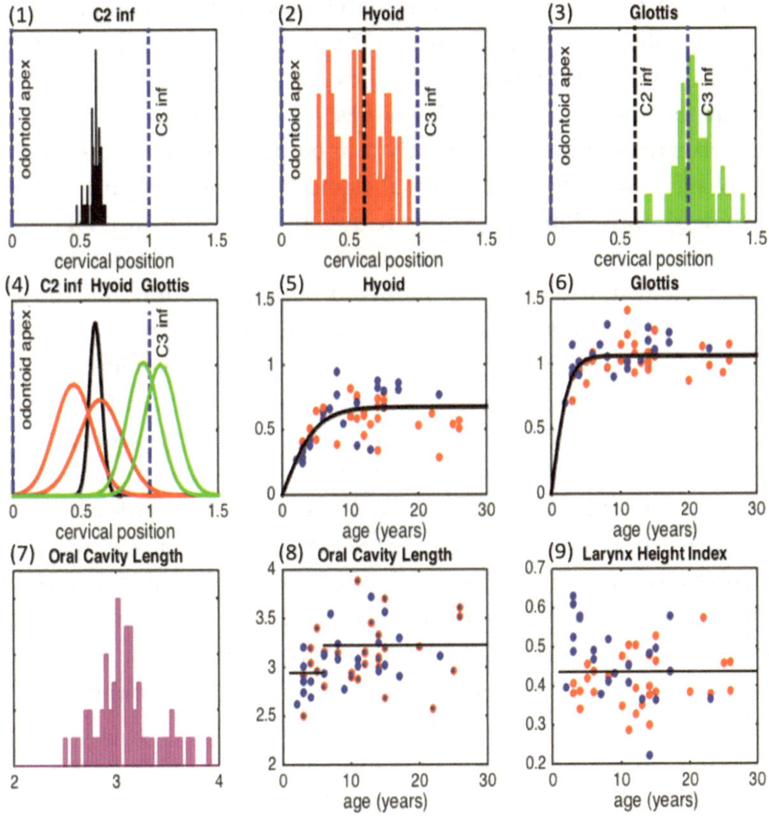

After normalization, the distribution of C2 inf position is narrow (Figure 12.1), providing a precise landmark. The distance between C2 inf and C3 inf (mean = 0.39) is useful for defining a metric expressed in fraction of a vertebra, knowing that one vertebra thus defined includes both the body and the intervertebral space. The distribution of (normalized) vertebral projections of the hyoid is bimodal with the main peak around C2 inf and a second smaller peak somewhat higher on C2 (Figure 12.2). The distribution according to age shows a clear increase which is well fitted with a sigmoid function (Figure 12.5) and the smaller peak apparently represents the young baboons. This is well shown thanks to Gaussian modeling (Figure 12.4) applied after decomposition in two age groups (<= 6 years and > 6 years): the main peak corresponds to adults (mean = 0.64) and the smaller to young baboons (mean = 0.44), the narrow Gaussian indicating the position of C2 inf at 0.61. Using these means for the 2 age groups and our estimate above of a standard vertebra, we can estimate the amount of hyoid descent as (0.64–0.44)/0.39=0.51 vertebra. The glottis histogram is apparently monomodal, but we see some variation by age that we fit with a partial sigmoid function (Figure 12.6). This is also decomposed in two groups with Gaussian modeling (Figure 12.4) having means at 0.97 (<= 6 years) and 1.1 (> 6 years). This is around C3 inf, and the distance between the hyoid bone projection and the glottis projection is approximately equal to one vertebra for both groups. The glottis descent, (1.1–0.97)/0.39=0.33 vertebra, is less than half a vertebra. Note that our determination of the location of the hyoid bone position is more precise than for the glottis, which is, ultimately, an empty space. Moreover, we do not apply any correction to the glottis position. Thus, the hypothesis of having approximately the same descent for hyoid bone and glottis, about half a vertebra, is reasonable. In summary, we quantify the process of laryngeal descent from baboon childhood to adulthood in the following manner: from the middle of C2 to C2 inf for the hyoid, and from the middle of C3 to C3 inf for the glottis, with a constant distance of one vertebra between hyoid and glottis. We have also found that dOC3 does not vary much with age (data not shown).

The oral cavity length (OCL) is normalized similarly in order to see if developmental laryngeal descent in baboons is associated with an increased prognathism. The histogram of OCL is monomodal with a peak at about 3 (Figure 12.7). In other words, the distance between the prosthion and the pharyngeal point (resp. Pr and PPh in Figure 11) is on average about 3 times the distance between the glottis projection (centered on C3 inf) and the odontoid. The distribution of OCL according to age is divided in two groups (young and adults with limit at 6 years) to show a small increase in prognathism (Figure 12.8). Complementarily, the larynx height index (LHI) (Honda and Tiede, 1998) is defined as the ratio of the OCL and laryngeal height, itself defined as the pharyngeal-glottal distance (Figure 11). It averages a constant 0.43 across age in baboons (Figure 12.9), because it is a ratio between two similarly increasing values. In contrast, LHI varies from 0.5 at birth to 1.0 at adulthood for humans, since there is no increase in OCL while over time there is laryngeal descent.

3.2 Vocal tract biometry of human children, from radiography

The position of the hyoid bone and larynx in children has been reported in various studies (Amayeri et al., 2014; Coelho-Ferraz et al., 2006, 2007; Grant, 1965; Westhorpe, 1987). Our own radiographs of children were obtained as part of the SkullSpeech project (Perrier and Boë, 2009–2012). We also obtained radiographic data from M. J. Deshayes (127 children, girls and boys from 3 to 7.5 years, mean 5.2 years, standard deviation 0.95 year), an age range in which there is no sexual dimorphism.

We analyzed the radiographs with the same general procedure as for baboons, adding C4 inf, but eliminating the glottis, which was not visible. We applied the same normalization with dOC3. It appears that the distributions of both C2 inf and C4 inf are narrow. The hyoid projection is between the two, near C3 inf. The mean OCL value is 1.85, and the ratio of OCL between young baboons and infants is 3/1.85=1.62 indicating the high degree of prognathism in young baboons. The variations of both hyoid position and OCL seem minimal around 5 years of age. LHI cannot be measured since it requires glottis position, and the glottis was not visible.

Figure 13: Biometric results for human children. (See text for discussion of normalized measures.) (1) Positions of C2 inf and C4 inf, while C3 inf is at 1 by definition; (2) Hyoid bone vertebral projection, Hyp, mean = 1.08 centered on the space below C3 inf; (3) Gaussian models of distributions of the positions of C2 inf and C4 inf and of the hyoid projection, Hyp; (4) Oral Cavity Length (OCL), mean=1.85; (5) Hyoid projection, Hyp, across age, with no significant change; (6) OCL across age, with no significant change.

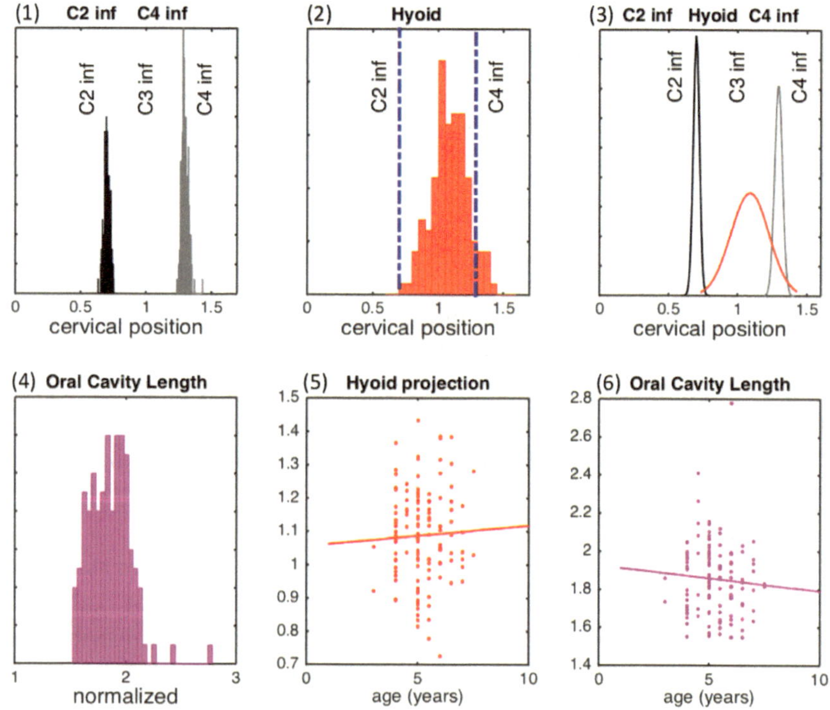

Figure 14: Position of hyoid and glottis compared to cervical vertebrae for young (≤ 6 years) and adult baboons, and in humans, for newborns (Westhorpe, 1987; Barbier, 2010) 5 year-olds (present data; Barbier, 2010), and adult females (Barbier, 2010) and males (Westhorpe, 1987; Barbier, 2010).

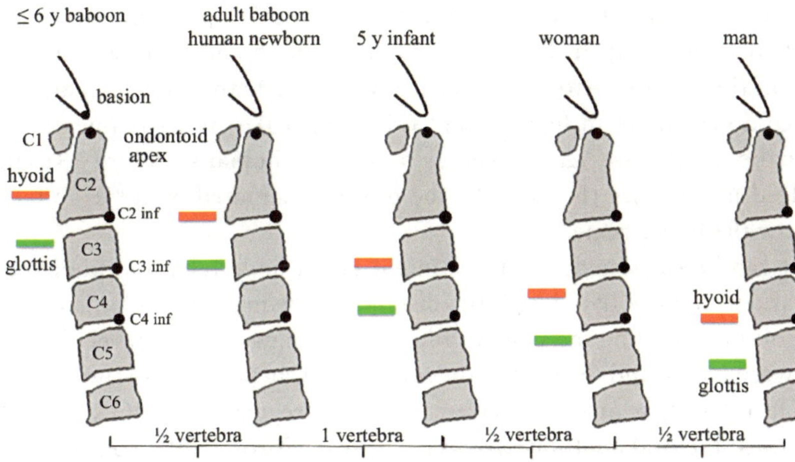

Though we detail here only data for human children, our previous studies address growth patterns of the glottis, hyoid, and vertebrae in humans from childhood through adulthood (Barbier, 2010; Barbier et al., 2012, 2015). These studies show that in both males and females, there is considerable vertical growth during the first two years of life, then the growth of C3 and the hyoid position stabilize between 3 and 8 years of age. Glottis descent appears highly correlated with those of C3 and the hyoid. At around 10 years there begins a second surge of vertical growth, which only affects the hyoid and the glottis, C3 being extremely stable after the 8th year. This second surge, greater for the glottis than for the hyoid, seems to stabilize for the hyoid at about 15 years for women, but continues for the glottis until nearly 20 years for men (Barbier, 2010; Barbier et al., 2012, 2015). Figure 14 summarizes all these data in five diagrams comparing young baboons, adult baboons, and human children and adults, female and male. In all these diagrams we observe an offset of about one vertebra between the vertebral projections of the hyoid and the glottis.

We showed in the previous section that the hyoid in the young baboon (less than 6 years old) projects to the level of the body of C2, and for the adult baboon to the level of C2 inf. Our data show that lowering in baboons takes place in a single step, without any clear sexual dimorphism. For comparison, the hyoid bone projects to the same level in the adult baboon, as in the human newborn (Barbier et al., 2012, 2015). However, the descent of the hyoid in humans takes place in two stages, and we found that in the 5-year-old child, the hyoid bone was around C3 inf (Figure 13). After adolescence, we estimate that the descent takes place down to C4 in females and C4 inf in males. It also appears that in baboons (Figure 12) as well as 5-year-old humans (Figure 13), and indeed in all cases (Figure 14), the glottis is situated about 1 vertebra below the hyoid bone.

It can be suggested that there might be an underlying morphological invariant, namely hyoid – glottis distance, allowing the epiglottis to play its role of protecting the airways while maintaining a constant relationship between its top and the base of the tongue. This hypothesis is reinforced by the fact that these anatomical elements are all regulated by HOX genes. In contrast, the oral anatomy would have escaped from HOX control (Chai and Maxson, 2006), leading to greater changes including a decrease in prognathism associated with a caudal displacement of the tongue and an increase in verticality. The consequence would be a caudal hyoid – glottis translation relative to the spinal column, maintaining the distance between them to enable swallowing and to protect the airways.

3.3 Vocal tract growth in humans and baboons

In a previous study (Barbier et al., 2012, 2015) a grouping of four American Association of Orthodontists (AAO) archives was used to quantify (human) vocal tract growth. These records contain 966 sagittal X-rays of the head and neck for 68 white North American subjects (33 women and 35 men), obtained approximately every year between 1 month and 25 years in order to study longitudinal growth of the dentition. For baboons, we were able to retain only 25 of our 56 subjects, the lips being sometimes absent from the MRI.

Figure 15: Vocal tract length (VTL) measurement from glottis to lips: 10 landmarks positioned by hand and then joined with a spline curve. Baboon Papio anubis, male, 15 years, VTL = 13.2 cm.

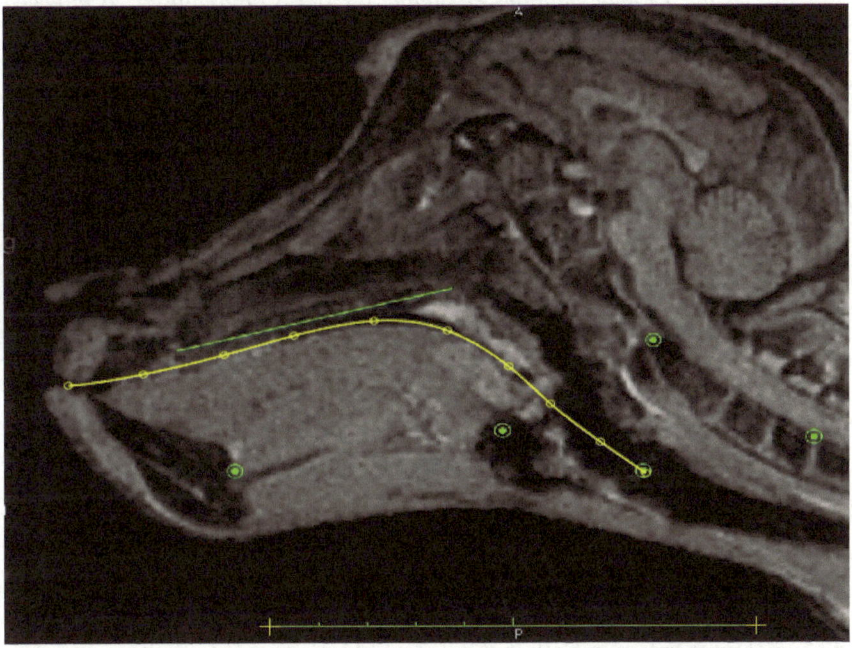

Figure 15 shows the method of VTL measurement used for the 25 baboons. The distance from lips to glottis is the length of a spline curve defined with ten points manually placed on the tongue surface. Figure 16 shows VTL growth obtained by fitting data with a double sigmoid function for human females and males (Barbier et al., 2012, 2015) and, for comparison, measurements modeled by a simple sigmoid for female and male baboons. The choice between simple and double sigmoids stems from the fact that baboons reach their maturity around 6–7 years without an adolescent phase. For adult baboons, as for humans, there is sexual dimorphism, so the growth profiles are not homogeneous and diverge after few years (Figure 16).

Figures 16: Development of vocal tract length from fertilization (assuming 9 months gestation) to adulthood for female and male humans (Barbier et al., 2012, 2015), as fitted with a double sigmoid, and for 12 female and 13 male baboons (assuming 6 months of gestation), as modeled with a simple sigmoid.

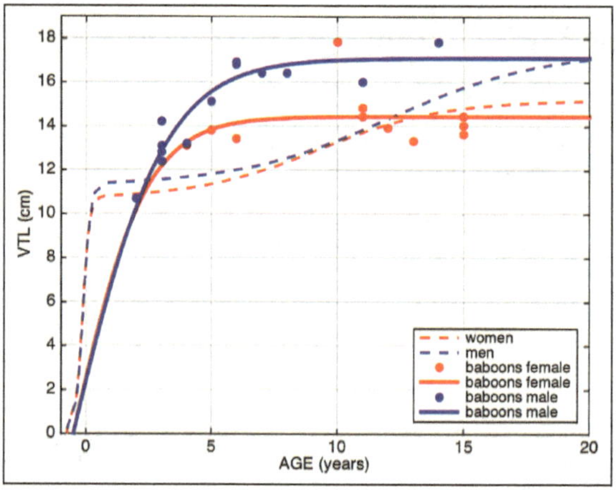

Length analysis of the vocal tract's oral segment as normalized to dOC3 shows that there is additional growth linked to prognathism in baboons from youth into adulthood (Figure 12.8, Oral Cavity Length). The fact that the vocal tract grows in both its OCL and LH dimensions results in stability of the Larynx Height Index (Figure 12.9). In addition, we measured VTL (Figure 15) using points manually placed on the back of the tongue and including the two oral and pharyngeal segments, which showed a one-step variation during baboon growth (Figure 16). In contrast, human data show a two-step VTL growth that is consistent with laryngeal descent, which takes place in two stages as well. We conclude that vocal tract growth in the baboon is less than humans in its pharyngeal region, but much greater than humans in the oral region (Goldstein, 1980). This results in a large difference in LHI between baboons and humans: 0.43 for baboons vs. 0.75 for female and 1.0 for male humans. Remarkably, laryngeal descent in humans compensates for the lack of prognathism, with the ultimate effect of preserving vocal tract length.

4. Conclusion

In this study, we propose a series of new qualitative and quantitative biometrical analyses of oropharyngeal anatomy adapted and normalized for the comparison between baboons and humans. Using the vertebral column as a fixed phylogenetic reference, we derive a measure of larynx height (the LHI), and also of variations of the oral cavity length expressed with a new metric. We show that the distance of one vertebra separating the hyoid bone and the glottis appears to be invariant, despite the great morphological differences illustrated in Figure 4. A new representation of the laryngeal descent process is summarized on Figure 14. In comparison to the young baboon, we find that the adult baboon has a single-stage laryngeal descent of only ½ vertebra. The oral cavity then grows an equivalent length, thus keeping LHI constant. The human newborn is at the level of the adult baboon, and humans undergo two descent stages, cumulatively amounting to 1½ vertebra for females and 2 for males. Since this is realized without the increased oral length from prognathism, it results in a large LHI increase.

Functionally, the distance of 1 vertebra between the hyoid bone and the glottis is highly constrained by the mechanical requirements of swallowing. As baboon tongue musculature is similar to that of humans, if we consider the hyoid bone as the tongue root, laryngeal descent in humans corresponds to a tongue shift toward the back, without modification of the relationship between the tongue and the oropharyngeal cavity because the length of the oral cavity is preserved. We assume this is also constrained by the swallowing function, because the role of the tongue is to drive the food, solid or liquid, from the lips to the esophagus. In baboons, the tongue tip appears to receive no fibers from GGa. This suggests that in humans, laryngeal descent divided the vocal tract in two cavities with a specialization of the anterior part in chewing and preparation of the food for ingestion. The corollary of this specialization was to acquire a better musculature of the tongue tip.

All these observations are compatible with the hypothesis of speech exaptation from feeding gestures. To begin with, tongue musculature in baboons and humans is similar, with the same two control axes enabling constrictions and cavities to form. Swallowing actually involves vocal tract constrictions, and even if the tongue is flat in baboons, the relationship between tongue and oropharyngeal cavity remains similar to that of humans.

Fitch (2010) recapitulates the argument that laryngeal descent is required to form the posterior cavity to produce the vowel /u/, and more generally, the human vowel triangle. While baboons can probably not produce a vocalization close to /i/ because their tongue tip musculature is not developed, a high larynx is not an obstacle preventing vocalizations with formants diverging significantly from the central vowel. More precisely, we consider that laryngeal descent is not required for the development of a posterior cavity. In fact, the stability of vocal tract length and the constant relationship of tongue to oropharynx ensure that the production of an /u/-like vocalization is possible for baboons through contraction of the styloglossus, the same muscle as in humans, allowing a retraction of the tongue and forming a posterior cavity (Figure 8; also see Boë et al., 2017). Constriction control, already present to allow swallowing, is the crucial factor in production of distinct vowel qualities, and this is clearly an example of exaptation from feeding gestures for speech.

Acknowledgements

Thanks to Mathieu Labrunie who transcoded the files (*.nii in *.dcm) of the original MRIs so that we could process them with the OSIRIX software; to Guillaume Barbier for furnishing his data for children growth; to Pierre Badin for providing us with radiographs for the /i a u/ vowels with different inclinations of the head; and to Daniel Lieberman for access to the AAO Denver radiographic database (see Lieberman and McCarthy, 2001).

References

Alemseged, Z., Spoor, F., Kimbel, W. H., Bobe, R., Geraads, D, Reed, D., and Wynn, J.G. (2006). A juvenile early hominin skeleton from Dikika, Ethiopia. *Nature, 443*, 296–301.

Amayeri, M., Saleh, F., and Daleh, M. (2014). The position of hyoid bone in different facial patterns: A lateral cephalometric study. *European Scientific Journal, 10*, 15, 19–34.

Andrew, R.J. (1976). Use of formants in the grunts of baboons and other nonhuman primates. In S. R. Harnad, H. D. Steklis, and J. Lancaster (Eds.), *Origins and evolution of language and speech* (pp. 673–693). New York: New York Academy of Sciences.

Arensburg, B., Schepartz, L.A., Tillier, M.A., Vandermeersch, B., and Rak, Y. (1990). A reappraisal of the anatomical basis for speech in middle paleolithic hominids. *American Journal of Physical Anthropology*, *83*, 137–146.

Arensburg, B., Tillier, A.M., Vandermeeersch, B., Duday, H., Schepartz, L.A., and Rak, Y. (1989). A middle palaeolithic human hyoid bone. *Nature*, 338, 558–760.

Badin, P., and Serrurier, A. (2006). Three–dimensional linear modeling of tongue: Articulatory data and models. In *Proceedings of the 7h International Seminar on Speech Production*, 395–402.

Ballard, K., Robin, D., and Folkins, J. (2003). An integrative model of speech motor control: A response to Ziegler. *Aphasiology*, *17*, 37–48.

Barbier, G. (2010). *Croissance du conduit vocal de la naissance à l'âge adulte. Étude radiographique longitudinale*. Mémoire de Master 1 Recherche. Université Stendhal, Grenoble.

Barbier, G., Boë, L.J., and Captier, G. (2012). La croissance du conduit vocal du fœtus à l'âge adulte : une étude longitudinale. *Biométrie humaine et Anthropologie*, *1–2*, 11–22.

Barbier, G., Boë, L.J., Captier, G., and Laboissière, R. (2015) Human vocal tract growth: A longitudinal study of the development of various anatomical structures. In *Proceedings 16th Interspeech*, 364–368.

Boë, L.J., Berthommier, F., Legou, T., Captier, G., Kemp, C., Sawallis, T.R., Becker, Y., Rey, A., and Fagot, J. (2017). Evidence of a vocalic proto-system in the baboon (*Papio papio*) suggests pre–hominin speech precursors. *PLoS One*, *12*, e0169321.

Boë, L.J., Perrier P., Guérin, B., and Schwartz, J.L. (1989). Maximal vowel space. In *Proceedings Eurospeech*, 2, 281–284.

Bonder, L. (1983). Equivalency of lossless n–tubes. *Acustica, 53*, 193–200.

Buchaillard, S., Perrier, P., and Payan, Y. (2010). A biomechanical model of cardinal vowel production: muscle activations and the impact of gravity on tongue positioning. *The Journal of the Acoustical Society of America*, *126*, 2033–2051.

Bunton, K. (2008). Speech versus nonspeech: different tasks, different neural organization. In *Proceedings Seminar Speech Language*, *29*, 267–275.

Chai, Y, and Maxson, R.E. Jr. (2006). Recent advances in craniofacial morphogenesis. *Developmental Dynamics*, 9, 2353–2375.

Clark, W. E. L., Cooper D. M., and Zuckerman, S. (1936). The endocranial cast of the chimpanzee. *Journal of the Royal Anthropological Institute of Great Britain and Ireland, 66*, 249–268.

Coelho–Ferraz, M. J. P., Nouer, D. F., Bérzin, F., de Sousa, M. A., and Romano, F. (2006). Cephalometric appraisal of the hyoid triangle in Brazilian people of Piracicaba's region. *Brazilian Journal of Oral Science*, *5*, 1001–106.

Coelho–Ferraz, M. J. P , Nouer, D. F., Teixeiras, J. R., and Bérzin, F. (2007). Cephalometric assessment of the hyoid bone position in oral breathing children. *Brazilian Journal of Otorhinolaryngology*, *73*, 47–52.

Couly, G., and Bennaceur, R. (1998). Biologie du développement de la face et du cou. Acquisitions récentes d'embryologie génétique. *Encyclopédie Médicale de Chirurgie et de Stomatologie*, 22–00l-A-l0, 7p.

Couly, G., Coltey, P., and Le Douarin, N. (1993). The triple origin of skull in higher vertebrates: A study in quail-chick chimeres. *Development, 117*, 409–429.

Couly, G., Creuzet, S., Bennaceur, S., Vincent, C., and Le Douarin, N. M. (2002). Interactions between Hox-negative cephalic neural crest cells and the foregut endoderm in patterning the facial skeleton in the vertebrate head. *Development*, *120*, 1061–1073.

Crompton, H. A. F, and German, R.Z. (1984). Mechanism of intraoral transport in macaques. *American Journal of Anthropology, 65*, 275–282.

D'Anastasio, R., Wroe, S., Tuniz, C., Mancini, L., Cesana, D. T., Dreossi, D., Ravichandiran, M., Attard, M., Parr, W. C. H., Agur, A., and Capasso, L. (2013). Micro-biomechanics of the Kebara 2 hyoid and its implications for speech in Neanderthals. PLoS One, 8, e82261.

Denny, M., and McGowan, R. S. (2012). Implications of peripheral muscular and anatomical development for the acquisition of lingual control for speech production: A review. *Folia Phoniatrica et Logopaedica*, *64*, 105–115.

Falk, D. (2014). Interpreting sulci on hominin endocasts: old hypotheses and new findings. *Frontiers in Human Neuroscience*, *8*, 134.

Fant, G. (1960). *Acoustic theory of speech production.* The Hague: Mouton and Co.

Fischer, J., Hammerschmidt, K., Cheney, D., and Seyfarth R. (2002). Acoustic features of male baboon loud calls: influences of context, age, and individuality. *The Journal of the Acoustical Society of America, 111*, 1465–1474.

Fitch, W.T. (2010). *The evolution of language*. Cambridge University Press: Cambridge.

Folkins, J., Moon, J., Luschei, E., Robin, D., Tye-Murray, N., and Moll, K. (1995). What can nonspeech tasks tell us about speech motor disabilities? *Journal of Phonetics*, *23*, 139–147.

Goldstein, U.G. (1980). *An articulatory model for the vocal tract of the growing children*. Thesis of Doctor of Science, MIT, Cambridge, Massachusetts.

Gould, S.J., and Vrba, E. (1982). Exaptation – a missing term in the science of form. *Paleobiology*, *8*, 4–15.

Grant, L.E. (1965). *A radiographic study of hyoid bone position in Angle's Class I, II, and III malocclusions* [Master's thesis]. University of Kansas City.

Green, J.R., and Wang, Y.-T. (2003). Tongue-surface movement patterns during speech and swallowing. *The Journal of the Acoustical Society of America*, *113*, 5, 2820–2833

Hall, K. R. L., and DeVore, I. (1965). Baboon social behavior. In I. DeVore (Ed.) *Primate Behavior: Field Studies of Monkeys and Apes* (pp. 53–110). New York: Holt, Rinehart and Winston.

Harrison, D. H. (1995). *The anatomy and physiology of the mammalian larynx*. New York: Cambridge Univ. Press.

Hiiemae, K. M. (1967). Masticatory function in mammals. *Journal of Dental Research*, *46*, 883–893.

Hiiemae, K. M. (2000). Feeding in mammals. In K. Schwenk (Ed.), *Feeding in tetrapods* (pp. 411–448). San Diego: Academic Press.

Hiiemae, K. M., and Crompton, A. W. (1985). Mastication, food transport and swallowing. In M. Hildebrand, D. Bramble, K. Liem, and D. Wake (Eds.), *Functional vertebrate morphology* (pp. 262–290). Cambridge, MA: Belknap Press, Harvard University Press.

Hiiemae, K. M., and Palmer, J.B. (1999). Food transport and bolus formation during complete feeding sequences on foods of different initial consistency. *Dysphagia*, *14*, 31–42.

Hiiemae, K. M., and Palmer, J .B. (2003). Tongue movements in feeding and speech. *Critical Reviews in Oral Biology and Medecine, 14*, 413–429.

Hiiemae, K. M., Palmer, J. B., Medicis, S. W., Hegener, J., Jackson, B. S., and Lieberman, D. E. (2002). Hyoid and tongue surface movements in speaking and eating. *Archives of Oral Biology, 47*, 11–27.

Hiiemae, K. M., Reese, A., and Hayenga, S. (1995). Patterns of tongue and jaw movement: A cinefluorographic study of feeding in the macaque. *Archives of Oral Biology, 40*, 229–246.

Hofer, H. O., Meinel, W., and Sauer, E. (1990). Comparative anatomic studies of the tongue of Pan troglodytes (Blumenbach, 1799) and other primates. I. The chimpanzee tongue. *Gegenbaurs morphologisches Jahrbuch, 136*, 455–492.

Honda, K. (1996). Organization of tongue articulation for vowels. *Journal of Phonetics, 24*, 39–52.

Honda, K., and Tiede, M. K. (1998). An MRI study on the relationship between oral cavity shape and larynx position. In *Proceedings 5th International Conference on Spoken Language Processing, 2*, 437–440.

Kelemen, G. (1969). Anatomy of the larynx and the anatomical basis of vocal performance. In G. Bourne (Ed.), *The Chimpanzee, 1* (pp. 165–186). Basel: Karger.

Kuratani, S., Matsuo, A. I., and Aizawa, S. (1997). Developmental patterning and evolution of the mammalian viscerocranium: Genetic insights into comparative morphology. *Developmental Dynamics, 209*, 139–155.

Lieberman, D. E., McCarthy, R. C., Hiiemae, K. M., and Palmer, J. B. (2001). Ontogeny and postnatal hyoid and larynx descent in humans. *Archives of Oral Biology, 46*, 117–128.

Lieberman, P. (1975). *On the origins of language: An introduction to the evolution of speech*. New York: Macmillan.

Lieberman, P. (1984). *The Biology and Evolution of Language*. Cambridge, MA: Harvard University Press.

Lieberman, P. (1998). *Eve spoke: Human language and human evolution*. New York: W.W. Norton; London: Picador, Macmillan.

Lieberman, P. (2007). The evolution of human speech: Its anatomical and neural bases. *Current Anthropology, 48*, 39–66.

Lieberman, P. (2015). Evolution of language. *Journal of Anthropological Sciences*, *94*, 1–20.

Lieberman, P., Klatt, D. H., and Wilson, W. H. (1969). Vocal tract limitations on the vowel repertoires of rhesus monkey and other nonhuman primates. *Science*, *164*, 1185–1187.

Liljencrants, J., and Lindblom, B. (1972). Numerical simulations of vowel quality systems: The role of perceptual contrast. *Language*, *48*, 839–862.

MacNeilage, P. F. (1998). The frame/content theory of evolution of speech production. *Behavioural and Brain Sciences*, *21*, 499–546.

MacNeilage, P. F., and Davis, B. (2000a). Deriving speech from nonspeech: a view from ontogeny. *Phonetica*, *57*, 284–296.

MacNeilage, P. F., and Davis, B. (2000b). On the origin of internal structure of word forms. *Science*, *288*, 527–530.

Martin, R. E. (1991). *A comparison of lingual movement in swallowing and speech production*. PhD dissertation, Madison, University of Wisconsin.

Martìnez, I., Arsuaga, J. L., Quam, R., Carretero, J. M., Gracia, A., and Rodriguez, L. (2008). Human hyoid bones from the middle pleistocene site of the Sima de los Huesos (Sierra de Atapuerca, Spain). *Journal of Human Evolution*, *54*, 118–124.

McGinnis, W., Garber, R. L., Wirz, J., Kuroiwa, A., and Gehring, W. J. (1984). A homologous protein–coding sequence in Drosophila homeotic genes and its conservation in other metazoans. *Cell*, *37*, 403–408.

Nishimura, T. (2003a) *Comparative morphology of the hyo-laryngral complex*. Heineman, London.

Nishimura, T. (2003b) Comparative morphology of the hyo–laryngeal complex in anthropoids: two steps in the evolution of the descent of the larynx. *Primates*, *44*, 41–49.

Nishimura, T. (2005). Developmental changes in the shape of the supralaryngeal vocal tract in chimpanzees. *American Journal of Physical Anthropology*, *126*, 193–204.

Nishimura, T. (2006). Descent of the larynx in chimpanzees : Mosaic and multiple steps evolution of the foundation for human speech. In T. Matsuzawa, M. Tomonaga, and M. Tanaka (Eds.), *Cognitive development in chimpanzees* (pp. 75–95). Tokyo: Springer–Verlag.

Nishimura, T., Mikami, A., Suzuki, J., and Matsuzawa (2003). Descent of the larynx in chimpanzee infants. *Proceedings of the National Academy of Sciences*, *100*, 6930–6933.

Nishimura, T., Oishi, T., Mikami, A., Suzuki, J., Matsuda, K., and Takahashi, T. (2008). Development of the supralyngeal vocal tract in Japanese macaques: Implication for the evolution of the descent of the larynx. *American Journal of Physical Anthropology*, *135*, 1882–194.

Owren, M. J., Seyfarth, R. M., and Cheney, D. L. (1997). The acoustic features of vowel-like grunt calls in chacma baboons (*Papio cyncephalus ursinus*): Implications for production processes and functions. *The Journal of the Acoustical Society of America*, *101*, 2951–2963.

Perrier, P., and Boë, L.J. (2009–2012). *SkullSpeech –Acquisition et contrôle articulatoire de la parole, croissance de l'extrémité céphalique du nouveau–né jusqu'à l'âge adulte.* ANR project. http://www.gipsa–lab.grenoble–inp.fr/crissp/projets.php?id_projet=125.

Pisanski, K., Cartei, V., McGettigan, C., Raine, J., and Reby, D. (2016). Voice modulation: A window into the origins of human vocal control? *Trends in Cognitive Sciences*, *20*, 304–318.

Reidenberg, J. S., and Laitman, J. T. (1991). Effect of basicranial flexion on larynx and hyoid position in rats: An experimental study of skull and soft tissue interactions. *The Anatomical Record*, *230*, 557–569.

Rendall, D., Kollias, S., Ney, C., and Lloyd , P. (2005). Pitch (F0) and formant profiles of human vowels and vowel-like baboon grunts: The role of vocalizer body size and voice-acoustic allometry. *The Journal of the Acoustical Society of America*, *117*, 944–955.

Riede, T., Bronson, E., Hatzikirou, H., and Zuberbühler, K. (2005). Vocal production mechanisms in a non–human primate: Morphological data and a model. *Journal of Human Evolution*, *48*, 85–96.

Roers, F., Mürbe, D., and Sundberg, J. (2009). Predicted singers' vocal fold lengths and voice classification: A study of x–ray morphological measures. *Journal of Voice*, *23*, 408–413.

Rodrìguez, L., Cabo, L. L., and Egocheaga, J. E. (2003). Breve nota sobre el hioides neandertalense de Sidron (Pilona, Asturias). In M. P. Aluja, A. Malgosa, and R. M. Nogues (Eds.), *Antropologìa y Diversidad (vol. 1)* (pp. 484–493). Barcelona : Edicions Bellaterra.

Scammon, R. E. (1930). The measurement of the body in childhood. In J. A. Harris, C. M. Jackson, D. G. Paterson, and R. E. Scammon (Eds.), *The Measurement of Man* (pp. 173–215). Minneapolis, USA: University of Minnesota Press.

Serrurier, A., Badin, P., Barney, A., Boë, L.J., and Savariaux, C. (2012). Comparative articulatory modelling of the tongue in speech and feeding. *Journal of Phonetics, 40*, 745–763.

Swindler, D. R., and Wood, C. D. (1973). *An atlas of primate gross anatomy. baboon, chimpanzee and man.* Seattle and London: University of Washington Press. 1982 Malabar R.E. Krieger Publ. Co.

Takemoto, H. (2008). Morphological analyses and 3D modeling of tongue musculature of the chimpanzee (*Pan troglodytes*). *American Journal of Primatology, 70*, 965–975.

Testut, L. (1897). *Traité d'anatomie humaine. Anatomie descriptive–histologie–développement.* Paris: Octave Doin.

Tominaga, T., Dickman, C. A., Sonntag, V. K., and Coons, S. (1995). Comparative anatomy of the baboon and the human cervical spine. *Spine, 20*, 131–137.

Vrolik, V. (1841). *Recherches d'anatomie comparée sur le Chimpanzé*. Amsterdam : J. Mueller.

Westhorpe, R. N. (1987). The position of the larynx in children and its relationship to the ease of intubation. *Anaesthesia and Intensive Care, 15*, 384–388.

Walker, E., and Fulton, J. F. (1936). The external configuration of the cerebral hemispheres of the chimpanzee. *Journal of Anatomy, LXXI*, 105–116.

Youatt, T. W. (1835–1836). Account of the habits and illness of the late chimpanzee. *The Lancet,* 202–206.

Ziegler, W. (2003). Speech motor control is task-specific: Evidence from dysarthria and apraxia of speech. *Aphasiology, 17*, 3–36.

Zuberbühler, K. (2012). Primate communication. *Nature Education Knowledge, 3*, 83.

Veena Kumar[1] and Kristina Simonyan[2,3]
[1]*Department of Neurology, Mount Sinai School of Medicine, New York, NY*
[2]*Department of Otolaryngology, Massachusetts Eye and Ear Infirmary, Boston, MA*
[3]*Department of Neurology, Massachusetts General Hospital, Boston, MA*

Evolution of the Laryngeal Motor Cortex for Speech Production

Abstract: Considerable progress has been recently made in understanding the brain mechanisms underlying speech and language control. One of the important but oftentimes overlooked aspects of speech production is the ability of the primary motor cortex to control fine movements of the laryngeal muscles for the production of learned vocalizations. In this respect, the laryngeal motor cortex is indispensable not only for the development of novel articulatory sequences but also for coordination of sensorimotor interactions for smooth speech motor output. In this chapter, we discuss the comparative organization and function of the laryngeal motor cortex in humans and non-human primates and provide some insights into the evolutionary importance of this cortical region in shaping human speaking.

Keywords: laryngeal control, non-human primates, laryngeal motor cortex

1. Introduction

One of the most puzzling questions in evolutionary biology is that of the unique human capacity for speech and song. Singers in particular exhibit a remarkable control of laryngeal muscles through prolonged expiration, phonation, and pitch. During phonation for speech and song, voice onset is precisely timed, which allows linguistic distinctions between voiced and voiceless consonants. Changes in the subglottal pressure due to changes in lung volume, the elastic properties of the chest wall, and the active contraction of the intercostal and abdominal muscles lead to modulations of voice intensity, whereas the resonance characteristics of the supraglottal region (e.g., oral and pharyngeal cavities) influence the spectral properties of the sound. Vocal fold movements are controlled by intrinsic and extrinsic laryngeal muscles. The intrinsic laryngeal muscles are confined to the larynx and participate in vocal fold closure (thyroarytenoid, TA, lateral cricoarytenoid,

LCA, and interarytenoid muscles, IA), opening (posterior cricoarytenoid muscle, PCA), and lengthening (cricothyroid muscle, CT) (Fig. 1C,D). The extrinsic muscles connect the larynx with surrounding structures, such as the hyoid bone, sternum and pharynx, and raise or lower the larynx within the neck relative to the spine to modulate vocal fold length, fundamental frequency, oro-pharyngeal resonance frequencies and formant structure.

In humans, fine movements of the laryngeal muscles are under control of the laryngeal motor cortex (LMC). Bilateral lesions of human LMC abolish speech and song production but preserve innate vocalizations such as laughing and crying (Amassian et al., 1933; Groswasser et al., 1988; Jurgens et al., 1982; Mao et al., 1989), which are types of non-verbal vocalizations controlled by subcortical structures and present even in infants (Jurgens, 2002; Simonyan and Horwitz, 2011). On the other hand, nonhuman primates, such as the rhesus monkey and squirrel monkey, produce innate vocalizations and have a very limited, if any, ability to learn new vocalizations. Laryngeal muscles in nonhuman primates, too, are controlled by the LMC, albeit to a lesser extent. However, bilateral lesions of LMC in monkeys have no effect on their vocalization, such as the LMC destruction does not affect their ability to produce species-specific calls (Jurgens et al., 1982; Kirzinger and Jurgens, 1982; Sutton et al., 1974). As we discuss below, these behavioral discrepancies may be explained by the location and structural organization of the LMC in humans vs. nonhuman primates, potentially contributing to the evolution of our ability to speak.

2. Specification of LMC localization and organization

One of the factors contributing to LMC developmental maturation in humans lays in its localization. Electrical stimulation in rhesus macaque and squirrel monkeys has long localized an isolated region of laryngeal muscle representation within the ventral premotor cortex (Brodmann area 6) between the inferior arcuate sulcus and subcentral dimple (Hast et al., 1974; Jurgens, 1982; Simonyan and Jurgens, 2002, 2003, 2005a, b) (see Figure 1). Stimulation of this region produces isolated and symmetrical adduction of the vocal folds but no vocalization. More posteriorly in the primary motor cortex (Brodmann area 4), the laryngeal movements are elicited only in combination with orofacial movements, suggesting that an isolated and segregated representation of the LMC in nonhuman primates is limited to the premotor cortex.

In humans, Penfield and colleagues have mapped the "vocalization" area within the primary motor cortex, stimulation of which elicited speech-like sounds (Penfield and Boldrey, 1937) (Figure 2). However, since these first studies in 1930s, the exact localization of the laryngeal muscles within this "vocalization" region (which is a behavior and not a body part representation) remained largely unknown. First systematic exploration of the primary motor cortex for the localization of the LMC in humans was performed by Rödel and colleagues in 2004, who used transcranial magnetic stimulation to selectively stimulate the intrinsic laryngeal muscles (cricothyroid and thyroarytenoid) (Rodel et al., 2004).

Figure 1: (A) Schematic drawing of the sites of vocalization elicitation during direct electrical stimulation of the primary motor cortex in humans (Penfield & Rasmussen, 1950) and the sagittal section of the brainstem depicting the distribution of degenerating fibers (small dots) in the nucleus ambiguus (Amb) and surrounding reticular formation (Kuypers, 1958). The arrows represent the direct (monosynaptic) connections from the LMC to the reticular formation and nucleus ambiguus, the site of laryngeal motoneurons, which project to the laryngeal muscles. (B) Schematic drawing of topographic representation of the intrinsic and extrinsic laryngeal muscle in the premotor cortex (Hast et al., 1974). Sca — subcentral dimple; right-angled triangle — cricothyroid muscle; circle — thyroarytenoid muscle; encircled right-angled triangle — combination of the cricothyroid and thyroarytenoid muscles; square — extrinsic laryngeal muscles. (bottom) The cross-section of the brainstem and photomicrographs show terminal fields of the laryngeal motor cortical projections in the reticular formation (RF) but not nucleus ambiguus in the rhesus monkey, which was injected with the anterograde tracer, biotin dextranamine, into the LMC (Simonyan and Jurgens, 2003). The arrows show indirect connection of the LMC with the nucleus ambiguus via the surrounding reticular formation. The scale bar corresponds to 50 mm. Adapted from Simonyan (2014).

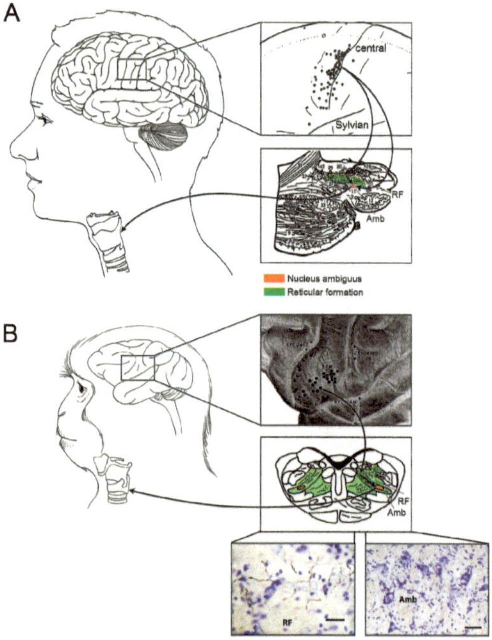

This study has not only provided the first description of the isolated laryngeal muscle representation in the motor cortex but also localized it to Brodmann area 4 in humans as opposed to Brodmann area 6 area in nonhuman primates. More recent functional MRI and electrocorticography (ECoG) studies have confirmed this initial finding and provided a more detailed characterization of the LMC somatotopy within the speech motor cortex and connectivity. Specifically, these studies localized the LMC to the subdivision of primary motor cortex, namely area 4p (Bouchard et al., 2013; Kumar et al., 2016; Simonyan, 2014) , as well as identified another segregated LMC location in area 6 of the premotor cortex (Simonyan, 2014), which is similar to the LMC location in nonhuman primates (Simonyan and Jurgens, 2002) (see Figure 2).

Figure 2: (A) The 'Motor sequence' within the primary motor cortex with the extensive vocalization region in the inferior portion of the precentral gyrus (Penfield and Jasper, 1954). (B) Meta-analysis of 19 fMRI studies in humans between 2000 and 2013 using activation likelihood estimation (ALE) of brain function during voice production. Bilateral peaks of LMC activation are found in the area 4p with an additional peak of activation in the left area 6 (Simonyan, 2014). Adapted from Simonyan (2014).

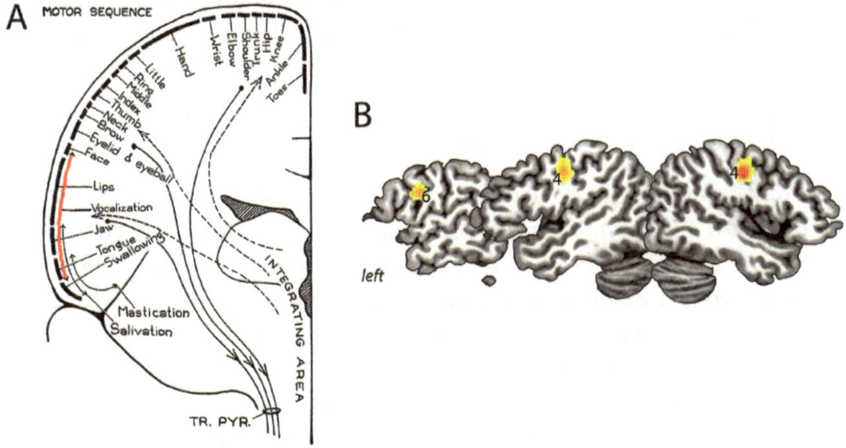

Nevertheless, the LMC in area 4p in humans and area 6 in nonhuman primates are presumed to be functionally homologues because stimulation of both regions evokes isolated and segregated intrinsic laryngeal movements (Kumar et al., 2016). It is plausible that the dual representation of this region in both primary motor and premotor cortices in humans as well as the shift of its functionality in the control of the laryngeal muscles to the primary motocortical location in humans may have played a role in its ability to coordinate more complex vocal tasks, such as production of speech and song. As this premotor position is present in both rhesus and squirrel monkeys, separated by 35 million years, it is possible that this shift to and involvement of the primary motor cortex evolved de novo in humans (Simonyan, 2014). The position of LMC within the primary motor cortex have presumably allowed for more enhanced coupling of expiration, phonation, and articulation, which are the three essential aspects of voice production for speech and song, within the same region. Moreover, this 'new' LMC in the primary motor cortex may have played a crucial role in the evolution of human speech, through expanded structural and functional connectivity with other brain regions involved in vocalization and improved regulation of speech-motor planning and feedback. The secondary 'old' localization of LMC in area 6, which is present in both humans and nonhuman primates, potentially controls more universal laryngeal functions present in both species, such as participation in breathing and other related function associated with straining of the laryngeal muscles like during jumping, lifting heavy weights, etc.

3. Connectivity of the laryngeal motor cortex

Another important factor contributing to the maturation and evolution of the LMC is the establishment of direct monosynaptic connections with laryngeal motoneurons located in the nucleus ambiguus of the brainstem. These neurons control intrinsic laryngeal muscles, which are responsible for a variety of laryngeal behaviors, including voice production. In contrast, the corticobulbar projections from the LMC in the monkey are indirect and first synapse in the surrounding reticular formation and brainstem phonatory sensory nuclei (Iwatsubo et al., 1990; Jurgens, 1976; Kuypers, 1958; Simonyan and Jurgens, 2003). Because of this direct in humans vs. indirect in monkeys control of brainstem laryngeal motoneurons, the LMC ability to finer modulation of laryngeal movements for learned vocalization is much more limited in monkeys compared to humans (see Figure 1).

Despite the lack of direct LMC-ambiguus connection, the neuroanatomical tract tracing studies in nonhuman primates, such as rhesus monkey and squirrel monkey, have established that the LMC has extensive connections to a number of cortical and subcortical regions (Jurgens, 1976; Simonyan and Jurgens, 2002, 2003, 2005a, b). Diffusion MRI studies in humans have revealed similar structural connectivity (Simonyan et al., 2009; Kumar et al., 2016). LMC in both species is highly connected to sensorimotor regions (primary somatosensory area, supplementary motor area), auditory processing and sensorimotor integration centers (superior temporal cortex, inferior parietal cortex, prefrontal gyrus, supramarginal gyrus) as well as limbic regions (cingulate gyrus, insula), basal ganglia and thalamus (see Figure 3). The overall similarity of LMC networks between humans and nonhuman primates might explain why monkeys are able to control the timing and duration of calls (Coude et al., 2011), while there are not able to learn finer motor control of their laryngeal muscles for more complex voice production as during speaking. It also suggests that the evolution of human speech relied largely on the maturation and reorganization of neural pathways within this vast structural network that is already in place in nonhuman primates.

Figure 3: LMC cortical and subcortical networks. Block diagrams illustrate the reciprocal (left box), outgoing (middle box), and incoming (right box) connections of the LMC as defined using neuroanatomical tracing studies in nonhuman primates and diffusion tensor tractography in humans. Asterisk (*) indicates that the projection from the laryngeal motor cortex to the nucleus ambiguus exists only in humans but not in nonhuman primates. Adapted from Simonyan and Horwitz (2011).

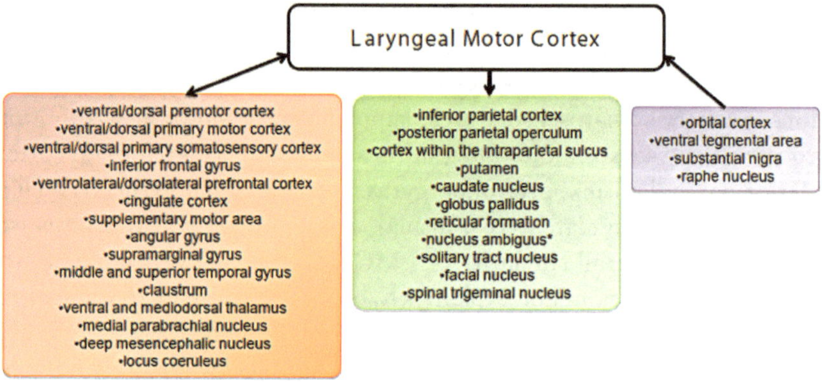

In addition, several other *cortical* LMC connections appear to be enhanced that have potentially allowed for the development of our ability to better control sensorimotor integration for speech production. A recent study examining LMC connectivity with probabilistic diffusion weighted tractography has revealed a remarkable 7-fold increase in LMC-parietal/temporal connectivity strength in humans compared to nonhuman primates (rhesus monkey) (Kumar et al., 2016) (see Figure 4). Within these connection is the supramarginal gyrus (SMG) of the inferior parietal lobule (IPL). Connections of the IPL with Broca's area and auditory areas suggest a key role of this region in language development, auditory language processing, and integration of auditory word forms to generate speech (Caspers et al., 2011; Price, 2000). Functional studies also implicate left IPL in the semantic and phonological processing network (Vigneau et al., 2006), while its lesions are known to produce receptive aphasia (Alexander et al., 1989; Hart and Gordon, 1990; Kertesz et al., 1982).

Figure 4: Quantitative distribution of LMC connections in human (A) and nonhuman primates (B). Connectivity fingerprints show the proportion (%, in logarithmic scale) of LMC tracts reaching each target region in humans and macaques. The corresponding percentage of each tract contribution is given on the right. S1 – primary somatosensory cortex, IFG – inferior parietal gyrus, STG – superior temporal gyrus, IPL – inferior parietal lobule, SMA – supplementary motor area, ACC – anterior cingulate cortex, MCC – middle cingulate cortex, Put – putamen, Cd – caudate nucleus, Gp – globus pallidus, Thal – thalamus.

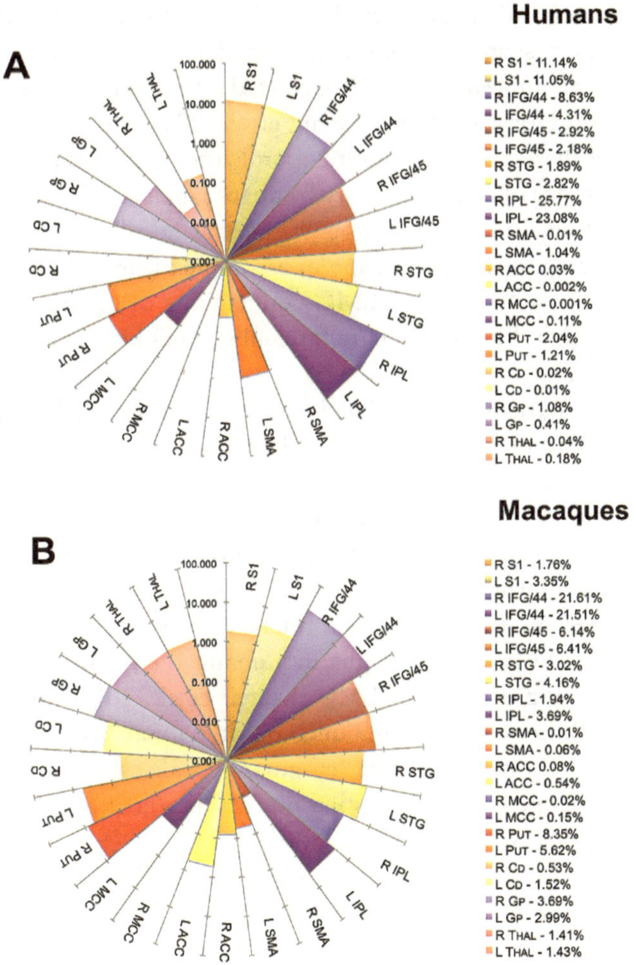

Other diffusion tractography studies have illustrated that dorsal arcuate fasciculus projecting to IPL was present in chimpanzees as well as humans as opposed to rhesus monkeys (Petrides and Pandya, 2009; Rilling et al., 2008, 2011, 2012). The arcuate fasciculus, which connects Broca's and Wernicke's area (Catani et al., 2005), is crucial for word-learning in humans (Lopez-Barroso et al., 2013). As a relay between frontal and temporal regions, IPL maps auditory stimuli into lexical and articulatory-gestural representations for speech production (Bohland and Guenther, 2006; Guenther, 2006; Jardri et al., 2007; McNamara et al., 2008). It is one of only a few sensorimotor centers, which coordinates speech production as well as comprehension (Fiebach et al., 2007; Hocking et al., 2009; Simonyan and Horwitz, 2011; Zheng et al., 2010). Taken together, the expansion of these LMC-parietal/temporal connections in humans points to the importance of somatosensory feedback, such as proprioceptive and tactile sensation, in the perception of speech and modulation of motor activities leading to the production of a complex learned vocal behavior. Whether the new IPL/STG – LMC connections originated de novo in humans or rather, like the arcuate fasciculus, in chimpanzees remains to be investigated.

While LMC-parietal/temporal pathway emerged as most dominant, comparison of the broader LMC network in humans and macaques points to the importance of many other regional hubs in higher-order processing centers. Subcortical connections to LMC appear to be bilateral and structurally similar in both species. Basal ganglia circuitry dysfunction is implicated in a variety of motor disorders, such as Parkinson's disease, tremor, and dystonia. Based on neuroanatomical tract tracing studies in macaques and probabilistic diffusion tractography in humans, the putamen is the major basal ganglia output target of LMC (Kumar et al., 2016; Simonyan and Jurgens, 2003). Lesions in the putamen result in dysarthria and dysphonia in humans but do not alter monkey innate vocalization (Jurgens, 2002; Jurgens and Ploog, 1970), suggesting that it has a greater role in the production of learned vs. innate vocalizations.

Tracts from LMC to superior temporal gyrus (STG) are more widely distributed in humans and left-lateralized when compared to macaques (Kumar et al., 2016; Simonyan et al., 2009). Bilateral isolated lesions in STG cause 'word deafness,' or inability to comprehend heard speech (Alexander et al., 1989; Buchman et al., 1986; Hart and Gordon, 1990; Kertesz et al., 1982). The diversification of left LMC-STG tracts in humans represents the importance of auditory feedback in rapid modification of speech production as opposed to innate vocalization. STG is activated during speech perception but also vocalization. Furthermore, fMRI studies show that LMC is activated mostly during syllable production, but also in speech perception (Wilson et al., 2004). The coupling of these regions points to an important role for LMC in auditory processing and integration into motor movements.

4. Conclusion

Although investigation into the role of LMC in speech has undergone great developments in the past decade, much is yet to be explored. We can assert that humans have undergone three important evolutionary modifications with respect to the LMC organization. First, the development of direct neuronal connections to nucleus ambiguus allowed pathways to bypass other relay stations, resulting in direct control of laryngeal motoneurons in the brainstem. Second, the shift of functionally active LMC from premotor to primary motor cortex facilitated a more precise control of the laryngeal muscles for production of complex vocal tasks. Third, this shift allowed for expansion of structural and functional LMC cortical connectivity with the parieto-temporal regions in addition to subcortical pathways.

References

Alexander, M. P., Hiltbrunner, B., and Fischer, R. S. (1989). Distributed anatomy of transcortical sensory aphasia. *Archives of Neurology, 46*, 885–892.

Amassian, V., Anziska, B., Cracco, J., Cracco, R., and Maccabee, P. (1933). Cerebral facial diplegia – cortical form of the pseudobulbar paralysis (contribution to the study on the dissociation of voluntary and reflex activities). *Revista de Neurologia, 60*, 441–458.

Bohland, J. W., and Guenther, F. H. (2006). An fMRI investigation of syllable sequence production. *Neuroimage, 32*, 821–841.

Bouchard, K. E., Mesgarani, N., Johnson, K., and Chang E. F. (2013). Functional organization of human sensorimotor cortex for speech articulation. *Nature, 495*, 327–332.

Buchman, A. S., Garron, D. C., Trost-Cardamone, J. E., Wichter, M. D., and Schwartz, M. (1986). Word deafness: one hundred years later. *Journal of Neurology, Neurosurgery, and Psychiatry, 49*, 489–499.

Caspers, S., Eickhoff, S. B., Rick, T., von Kapri, A., Kuhlen, T., Huang, R., Shah, N. J., and Zilles, K. (2011). Probabilistic fibre tract analysis of cytoarchitectonically defined human inferior parietal lobule areas reveals similarities to macaques. *Neuroimage, 58*, 362–380.

Catani, M., Jones, D. K., and ffytche, D. H. (2005). Perisylvian language networks of the human brain. *Annals of Neurology, 57*, 8–16.

Coude, G., Ferrari, P. F., Roda, F., Maranesi, M., Borelli, E., Veroni, V., Monti, F., Rozzi, S., and Fogassi, L. (2011). Neurons controlling voluntary vocalization in the macaque ventral premotor cortex. *PLoS One, 6*, e26822.

Fiebach, C. J., Friederici, A. D., Smith, E. E., and Swinney, D. (2007). Lateral inferotemporal cortex maintains conceptual-semantic representations in verbal working memory. *Journal of Cognive Neuroscience, 19*, 2035–2049.

Groswasser, Z., Korn, C., Groswasser-Reider, I., and Solzi, P. (1988). Mutism associated with buccofacial apraxia and bihemispheric lesions. *Brain and Language, 34*, 157–168.

Guenther, F. H. (2006). Cortical interactions underlying the production of speech sounds. *Journal of Communication Disorders, 39*, 350–365.

Hart. J., and Gordon, B. (1990). Delineation of single-word semantic comprehension deficits in aphasia, with anatomical correlation. *Annals of Neurology, 27*, 226–231.

Hast, M. H., Fischer, J. M., Wetzel, A. B., and Thompson, V. E. (1974). Cortical motor representation of the laryngeal muscles in Macaca mulatta. *Brain and Research, 73*, 229–240.

Hocking, J., McMahon, K. L., and de Zubicaray, G. I. (2009). Semantic context and visual feature effects in object naming: an fMRI study using arterial spin labeling. *Journal of Cognive Neuroscience, 21*, 1571–1583.

Iwatsubo, T., Kuzuhara, S., Kanemitsu, A., Shimada, H., and Toyokura, Y. (1990). Corticofugal projections to the motor nuclei of the brainstem and spinal cord in humans. *Neurology, 40*, 309–312.

Jardri, R., Pins, D., Bubrovszky, M., Despretz, P., Pruvo, J.P., Steinling, M., and Thomas, P. (2007). Self awareness and speech processing: an fMRI study. *Neuroimage, 35*, 1645–1653.

Jurgens, U. (1976). Projections from the cortical larynx area in the squirrel monkey. *Experimental Brain Research, 25*, 401–411.

Jurgens, U. (1982). Afferents to the cortical larynx area in the monkey. *Brain Research, 239*, 377–389.

Jurgens, U. (2002). Neural pathways underlying vocal control. *Neuroscience & Biobehavioral Reviews, 26*, 235–258.

Jurgens, U., Kirzinger, A., and von Cramon, D. (1982). The effects of deep-reaching lesions in the cortical face area on phonation. A combined case report and experimental monkey study. *Cortex, 18*, 125–139.

Jurgens, U., and Ploog, D. (1970). Cerebral representation of vocalization in the squirrel monkey. *Experimental Brain Research, 10*, 532–554.

Kertesz, A., Sheppard, A., and MacKenzie, R. (1982). Localization in transcortical sensory aphasia. *Archives of Neurology, 39*, 475–478.

Kirzinger, A., and Jurgens, U. (1982). Cortical lesion effects and vocalization in the squirrel monkey. *Brain Research, 233*, 299–315.

Kumar, V., Croxson, P. L., and Simonyan, K. (2016). Structural organization of the laryngeal motor cortical network and its implication for evolution of speech production. *Journal of Neuroscience, 36*, 4170–4181.

Kuypers, H. G. (1958). Corticobular connexions to the pons and lower brain-stem in man: an anatomical study. *Brain, 81*, 364–388.

Lopez-Barroso, D., Catani, M., Ripolles, P., Dell'Acqua, F., Rodriguez-Fornells, A., and de Diego-Balaguer, R. (2013). Word learning is mediated by the left arcuate fasciculus. *Proceedings of the National Academy of Sciences of the United States of America, 110*, 13168–13173.

Mao, C. C., Coull, B. M., Golper, L. A., Rau, M. T. (1989). Anterior operculum syndrome. *Neurology, 39*, 1169–1172.

McNamara, A., Buccino, G., Menz, M. M., Gläscher, J., Wolbers, T., Baumgärtner, A., and Binkofski, F. (2008). Neural dynamics of learning sound-action associations. *PLoS One, 3*, e3845.

Penfield, W., and Boldrey, E. (1937). Somatic motor and sensory representation in the cerebral cortex of man as studied by electrical stimulation. *Brain, 60*, 389–443.

Penfield, W., and Jasper, H. H. (1954). *Epilepsy and the functional anatomy of the human brain, 1st Edition*. Boston: Little.

Penfield, W., and Rasmussen, T. (1950). *The cerebral cortex of man; a clinical study of localization of function*. New York: Macmillan.

Petrides, M., and Pandya, D. N. (2009). Distinct parietal and temporal pathways to the homologues of Broca's area in the monkey. *PLoS Biol*, 7, e1000170.

Price, C. J. (2000). The anatomy of language: contributions from functional neuroimaging. *Journal of Anatomy, 197, 335–359.*

Rilling, J. K., Glasser, M. F., Jbabdi, S., Andersson, J., and Preuss, T. M. (2011). Continuity, divergence, and the evolution of brain language pathways. *Frontiers in Evolutionary Neuroscience, 3*, 11.

Rilling, J. K., Glasser, M. F., Preuss, T. M., Ma, X., Zhao, T., Hu, X., and Behrens, T. E. (2008). The evolution of the arcuate fasciculus revealed with comparative DTI. *Nature Neuroscience, 11*, 426–428.

Rilling, J. K., Scholz, J., Preuss, T. M., Glasser, M. F., Errangi, B. K., and Behrens, T. E. (2012). Differences between chimpanzees and bonobos in neural systems supporting social cognition. *Social Cognitive and Affective Neuroscience, 7, 369–379.*

Rodel, R. M., Olthoff, A., Tergau, F., Simonyan, K., Kraemer, D., Markus, H., and Kruse, E. (2004). Human cortical motor representation of the larynx as assessed by transcranial magnetic stimulation (TMS). *Laryngoscope, 114*, 918–922.

Simonyan, K. (2014). The laryngeal motor cortex: its organization and connectivity. *Current Opinion in Neurobiology, 28*, 15–21.

Simonyan, K., and Horwitz, B. (2011). Laryngeal motor cortex and control of speech in humans. *Neuroscientist, 17*, 197–208.

Simonyan, K., and Jurgens, U. (2002). Cortico-cortical projections of the motorcortical larynx area in the rhesus monkey. *Brain Research, 949*, 23–31.

Simonyan, K., and Jurgens, U. (2003). Efferent subcortical projections of the laryngeal motorcortex in the rhesus monkey. *Brain Research, 974*, 43–59.

Simonyan, K., and Jurgens, U. (2005a). Afferent subcortical connections into the motor cortical larynx area in the rhesus monkey. *Neuroscience, 130*, 119–131.

Simonyan, K., and Jurgens, U. (2005b). Afferent cortical connections of the motor cortical larynx area in the rhesus monkey. *Neuroscience, 130*, 133–149.

Simonyan, K., Ostuni, J., Ludlow, C. L., Horwitz, B. (2009). Functional but not structural networks of the human laryngeal motor cortex show left hemispheric lateralization during syllable but not breathing production. *Journal of Neuroscience, 29*, 14912–14923.

Sutton, D., Larson, C., and Lindeman, R. C. (1974). Neocortical and limbic lesion effects on primate phonation. *Brain Research, 71*, 61–75.

Vigneau, M., Beaucousin, V., Hervé, P. Y., Duffau, H., Crivello, F., Houdé, O., Mazoyer, B., and Tzourio-Mazoyer, N. (2006). Meta-analyzing left hemisphere language areas: phonology, semantics, and sentence processing. *Neuroimage, 30*, 1414–1432.

Wilson, S. M., Saygin, A. P., Sereno, M. I., and Iacoboni, M. (2004). Listening to speech activates motor areas involved in speech production. *Nature Neuroscience, 7*, 701–702.

Zheng, Z. Z., Munhall, K. G., and Johnsrude, I. S. (2010). Functional overlap between regions involved in speech perception and in monitoring one's own voice during speech production. *Journal of Cognitive Neuroscience, 22*, 1770–1781.

William D. Hopkins
Neuroscience Institute and Language Research Center,
Georgia State University &
Division of Developmental and Cognitive Neuroscience, Atlanta

Motor and Communicative Correlates of the Inferior Frontal Gyrus (Broca's Area) in Chimpanzees

Abstract: The inferior frontal gyrus (IFG) encompasses Broca's area, a brain region implicated in a variety of cognitive and linguistic functions. For instance, clinical and experimental data suggest that the left IFG plays an important role in language and speech. In this paper, I briefly summarize data on the sulci and morphological landmarks that define the IFG in humans, chimpanzees and monkeys. I also present some preliminary data on the surface area, mean depth and gray matter thickness of the three primary sulci that comprise the IFG in the chimpanzee brain including the fronto-orbital, precentral inferior and inferior frontal sulci. I further present data on associations between individual variation in asymmetries of each sulcus with measures of oro-facial motor control and tool use skill. The implications of these findings for different theories on the evolution of language and higher order motor and cognitive functions in primates are discussed.

Keywords: Inferior frontal gyrus, chimpanzees, oro-facial motor control, language evolution

1. Introduction

A portion of the inferior frontal gyrus (IFG) in the human brain constitutes Broca's area, named for the famous French physician Paul Broca who originally described this brain regions role in language and speech production in patients that suffered damage to this area (Broca, 1861). Specifically, Broca obtained the post-mortem brains of two of his patients that showed deficits in speech production and found that both individuals had pronounced damage to the left inferior frontal gyrus (Dronkers et al., 2007). From these results, Broca described the critically important role of the inferior frontal gyrus for the faculty of language in humans, particularly in the left hemisphere. More than 150 years later, the conclusions drawn by Broca have largely been confirmed in both the clinical and experimental literature. For instance, an abundance of data has shown that deficits in speech (aphasia) and praxic functions (motor planning) are significantly more prevalent when damage occurs in the left compared to right hemisphere (Goldenberg and Randerath, 2015; Meador et al., 1999). Findings from the Wada test and more recently using transcranial doppler sonography have shown convincingly that a majority of individuals, particularly right-handed subjects, are left lateralized for language (Knecht et al., 2000; Rasmussen and Milner, 1977). More recent functional imaging studies have shown that Broca's area (along with other cortical areas) play an important role in a variety of linguistic and other cognitive functions (Cooper, 2006; Fazio et al., 2009; Horwitz et al., 1999; Makuuchi, 2005; Nishitani et al., 2005; Passingham, 1981). Lastly, the evidence that lesions to the left but not right hemisphere affected language functions spurred considerable interest in the concept of hemispheric specialization. Indeed, given the robust nature of lateralization for language, it is difficult to discuss the evolution of language without consideration of the topic of hemispheric specialization (Bradshaw and Rogers, 1993; Corballis, 1992, 2003).

Given the historic and more recent significance placed on the role of Broca's area in language, higher order cognition and motor control, there has naturally been considerable interest in the evolution of this brain region, particularly within primates and specifically among great apes (Loh et al., 2016). The motivation for studies on the evolution of the IFG has no doubt been stimulated by the growing body of evidence of sophisticated cognitive and some language-like abilities displayed by primates in comparison to more distantly related mammalian species (Crockford et al., 2004; Crockford et al., 2012; Schel et al., 2013; Seyfarth and Cheney, 2010; Slocombe and Zuberbühler, 2005; Slocombe and Zuberbühler, 2007). For instance, more than 50 years of so-called ape language research has demonstrated that apes can acquire and use augmentative or alternative communication systems for interspecies communication (Lyn, 2012, in press). When considering their natural communication, there is recent evidence of intentional vocal communication in both captive and wild chimpanzees (Crockford et al., 2012; Hopkins et al., 2007a; Hopkins et al., 2011; Leavens et al., 2014a; Schel et al., 2013) suggesting that they have voluntary control of their vocalizations, findings that challenges many historical and contemporary views of primate vocalizations (Premack, 2004; Seyfarth and Cheney, 2010). There is also evidence that nonhuman primates, and particularly great apes, produce intentional, referential gestures, sometimes in sequences, during inter- and intra-specific communicative events (Cartmill and Byrne, 2007; Gentry and Byrne, 2010; Hobaiter and Byrne, 2011a; Hobaiter and Byrne, 2011b; Leavens et al., 1996; Leavens et al, 2004b; Leavens et al., 2015; Leavens et al., 2005a; Liebal et al., 2004; Pika et al., 2003; Pika and Mitani, 2006; Pollick and de Waal, 2007), which some believe supports a gestural origins view of language evolution (Arbib et al., 2008; Corballis, 2003; Corballis et al., 2012). Interestingly, studies in captive and to a lesser degree wild apes as well as baboons have demonstrated consistent evidence of population-level right handedness during intra- and inter-specific gestures, suggesting a left lateralized system for gestural communication (Hobaiter and Byrne, 2013; Hopkins et al., 2012a; Hopkins et al., 2005; Meguerditchian and Vauclair, 2006; Prieur et al., 2016a, 2016b).

The purpose of this paper is two-fold. First, a summary of the available literature on the morphology and cellular organization of Broca's area in primates with a particular emphasis on data from chimpanzees is presented in the first part of the chapter. Second, I present some new descriptive data on different dimensions of the sulci comprising Broca's area in chimpanzees as well as their association with communicative, cognitive and motor functions. I emphasize throughout the paper what information seems well established and what limitations exist within the literature.

2. Morphology and cellular organization of Broca's area in primates

2.1 Broca's Area: Morphology

The morphology of Broca's area in the human and chimpanzee brain has been eloquently described in a series of papers by Keller and colleagues (Keller et al., 2012; Keller et al., 2007; Keller et al., 2009). In the human brain, morphologically, Broca's area is comprised of two primary regions, the Pars opercularis (ParsO) and Pars triangularis (ParsT) (see Figure 1a). The ParsO is defined posteriorally by the precentral inferior sulcus, superiorly by the inferior frontal sulcus and anteriorly by the ascending ramus. The ParsT is the gyrus that lies between the horizontal sulcus or rami and the anterior ramus. Connecting the anterior terminal points of the horizontal and ascending rami closes the formation of the "triangularis" gyrus. In humans, within the ParsO, there is also the presence of a small, shallow fold in some brains called the dimple. Though the volume of Pars opercularis and Pars triangularis have historically been considered anatomically larger in the left compared to right hemisphere, according to Keller et al. (2007) and others (Knaus et al., 2007; Tomaiuolo et al., 1999), the evidence is not as compelling as once thought. For instance, Keller et al. (2007) reported that a left hemisphere asymmetry in the gray matter volume of the Pars opercularis was contingent upon the presence of a dimple (a small fold) within the gyrus. In those brains that lacked the dimple, no interhemispheric differences in gray matter volume were evident.

In great apes, Broca's area is less well developed but there is some homology with respect to the ParsO in humans (Keller et al., 2009) (see Figure 1b). Specifically, in chimpanzees, bonobos, gorillas and orangutans, the ParsO can be anatomically defined using essentially the same landmarks as used in human brains. The posterior border is the precentral inferior (PCI) sulcus while the superior border is the inferior frontal sulcus (IFS). The anterior border is the fronto-orbital (FO) sulcus, which is the same sulcus as the anterior ramus in the human brain. Thus, the same sulci can be used to define the ParsO. In contrast, the great ape brains all lack a horizontal ramus which therefore prevents the measurement of a homolog to ParsT.

In more distantly related Old and New World monkeys, from the standpoint of morphology, there are no common sulci landmarks with either chimpanzees or humans that can be used to define either ParsO or ParsT. To be clear, there are sulci landmarks within the inferior portion of the frontal lobe in monkeys that can be quantified and arguably could be used as proxy measures to Broca's. For example, the arcuate and principalis are two sulci that upon appearance of their spatial location seem to be analogous to the inferior frontal sulcus and precentral inferior sulcus (Figure 1c). The challenge in defining a morphological homolog to either the ParsO or ParsT is the absence of a sulcus that can serve as an anterior border for either region.

Figure 1: 3D rendering of (a) human brain (b) chimpanzee brain and (c) rhesus monkey brain with the sulcal landmarks used to define the inferior frontal gyrus or ventrolateral prefrontal region.

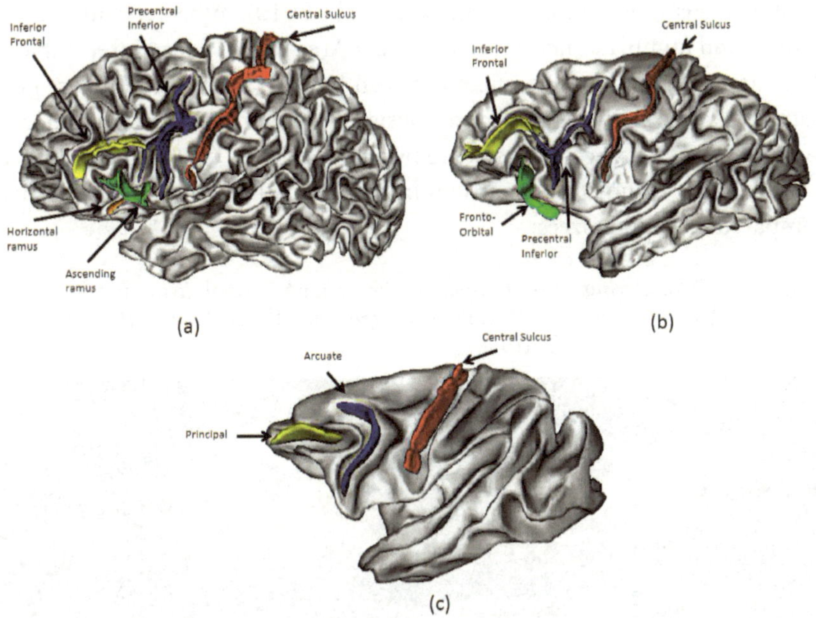

2.2 Broca's Area: Cytoarchitectonics

Though there has been tremendous historical and contemporary interest in the cellular organization and morphology of Broca's area (Zilles and Amunts, 2010), there are remarkably few studies on the cellular organization of this region in human and nonhuman primates (Schenker et al., 2007). Brodmann (1909) was the first to systematically describe the cytoarchitectonic regions of Broca's area subsequently labeled Area44 and Area45. In the functional neuroimaging literature, Area44 and Area45 are often used or considered synonymous with the ParsO and ParsT but the relationship between the cellular and morphological definition of Broca's area is not perfect (Amunts et al., 2007; Zilles and Amunts, 2010). In humans, generally, Area44 cells are found at a greater probability within the ParsO while Area45 cells are found within ParsT (Amunts et al., 1999; Uylings et al., 2006). For chimpanzees, like in humans, the spatial distribution of Area44 cells are more consistently found within the ParsO whereas Area45 cells are found in greatest probability in the gyrus and cortical fold immediately anterior and superior to the frontal orbital sulcus (see Figure 2) (Schenker et al., 2010). When considering the volume and number of neurons comprising Area44 and Area45 (see Table 1), there are two notable differences between humans and chimpanzees. First, humans tend to show a much more consistent leftward asymmetry compared to the chimpanzees. Second, as noted by Schenker et al. (2010), there has been a 6- to 7-fold volumetric expansion in the Area 44 and Area 45 in humans compared to chimpanzees.

Figure 2: 3D rendering of a chimpanzee brain with probabilistic distribution of Area 44 and Area 45 cells projected onto the surface of the cortex (see Schenker et al., 2010).

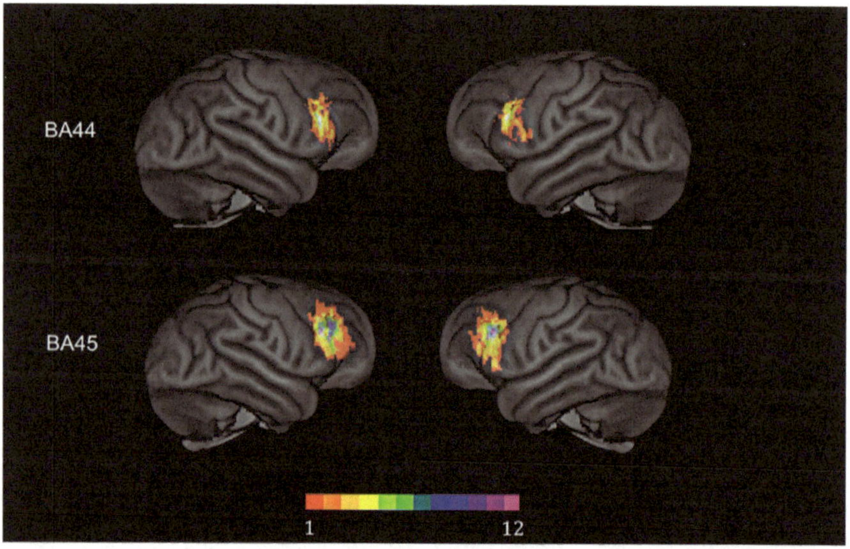

There are two additional sets of data and observations regarding Area 44 and Area 45 in the human and chimpanzee brain. First, Schenker et al. (2010) created probabilistic maps of Area 44 and Area 45 in their chimpanzee sample by registering the cellular data to *ex vivo* MRI scans in the same stereotaxic space. Then they computed the percentage of overlap in the location of Area 44 and Area 45 voxels that were present in at least

50% of the chimpanzees. For Area 44, the volume of the left and right hemisphere probabilistic maps is respectively 642mm^3 and 349 mm^3. For Area 45, the volume of the left and right hemisphere probabilistic maps is respectively 280mm^3 and 249 mm^3. The average native space left and right volumes for Area 44 and Area 45 were 601 mm^3, 648 mm^3, 501 mm^3 and 633 mm^3, respectively. Thus, for Area 44, the average native left hemisphere volume was very similar to the probabilistic volume (601 mm^3 vs 642 mm^3) whereas for the right hemisphere, the probabilistic volume (349 mm^3) was nearly half that of the native space (648 mm^3). For Area 45, for both the left and right hemisphere there were significant reduction in volume between the native and probabilistic volume but it was much greater for the right (633 mm^3 vs 249 mm^3) compared to left (501 mm^3 vs 280 mm^3) hemisphere. One interpretation of these results is that the spatial location of Area 44 and Area 45 cells are much more consistent across subjects in the left compared to right hemisphere, particularly for Area 44. Second, in a more recent paper, Spocter et al. (2012) examined neuropil space in several brain regions in humans and chimpanzees including Area 45 (see Table 1). The proportion of neuropil within a region of the cerebral cortex represents a key aspect of neuroanatomical organization, indicating its functional capacity. In a Nissl stain, the unstained portion of the tissue comprises the space occupied by synapses, dendrites, and blood vessels. A simple calculation of the neuropil fraction (NF) of tissue can be obtained by converting a high-resolution image of Nissl-stained sections to binary and calculating the ratio of the tissue compartment that is stained (i.e., cell bodies of neurons, glia, and endothelial cells) versus unstained. For a majority of the regions, including Area 45, humans have a smaller NF value compared to the chimpanzees suggesting a higher proportion of synapses in this region.

Unfortunately, there are no available data on the volume, neuron number or cell density for Area 44 and Area 45 in other great apes. For macaque monkeys, Petrides and colleagues (Petrides et al., 2005; Petrides and Pandya, 1999, 2009) have done the most extensive analyses and these authors have identified two distinct Area 45 (a & b) regions within the gray matter comprising the anterior fold of the arcuate sulcus (see Figure 1c). Specifically, Area 44 cells are found in the caudal bank and Area 45 cells are found in the rostral of the inferior portion of arcuate sulcus. Interestingly, electrical stimulation of the BA44 region elicits oro-facial and fingers movements in macaque monkeys (Petrides et al., 2005).

2.3 Sulci Surface Area, Depth and Gray Matter Thickness

Figure 3: Pipeline process for sulci extraction in BrainVisa. a. T1-scan; b. inhomogeneity correction; c. mask of tissue from skull; d. split brain mask and segmentation of the cerebellum; e. gray and white matter delineation; f. white matter mould; g. gray matter mould.

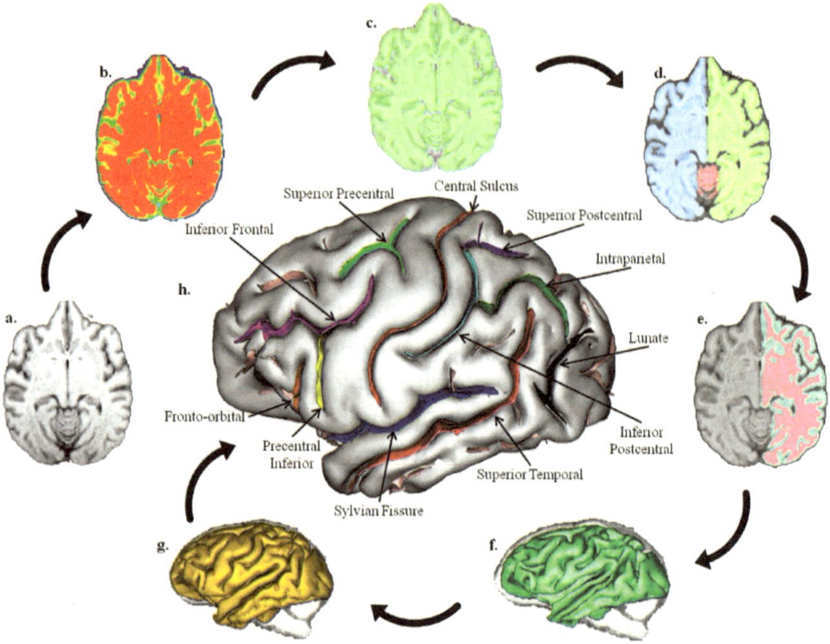

As noted above, the Pars opercularis of the chimpanzee brain is defined by three sulci including the fronto-orbital, precentral inferior and inferior frontal. Rather than focus on the volume, surface area and cortical thickness of the entire gyrus, it is also possible to quantify the surface area, mean depth and gray matter thickness of each of these folds using software programs such as BrainVisa (see Figure 3). BrainVisa (BV) is a software program that allows for extraction of the cortical folds or sulci and enables quantification of variability in their surface area, maximum and mean depth, length, gray matter thickness and fold opening (see Figure 4). We have previously used

BV to quantify individual and phylogenetic variability in the surface area and mean depth of the central sulcus (Cykowski et al., 2008; Hopkins et al., 2010; Hopkins et al., 2014; Hopkins, et al., 2015a). Here, I present data on the surface areas, mean depth and gray matter thickness for the three sulci used to define the Pars opercularis in the chimpanzee brain including the precentral inferior (PCI), inferior frontal (IFS) and fronto-orbital (FO) sulci. Specifically, my laboratory has recently quantified the surface area, mean depth and gray matter thickness of each sulcus in 271 *in vivo* and post-mortem chimpanzee MRI scans and the descriptive data for each sulcus and measure are shown in Table 2. Chimpanzees showed a significant leftward asymmetry in the surface area for FO and a rightward asymmetry for PCI. In terms of mean depth, significant leftward asymmetries were evident for the FO and IFS. Lastly, a significant leftward asymmetry was found for gray matter thickness in the PCI sulcus.

Figure 4: a. 3D rendering of chimpanzee brain with sulci boundaries of Broca's area indicated in green, yellow and blue; b. extraction of example sulcus, in this case, the central sulcus and how the surface area, depth and gray matter thickness measures are derived.

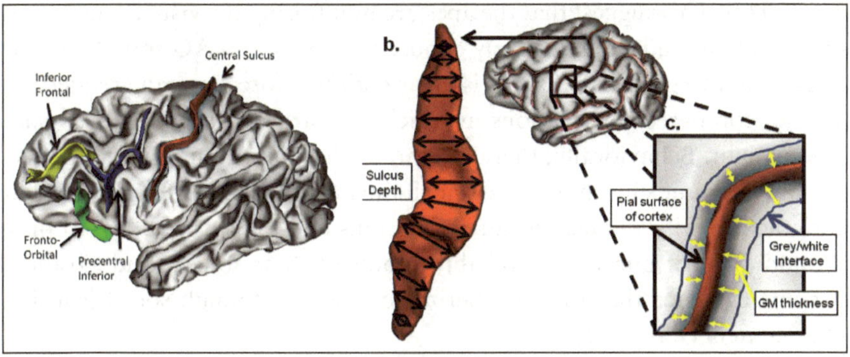

3. Behavioral Associations with Broca's Area in Chimpanzees

As noted above, Broca's area is linked to a variety of praxic, linguistic, and cognitive functions in the human brain. In our laboratory, we have been particularly interested in the potential role that Broca's area plays in praxic and communication functions. Thus, in the following sections, I present an overview of findings on the associations between praxic and communicative functions and Broca's area in chimpanzees. I also present some new data on the relationship between sulci variablity within the inferior frontal gyrus in relation to communication and praxic skill in tool use actions.

3.1 Gestural and Vocal Communication

Studies in our group have previously demonstrated that chimpanzees use manual gesture intentionally and referentially, and also show a prominent right-hand preference during both inter- and intra-specific interactions (Hopkins et al., 2005; Leavens et al., 1996; Leavens et al., 2004b). Additionally, our group has previously reported that chimpanzees produce what we define as attention-getting (AG) sounds. As the name implies, AG sounds are produced to capture the attention of an otherwise inattentive audience. For instance, as their first communicative response, captive chimpanzees are more likely to produce an AG sound when a food is visible to them but a human experimenter is looking away from them compared to when the experimenter is looking at or facing them (Hostetter et al., 2001; Leavens et al., 2004c). This suggest that the apes are monitoring the visual orientation of the human and can selectively choose to produce an AG sound when it is necessary to get the human's visual attention before manually requesting a food, compared to situations in which they are sharing the same visual line of sight. Behaviorally, there are three other important findings on the use of AG sounds by chimpanzees including (1) they are often produced in conjunction with manual gestures (Hopkins and Cantero, 2003) (2) only around 50% of our sample reliably produces these sounds and (3) there is some evidence that they are heritable, possibly though social learning (Taglialatela et al., 2012).

With respect to the neural correlates of hand use for gestural as well as differences between AG+ and AG- chimpanzees, several previous studies have revealed significant associations. AG+ and AG- chimpanzees refers to those individuals who were classified as reliably producing attention-getting sounds or not. For example, using a region-of-interest approach, we have previously found that right-handed chimpanzees for pointing or other manual gestures such as clapping show larger leftward asymmetries in the volume of the IFG and, to a lesser extent the planum temporale, compared to non-right-handed individuals and these handedness effects were not found in other brain regions (Hopkins and Nir, 2010; Meguerditchian et al., 2012; Taglialatela et al., 2006). More recently, Bianchi et al. (2016) using voxel-based morphometry analyses, have reported that AG+ chimpanzees have increased gray matter volume in the left ventrolateral premotor cortex and rightward asymmetries in dorsal prefrontal cortex. In a related study but using a different approach, Hopkins et al. (2017a) compared the depth of the central sulcus along the dorsal-ventral plane between AG+ and AG- chimpanzees and found significant differences in asymmetries in the ventral but not dorsal portion of this fold, particularly among males. The ventral portion of the central sulcus (CS) corresponds to the motor regions that control oro-facial and laryngeal movements and lies below the elbow that defines the inferior border of the motor hand area of the precentral gyrus (BA6) (Bailey et al., 1950).

3.2 Tool Use Handedness and Skill

One theory on the evolution of language and speech pertains to the emergence of tool manufacture and use (Bradshaw and Rogers, 1993; Gibson and Ingold, 1993; Greenfield, 1991; Stout and Chaminade, 2012). Basically, this theory proposes that human speech evolved by co-opting neural systems that were initially involved in praxic functions, notably the manufacture and use of tools. In humans, there is clinical and experimental data evidence for shared neural systems under lying both praxic and verbal functions including the inferior frontal gyrus, inferior parietal lobe and posterior and middle temporal gyri (Frey, 2008; Johnson-Frey, 2004; Lewis, 2006; Roby-Brami et al., 2012; Stout and Chaminade, 2012). Further, some clinical studies have shown both apraxia and aphasia exists when lesions occur within the IFG or adjacent cortical regions, particularly in the left hemisphere (Kobayashi and Ugawa, 2013).

Previous studies by our group have shown that individual variation in hand use and tool use skill is associated with different aspects of cortical organization in the IFG in chimpanzees (Hopkins et al., 2012b) . Our group has characterized hand preference and performance asymmetries on a probing tool use task that is designed to simulate termite fishing in wild chimpanzees (Hopkins et al., 2009). Although population-level handedness for this tool use task is not evident in our chimpanzee sample, we have found that right- and left-handed individuals differ in the linear length of the FO sulcus (Hopkins et al., 2007b). Our group has also found that performance asymmetries on the tool use tasks are associated with the volume of the IFG and the depth of the central sulcus, particularly in regions corresponding to the motor-hand area of the precentral gyrus or KNOB (Hopkins et al., 2017b). It is also important to note that in wild chimpanzees, there are several interesting observations regarding tool use as it may relate to brain asymmetries within the IFG and related areas. First, different types of tool use tasks seem to elicit strong individual hand preferences with a majority of individuals showing a clear bias (i.e., there are very few ambidextrous individuals), which differ somewhat from data in captive chimpanzees (Hopkins et al., 2017b; Hopkins et al., 2009). Second, though directional biases vary as a function the type of tool use task, population-level handedness is evident for several measures such as ant-dipping, termite-fishing, pestle-pounding, wadge-dipping but not nut-cracking (reviewed in Hopkins, 2013; Sanz et al., 2016). Thus, with respect to the theories linking the evolution of praxic function and functional asymmetries, extant data on tool use clearly support this view.

3.3 Some New Data and Findings

As a means of further examining the influence of handedness for tool use, manual gestures and the use of AG sounds on variation in cortical folding, here I present some new findings on analyses of asymmetries in surface area, mean depth and gray matter thickness in the three sulci that define the inferior frontal gyrus of chimpanzees, notably FO, PCI and IFS. For each sulcus, we used the pipeline procedures in BrainVisa to extract the folds as we have done in previous studies (Hopkins et al., 2010; Hopkins et al., 2017a; Hopkins et al., 2014). The sulci were manually labeled and we subsequently obtained measures of their surface area, mean depth and gray matter thickness for the left and right hemispheres. Asymmetry quotients for each sulci and measure were derived following the formula [AQ=(R – L)/ ((R + L) *.5)] where R and L reflect the right and left hemisphere values. Positive AQ values indicated rightward asymmetries and negative values indicate leftward.

3.3.1 Manual Gesture and AG Sound Production

Figure 5: a. Picture of a chimpanzee gesturing to another individual; b. Mean AQ scores (+/- s.e.) for left-, ambiguously- and right-handed chimpanzees for surface area, mean depth and gray matter (GM) thickness. Negative AQ values indicate leftward asymmetries and positive indicate rightward biases.

For this analysis, I examined the influence of AG sound production and handedness for manual gestures on AQ scores for each measure using a mixed model analysis of covariance. AQ scores for each sulcus were the repeated measures while handedness (left-handed, right-handed, ambiguously-handed) and AG group (AG+, AG-) were the between group factors. Age (in years) was the covariate. For surface area, a main effect for gesture handedness was found ($F(2, 238) = 3.06$, $p = .049$; see Figure 5). Right-handed individuals showed increased leftward asymmetries compared to the ambiguously-handed and left-handed apes. No other significant main effects of interactions were found. For mean depth, no significant main effects or interactions were found. For gray matter thickness, a significant main effect for handedness was found ($F(2, 238)=3.07$, $p = .048$) and, as was the case with the surface area measures, right-handed chimpanzees had greater leftward asymmetries than ambiguously-handed and left-handed individuals (see Figure 5). A three-way interaction between sex, vocal grouping and region was also found ($F(2, 474) = 9.327$, $p = .001$, Figure 6). Post-hoc analysis indicated that for FO, no significant differences were found between AG groups and sex. For IFS, females AG+ chimpanzees had significantly greater leftward asymmetries than AG- females and AG+ males. In contrast, for the PCI, AG+ males had significantly greater leftward asymmetries than AG- males. Thus, differences between sex and AG grouping asymmetries in gray matter thickness varied primarily between the PCI and IFS sulci, respectively.

Figure 6: Mean gray matter thickness AQ scores (+/- s.e.) for male and females AG+ (vocal+) and AG- (Vocal-) chimpanzees for the (a) FO (b) IFS and (c) PCI sulci. 3D rendering adjacent to each graph depict the sulci measured in each analysis. Negative AQ values indicate leftward asymmetries and positive indicate rightward biases

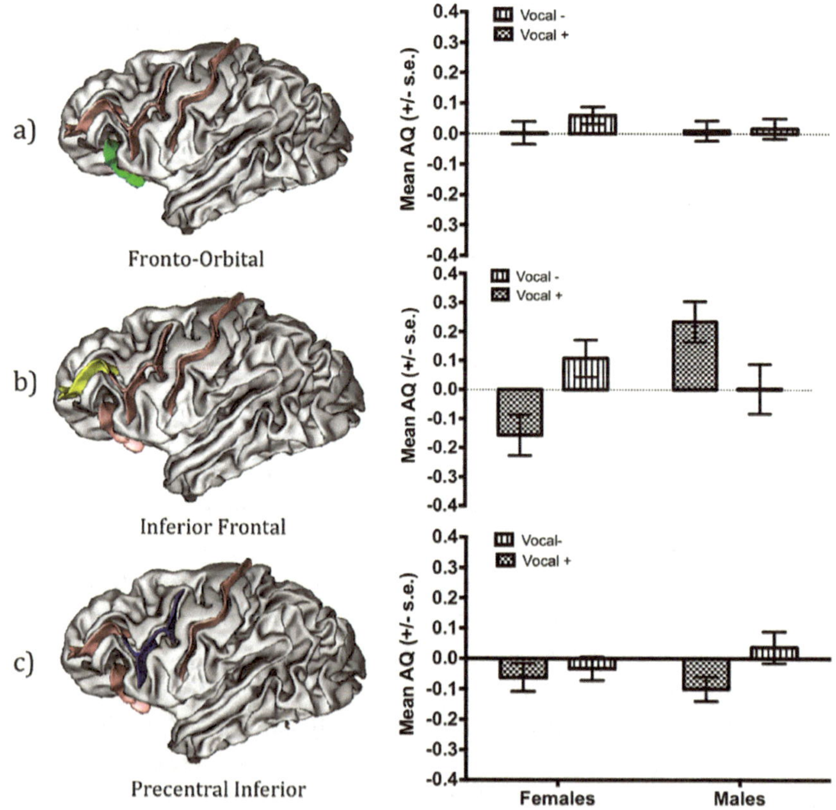

3.3.2 Tool Use Skill

For these analyses, our group quantified asymmetries in motor skill on a tool use task designed to simulate ant- or termite-fishing in wild chimpanzees. Briefly, and as shown in Figure 7, food (usually mustard, honey or salsa) is placed inside an opaque PVC pipe that is attached to the outside mesh of their home cages. On the end of the PVC pipe facing the chimpanzees, there is a small opening (1 cm in diameter) into which the chimpanzees can insert a lollipop stick in order to obtain the food. The lollipop stick is slightly smaller than the hole (9 mm in diameter) thereby increasing the motor and spatial demands of inserting the sticking the hole. In previous studies, we have recorded hand use and the latency from the initiation of a probing response to the successful insertion of the stick on 50 trials in more than 200 chimpanzees on this task. In these previous studies, the chimpanzees were free to use whichever hand they preferred; however, for the analysis presented here, I focused on assessment of asymmetries in hand skill in which we controlled for the number of dipping responses for each hand. Specifically, in 90 chimpanzees, we measured the average latency to successfully insert the stick on 30 trials for each hand (see Phillips et al., 2013). Thus, to more fairly assess motor skill asymmetries, we obtained an equal number of dipping responses for each hand and calculated a difference scores based on the mean latency scores for each hand. Based on the sign of the difference score, chimpanzees were classified as performing better with their right or left hand. We then compared the AQ scores between the two handedness groups (based on their asymmetries in performance) using a mixed model ANCOVA which included vocal group (AG+, AG-) and sex (male, female) as additional between group factors and age as a covariate.

Figure 7: (a) Mean gray matter thickness AQ scores (+/- s.e.) for PCI in male and females chimpanzees that perform better with their right or left hand on the tool use task (illustrated in the left panel) (b) Mean gray matter thickness AQ scores (+/- s.e.) for PCI in male and females AG+ (vocal+) and AG- (Vocal-) chimpanzees (behavior illustrated in left panel). Negative AQ values indicate leftward asymmetries and positive indicate rightward biases

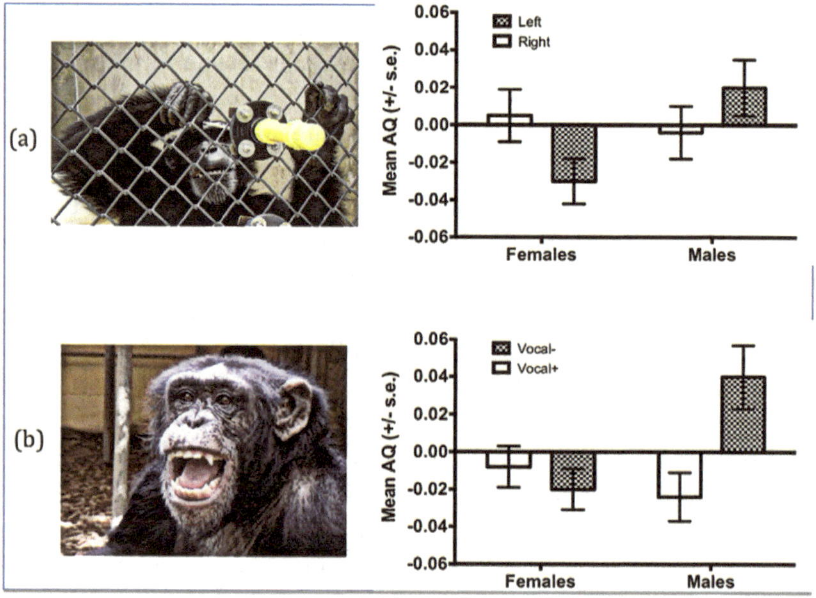

For surface area and mean depth AQ scores, no significant main effects of interactions were found. In contrast, for gray matter thickness, a borderline significant three-way interaction was found between sex, sulcus and performance asymmetry group ($F(2, 162) = 2.61$, $p < .07$) together, as we found before, with a significant three-way interaction between sex, sulcus and vocal group ($F(2, 162)=6.33$, $p < .003$). To explore these interactions more thoroughly, univariate analyses of covariance were performed on the gray matter AQ scores for each sulcus with vocal group, tool performance group and sex serving as between group factors while age was the covariate. No significant main effects or interactions were found for FO and IFS but, for PCI, significant two-way interactions were found between sex and performance group ($F(1, 82) = 4.586$, $p < .036$) and sex and vocal grouping

($F(1, 82) = 8.747, p < .005$). The mean gray matter thickness AQ scores for PCI in males and females within the tool performance asymmetry and vocal group can be seen in Figure 7a & b. As can be seen, the pattern of sex-dependent variation in AQ scores within PCI as a function of tool use and AG sound production are similar.

4. Discussion

From a purely gross anatomical standpoint, as has been articulated previously, there are clear differences in the cortical folding patterns of the inferior frontal gyrus between humans and chimpanzees and, indeed, all great apes. The sulcal landmarks used to define the Pars opercularis appear to be highly conserved between humans and other apes. Moreover, when comparing patterns of asymmetry in total or gray matter volume within the IFG between humans and chimpanzees, the data seems fairly consistent between species with neither showing a population-level bias. However, for the Pars triangularis, apes lack a horizontal ramus that serves as the inferior boundary for this anatomical region. Thus, there is some increased folding and gyrification within the ventral-lateral premotor and, indeed, the entire prefrontal cortex in humans compared to apes. From the cytoarchitectonic studies, the data also suggest that the volume of Area 44 and Area 45 in humans is between 6 and 7 times larger than the comparable regions in the chimpanzee brain (Schenker et al., 2010). In short, morphological and cellular changes in Broca's area have evolved in humans after the split from the common ancestor with chimpanzees and this likely reflects increased cortical expansion and connectivity in response to selection for motor and cognitive demands associated with praxic and communicative functions.

In chimpanzees, significant associations are found between motor and communicative functions and different aspects of cortical organization within the inferior frontal gyrus. With respect to communication, differences are found between individuals who prefer to gesture with their right or left hand in asymmetries in (1) gray matter volume (2) surface area and gray matter thickness of the FO, PCI and IFS sulci and (3) depth of the central and middle portions of the central sulcus. Associations between handedness and asymmetries in these same regions are not evident for actions that are not communicative in function (i.e., simple reaching), the exception being tool use (see below). At face value, these results are consistent with the gestural origins theory of language evolution, at least at the neurological level of analysis.

AG+ and AG- chimpanzees also differ in cortical organization within the IFG. Notably, Bianchi et al. (2016) found that AG+ chimpanzees had a small but significant difference in gray matter density within the left ventro-lateral prefrontal cortex and right dorsolateral prefrontal cortex compared to AG- apes. It is worth noting that the gray matter cluster within the left ventro-lateral prefrontal cortex reported by Bianchi et al. (2016) is located at the most medial point of the PCI. As reported here, AG+ and AG- chimpanzees, particularly males, show greater leftward asymmetries in gray matter thickness within the PCI fold. Recall that PCI sulci define the posterior border and include Area 44 cells (see Figure 2). Further, the gyrus immediately posterior to PCI is the ventral portion of the precentral gyrus. Previous studies by our group have found leftward asymmetries in the depth of the ventral portion of the central sulcus in AG+ but not AG- chimpanzees (Hopkins et al., 2017a). Thus, the emerging and convergent data strongly suggest that variability in cortical activation, cortical folding, gray matter thickness and volume are associated with AG sound production (and perception) by chimpanzees and possibly more distantly related Old World monkeys (Coude et al., 2011; Petrides et al., 2005; Romanski et al., 2004; Taglialatela et al., 2008, 2011).

One interesting observation of the PCI sulci that has been discussed in the literature is the degree of variability in folding across subjects (Sherwood et al., 2003). Specifically, there are sometimes more than one PCI branch that's extends vertically from the IFS and this can influence measurement of the volume of the IFS. In our labeling of PCI from the BrainVisa extractions, if PCI bifurcated, we included all these additional folds in the analysis. However, it would be interesting to assess whether the complexity of folding for PCI is associated with any of the communicative or praxic functions we have quantified in the chimpanzees. In humans, for some sulci such as the anterior cingulate, variability in the sulci patterns are predictive of localization of individual motor functions (Amiez and Petrides, 2014) and this type of approach could potentially provide some additional and novel findings that are not captured with our existing methods.

Like AG sound production, previous studies have found that chimpanzees that perform better on a tool use task with their right hand show greater leftward asymmetries in the IFG compared to those that perform better with their left hand, particularly among males (Hopkins et al., 2017b). Interestingly, hand preferences did not account for a significant proportion of variance in IFG asymmetry; thus, asymmetries in manual skill preference were more strongly linked to variation in IFG asymmetry. Analysis of the surface area, mean depth and gray matter (GM) thickness data for the FO, IFS and PCI reported here further reinforce these findings. When controlling for the number of right and left hand responses, we found that males who perform better with their right compared to left hand showed greater leftward asymmetries in PCI. This finding is quite consistent with the findings for AG sound production.

In my view, what links the neuroanatomical findings from gesture, tool use skill and AG sound production is that all these actions require voluntary motor control and planning. As noted earlier, behavioral studies clearly show that the use of manual gestures and AG sounds are intentional and referential and, in many ways, one can consider the function and use of AG sounds as a "gesture"; indeed, the use of AG sounds as gestures is context and modality specific but functions similarly to gestures (i.e., to draw the attention of a human experimenter to an object). It should be noted that AG sound production and gestures often co-occur, further suggesting that common neural systems likely underlie their expression (Hopkins and Cantero, 2003). The tool use testing we have done in our chimpanzees also requires hand-eye coordination and planned actions given the temporal and spatial constraints of the task; thus, this task also requires complex planning and execution of multiple motor systems. In addition, tool use, gesture and AG sound production share a common cognitive foundation in the form of means-ends reasoning (Hopkins et al., 2012a; Leavens et al., 2005b). For example, for tool use in wild chimpanzees, many forms are exhibited on the context of obtaining food that is otherwise unavailable (i.e., ants in a termite mound, meat inside a nut, water inside the truck of a tree, etc...). In captivity, most chimpanzees produce gestures or AG sounds to solicit the behavior of a human that has access to a food they see but is otherwise unavailable. Thus, in the captive setting, the chimpanzees are using the human experimenter effectively as a tool (i.e., social tool use) and, if a physical tool were available, they would likely use it instead of the human (see Volter et al., in press for discussion).

One important question that comes from finding on the association between Broca's area and communication and tool use skills is the extent to which common biological or genetic mechanisms account for unique or shared contributions to these brain-behavior associations. Specifically, our group has previously reported that manual gestures, AG sound production and tool use hand preference and skill are all heritable in chimpanzees (Hopkins, 2013; Hopkins et al., 2013; Hopkins et al., 2015b; Taglialatela et al., 2012). Similarly, we have found that gray matter volume of the IFG and planum temporale are significantly heritable in chimpanzees (Hopkins et al., 2015c). Whether common genes underlie both gray matter variation in the IFG and either tool use, manual gestures or AG sound production remains unclear but should be tested in future studies.

Another important convergent set of the results that seem to exist in our chimpanzee sample is the sex dependent brain-behavior associations, particularly as it relates to asymmetry. Males appear to show more consistent and robust brain-behavior associations compared to females. This is the case for the association between asymmetries in the PCI and AG sounds production and tool use performance asymmetries (Figures 7a & 7b). The stronger brain-behavior associations in males were also reported in relation to depth in the central sulcus and AG sound production (Hopkins et al., 2017a). Why this is the case is not clear but we have previously found that variation in corpus callosum morphology and fiber count differ between male and female chimpanzees in relation to neuroanatomical asymmetries in the planum temporale (Hopkins et al., 2016). Therefore, one possible explanation maybe that males, as a whole, show more pronounced asymmetries and these align themselves with individual differences in behavior more so than females.

Though Broca and many early investigators focused on the role of the inferior frontal gyrus in language functions, it is now clear that this region plays a role in a variety of cognitive and motor functions outside of the domain of language. For instance, the IFG is one brain region constituting the mirror neuron system, which has been implicated in the perception and production of imitation and related action-perception processes (Fazio et al., 2009; Kilner et al., 2009; Makuuchi, 2005). Further, there is evidence that the IFG, particularly within the right hemisphere plays an important function in self- and cognitive control (Kawashima et al., 1996; Konishi et al., 1999; Miller, 2000). There is also good evidence the apes exhibit imitation recognition and can be taught basic "do-as-i-do" imitation tasks (Custance et al., 1995; Haun and Call, 2008; Nielsen et al., 2005; Pope et al., 2015) and recent findings using DTI suggest important differences in connectivity between Broca's area and the parietal and temporal cortex may underlie phylogenetic differences in social learning, including imitation (Hecht et al., 2013a; Hecht et al., 2013b). Similarly, a recent paper by Hecht et al. (2017) found that individual difference in cortical connectivity within Broca's area was associated with individual differences in mirror self-recognition in chimpanzees. Thus, the extent that the IFG plays a role in cognitive functions other than tool use and communication warrants further investigation.

Finally, with respect to language, Broca took a localization or phrenology view with the notion that a single brain region played a specific and necessary role in a given function. We now know that a number of cortical and subcortical brains regions, in addition to the IFG, represent a circuit of connected areas that subserve language and speech as well as praxic functions such as tool use and tool making (Belton et al., 2003; Enard, 2011; Frey, 2008; Lewis, 2006; Lieberman, 2007; Stout and Chaminade, 2012; Vargha-Khadem et al., 2005). Here, we focused primarily on different aspects of cortical organization within the IFG but clearly future studies need to consider additional brain regions either alone or in conjunction with the IFG to gain a fuller understanding of the morphological correlates of tool use and communication skills in primates, including humans.

Acknowledgement

This research was supported in part by NIH grants NS-73134, NS-42867, HD-60563, NSF INSPIRE grant 1542848 and Human Frontiers Science Program grant RGP0044. American Psychological Association guidelines for the ethical treatment of animals were adhered to during all aspects of this study. Correspondence regarding this paper can be sent to: William D. Hopkins, Neuroscience Institute, Georgia State University, P.O. Box 5030, Atlanta, GA 30302. Email: whopkin@emory.edu or whopkins4@gsu.edu

References

Amiez, C., and Petrides, M. (2014). Neuroimaging evidence of the anatomo-functional organization of the human cingulate motor areas. *Cereb Cortex, 24(3)*, 563–578. doi:10.1093/cercor/bhs329.

Amunts, K., Schleicher, A., Bürgel, U., Mohlberg, H., Uylings, H.B., and Zilles, K. (1999). Broca's region revisited: Cytoarchitecture and intersubject variability. *The Journal of Comparative Neurology, 412(2)*, 319–341.

Amunts, K., Schleicher, A., and Zilles, K. (2007). Cytoarchitecture of the cerebral cortex—More than localization. *NeuroImage, 37*, 1061–1065.

Arbib, M.A., Liebal, K., and Pika, S. (2008). Primate vocalization, gesture and the evolution of human language. *Current Anthropology, 49(6)*, 1053–1076.

Bailey, P., von Bonin, G., and McCulloch, W.S. (1950). *The isocortex of the chimpanzee*. Urbana-Champaign: University of Illinois Press.

Belton, E., Salmond, C.H., Watkins, K.E., Vargha-Khadem, F., and Gadian, D.G. (2003). Bilateral brain abnormalities associated with dominantly inherited verbal and orofacial dyspraxia. *Human Brain Mapping, 18(3)*, 194–200. doi:10.1002/hbm.10093.

Bianchi, S., Reyes, L.D., Hopkins, W.D., Taglialatela, J.P., and Sherwood, C.C. (2016). Neocortical grey matter distribution underlying voluntary, flexible vocalizations in chimpanzees. *Scientific Reports, 6*, 34733.

Bradshaw, J.L., and Rogers, L.J. (1993). *The evolution of lateral asymmetries, language, tool use, and intellect.* San Diego: Academic Press, Inc.

Broca, P. (1861). Remarques sur le siège de la faculté du langage articulé; suivies d'une observation d'aphémie. *Bulletin de la Société Anatomique, 6*, 330–357.

Brodmann, K. (1909). *Vergleichende Lokalisationslehre der Grobhirnrinde in ihren Prinzipien dargestellt auf Grund des Zellenbaues*. Leipzig: Barth.

Cartmill, E., and Byrne, RW. (2007). Orangutans modify their gestural signaling according to their audience's comprehension. *Current Biology, 17*, 1–14.

Cooper, D.L. (2006). Broca's Arrow: Evolution, prediction and language in the brain. *The Anatomical Record (Part B: New Anatomy), 289B*, 9–24.

Corballis, M.C. (1992). *The lopsided brain: Evolution of the generative mind.* New York: Oxford University Press.

Corballis, M.C. (2003). From mouth to hand: gesture, speech, and the evolution of right-handedness. *Behavioral and Brain Sciences, 26*, 199–260.

Corballis, M.C., Badzakova-Trajkov, G., and Haberling, I.S. (2012). Right hand, left brain: genetic and evolutionary basis of cerebral asymmetries for language and manual action. *WIREs Cognitive Science, 3*, 1–17.

Coude, G., Ferrari, P.F., Rodà, F., Maranesi, M., Borelli, E., Veroni, V., Monti, F., Rozzi, S, and Fogassi, L. (2011). Neurons controlling voluntary vocalization in the macaque ventral premotor cortex. *PLoS One,* 6(11), e26822. doi:10.1371/journal.pone.0026822.

Crockford, C., Herbinger, I., Vigilant, L., and Boesch, C. (2004). Wild chimpanzees produce group-specific calls: a case for vocal learning? *Ethology, 110*, 221–243.

Crockford, C., Wittig, R.M., Mundry, R., and Zuberbühler, K. (2012). Wild chimpanzees inform ignorant group members of danger. *Current Biology, 22*, 1–5.

Custance, D., Whiten, A., and Bard, K. (1995). Can young chimpanzees (*Pan troglodytes*) imitate arbitrary actions? Hayes and Hayes (1952) revisited. *Behaviour, 132*, 837–850.

Cykowski, M.D., Coulon, O., Kochunov, P.V., Amunts, K., Lancaster, J.L., Laird, A.R., Glahn, D.C., and Fox, P.T. (2008). The central sulcus: an observer-independent characterization of sulcal landmarks and depth asymmetry. *Cerebral Cortex, 18*, 1999–2009.

Dronkers, N.F., Plaisant, O., Iba-Zizen, M.T., and Cabanis, A. (2007). Paul Broca's historic cases: high resolution MR imaging of the brains of Leborgne and Lelong. *Brain, 130(5)*, 1432–1441.

Enard, W. (2011). FOXP2 and the role of cortico-basal ganglia circuits in speech and language evolution. *Current Opinion in Neurobiology, 21*, 1–10.

Fazio, P., Cantagallo, A., Craighero, L., D'Ausilio, A., Roy, A.C., Pozzo, T., Calzolari, F., Granieri, E., and Fadiga, L. (2009). Encoding of human action in Broca's area. *Brain Research Reviews, 132*, 1980–1988.

Frey, S.H. (2008). Tool use, communicative gesture and cerebral asymmetries in the modern human brain. *Philosophical Transactions of the Royal Society B: Biological Sciences, 363*, 1951–1957.

Gentry, E., and Byrne, R.W. (2010). Why do gorillas make sequences of gestures? *Animal Cognition, 13(2)*, 287–301.

Gibson, K.R., and Ingold, T. (1993). *Tools, Language and Cognition in Human Evolution*. Retrieved from Cambridge:

Goldenberg, G., and Randerath, J. (2015). Shared neural substrates of apraxia and aphasia. *Neuropsychologia, 75*, 40–49.

Greenfield, P.M. (1991). Language, tools, and brain: The ontogeny and phylogeny of hierarchically organized sequential behavior. *Behavioral and Brain Sciences, 14(4)*, 531–550.

Haun, D.B.M., and Call, J. (2008). Imitation recognition in great apes. *Current Biology, 18(7)*, 288–290.

Hecht, E.E., Gutman, D.A., Preuss, T.M., Sanchez, M.M., Parr, L.A., and Rilling, J.K. (2013a). Process versus product in social learning: Comparative diffusion tensor imaging of neural systems for action execution–Observation matching in macaques, chimpanzees, and humans. *Cerebral Cortex, 23*, 1014–1024.

Hecht, E.E., Murphy, L.E., Gutman, D.A., Votaw, J.R., Schuster, D.M., Preuss, T.M., Orban, G.A., Stout, D., and Parr, L.A. (2013b). Differences in neural activation for object-directed grasping in chimpanzees and humans. *The Journal of Neuroscience, 33(35)*, 14117–14134.

Hecht, E.E., Mahovetz, L.M., Preuss, T.M., and Hopkins, W.D. (2017). A neuroanatomical predictor of mirror self-recognition in chimpanzees. *Soc Cogn Affect Neurosci., 12(1)*, 37–48. doi:10.1093/scan/nsw159.

Hobaiter, C., and Byrne, R.W. (2011a). The gestural repertoire of wild chimpanzees. *Animal Cognition, 14(5)*, 745–767.

Hobaiter, C., and Byrne, R. W. (2011b). Serial gesturing by wild chimpanzees: its nature and function for communication. *Animal Cognition, 14(6)*, 827–838.

Hobaiter, C., and Byrne, R.W. (2013). Laterality in gestrual communication of wild chimpanzees. *Annals of the New York Academy of Sciences, 1288(1)*, 9–16.

Hopkins, W.D. (2013). Behavioral and brain asymmetries in chimpanzees: A case for continuity. *Annals of the New York Academy of Sciences, 1288*, 27–35.

Hopkins, W.D., Adams, M., and Weiss, A. (2013). Genetic and environmental contributions to the expression of handedness in chimpanzees (*Pan troglodytes*) *Genes, Brain and Behavior, 12*, 446–452.

Hopkins, W.D., and Cantero, M. (2003). From hand to mouth in the evolution of language: the influence of vocal behavior on lateralized hand use in manual gestures by chimpanzees (*Pan troglodytes*). *Developmental Science, 6(1)*, 55–61.

Hopkins, W.D., Coulon, O., and Mangin, J.F. (2010). Observer-independent characterization of sulcal landmarks and depth asymmetry in the central sulcus of the chimpanzee brain. *Neuroscience, 171*, 544–551.

Hopkins, W.D., Coulon, O., Meguerditchian, A., Autrey, M.M., Davidek, K., Mahovetz, L.M., Pope, S., Mareno, M.C., and Schapiro, S.J. (2017a). Genetic factors and oro-facial motor learning selectively influence variability in central sulcus morphology in chimpanzees (*Pan troglodytes*). *Journal of Neuroscience, 37(22)*, 5475–5483.

Hopkins, W.D., Hopkins, A.M., Misiura, M., Latash, E.M., Mareno, M.C., Schapiro, S.J., and Phillips, K.A. (2016). Sex differences in the relationship between planum temporale asymmetry and corpus callosum morphology in chimpanzees (*Pan troglodytes*): A combined MRI and DTI analysis. *Neuropsychologia, 93(Pt B)*, 325–334. doi:10.1016/j.neuropsychologia.2016.04.003.

Hopkins, W.D., Meguerditchian, A., Coulon, O., Bogart, S.L., Mangin, J.F., Sherwood, C.C., Grabowski, M.W., Hof, P.R., and Vauclair, J. (2014). Evolution of the central sulcus morphology in primates. *Brain, Behavior and Evolution, 84*, 19–30.

Hopkins, W.D., Meguerditchian, A., Coulon, O., Misiura, M., Pope, S.M., Mareno, M.C., and Schapiro, S.J. (2017b). Motor skill for tool-use is associated with asymmetries in Broca's area and the motor hand area of the precentral gyrus in chimpanzees (*Pan troglodytes*). *Behavioural Brain Research, 318*, 71–81.

Hopkins, W.D., Misiura, M., Pope, S.M., and Latash, E.M. (2015c). Behavioral and brain asymmetries in primates: A preliminary evaluation of two evolutionary hypotheses. *Yearbook of Cognitive Neuroscience, 1359*, 65–83.

Hopkins, W.D., and Nir, T. (2010). Planum temporale surface area and grey matter asymmetries in chimpanzees (*Pan troglodytes*): The effect of handedness and comparison within findings in humans. *Behavioural Brain Research, 208(2)*, 436–443.

Hopkins, W.D., Pika, S., Liebal, K., Bania, A., Meguerditchian, A., Gardner, M., and Schapiro, S.J. (2012a). Handedness for manual gestures in great apes: A meta-analysis. In S. Pika and K. Liebal (Eds.), *Current developments in non-human primate gesture research*. Amsterdam: John Benjamins Publishing.

Hopkins, W.D., Reamer, L., Mareno, M.C., and Schapiro, S.J. (2015b). Genetic basis for motor skill and hand preference for tool use in chimpanzees (*Pan troglodytes*). *Proceedings of the Royal Society of London: Biological Sciences B, 782(1800)*, 20141223.

Hopkins, W.D., Russell, J., Freeman, H., Buehler, N., Reynolds, E., and Schapiro, S.J. (2005). The distribution and development of handedness for manual gestures in captive chimpanzees (*Pan troglodytes*). *Psychological Science, 16(6)*, 487–493.

Hopkins, W.D., Russell, J.L., and Cantalupo, C. (2007b). Neuroanatomical correlates of handedness for tool use in chimpanzees (*Pan troglodytes*): Implication for theories on the evolution of language. *Psychological Science, 18(11)*, 971–977.

Hopkins, W.D., Russell, J.L., and Schaeffer, J.A. (2012b). The neural and cognitive correlates of aimed throwing in chimpanzees: a magnetic resonance image and behavioural study on a unique form of social tool use. *Philosophical Transactions of the Royal Society B: Biological Sciences, 367(1585)*, 37–47.

Hopkins, W.D., Russell, J.L., Schaeffer, J.A., Gardner, M., and Schapiro, S.J. (2009). Handedness for tool use in captive chimpanzees (*Pan troglodytes*): sex differences, performance, heritability and comparison to the wild. *Behaviour, 146(11)*, 1463.

Hopkins, W D., Schaeffer, J., Russell, J.L., Bogart, S.L., Meguerditchian, A., and Coulon, O. (2015a). A comparative assessment of handedness and its potential neuroanatomical correlates in chimpanzees (*Pan troglodytes*) and bonobos (Pan paniscus). *Behaviour, 152(3–4)*, 461–492.

Hopkins, W.D., Taglialatela, J.P., and Leavens, D.A. (2007a). Chimpanzees differentially produce novel vocalizations to capture the attention of a human. *Animal Behaviour, 73*, 281–286.

Hopkins, W.D., Taglialatela, J.P., and Leavens, D.A. (2011). Do chimpanzees have voluntary control of their facial expressions and vocalizations? In A. Vilain, J.-L. Schwartz, C. Abry, and J. Vauclair (Eds.), *Primate communication and human language: Vocalisation, gestures, imitation and deixis in humans and non-humans* (pp. 71–90). Amsterdam: John Benjamins Publishing Company.

Horwitz, B., Amunts, K., Bhattacharyya, R., Patkin, D., Jeffries, K., Zilles, K., and Braun, A.R. (1999). Activation of Broca's area during the production of spoken and signed language: a combined cytoarchitectonic mapping and PET analysis *Neuropsychologia, 41(14)*, 1868–1876.

Hostetter, A.B., Cantero, M., and Hopkins, W.D. (2001). Differential use of vocal and gestural communication by chimpanzees *(Pan troglodytes)* in response to the attentional status of a human *(Homo sapiens)*. *Journal of Comparative Psychology, 115(4)*, 337–343.

Johnson-Frey, S.H. (2004). The neural basis of complex tool use in humans. *Trends in Cognitive Sciences, 8(2)*, 71–78.

Kawashima, R., Satoh, K., Itoh, H., Ono, S., Furumoto, S., Gotoh, R., Koyama, M., Yoshioka, S., Takahashi, T., Takahashi, K., Yanagisawa, T., and Fukuda, H. (1996). Functional anatomy of GO/NO-GO discrimination and response selection--a PET study in man. *Brain Research, 728(1)*, 79–89.

Keller, S.S., Deppe, M., Herbin, M., and Gilissen, E. (2012). Variabilty and asymmetry of the suclal contours defining Broca's area homologue in the chimpanzee brain. *Journal of Comparative Neurology, 520*, 1165–1180.

Keller, S.S., Highley, J.R., Garcia-Finana, M., Sluming, V., Rezaie, R., and Roberts, N. (2007). Sulcal variability, stereological measurement and asymmetry of Broca's area on MR images. *Journal of Anatomy, 211*, 534–555.

Keller, S.S., Roberts, N., and Hopkins, W. D. (2009). A comparative magnetic resonance imaging study of the anatomy, variability and asymmetry of Broca's area in the human and chimpanzee brain. *Journal of Neuroscience, 29*, 14607–14616.

Kilner, J.M., Neal, A., Weiskopf, N., Friston, K.J., and Frith, C.D. (2009). Evidence of mirror neurons in human inferior frontal gyrus. *The Journal of Neuroscience, 29(32)*, 10153–10159.

Knaus, T.A., Corey, D.M., Bollich, A.M., Lemen, L.C., and Foundas, A.L. (2007). Anatomical asymmetries of anterior perisylvian speech-language regions. *Cortex, 43*, 499–510.

Knecht, S., Drager, B., Deppe, M., Bobe, L., Lohmann, H., Floel, A., Ringelstein, E.-B, and Henningsen, H. (2000). Handedness and hemispheric language dominance in healthy humans. *Brain, 123(12)*, 2512–2518.

Kobayashi, S., and Ugawa, Y. (2013). Relationships between aphasia and apraxia. *Journal of Neurology and Translational Neuroscience, 2(1)*, 1028.

Konishi, S., Nakajima, K., Uchida, I., Kikyo, H., Kameyama, M., and Miyashita, Y. (1999). Common inhibitory mechanism in human inferior prefrontal cortex revealed by event-related functional MRI. *Brain, 122 (Pt 5)*, 981–991.

Leavens, D.A., Hopkins, W.D., and Bard, K.A. (1996). Indexical and referential pointing in chimpanzees *(Pan troglodytes)*. *Journal of Comparative Psychology, 110(4)*, 346–353.

Leavens, D.A., Hopkins, W.D., and Bard, K.A. (2005b). Understanding the point of chimpanzee pointing: Epigenesis and ecological validity. *Current Directions in Psychological Science, 14(4)*, 185–189.

Leavens, D.A., Hopkins, W.D., and Thomas, R. (2004b). Referential communication by chimpanzees *(Pan troglodytes)*. *Journal of Comparative Psychology, 118*, 48–57.

Leavens, D.A., Hostetter, A.B., Wesley, M.J., and Hopkins, W.D. (2004c). Tactical use of unimodal and bimodal communication by chimpanzees, *Pan troglodytes*. *Animal Behaviour, 67*, 467–476.

Leavens, D.A., Reamer, L.A., Mareno, M.C., Russell, J.L., Wilson, D.C., Schapiro, S.J., and Hopkins, W.D. (2015). Distal communication by chimpanzees (Pan troglodytes): Evidence for common ground? *Child Development, 86(5)*, 1623–1638.

Leavens, D.A., Russell, J.L., and Hopkins, W.D. (2005a). Intentionality as measured in the persistence and elaboration of communication by chimpanzees *(Pan troglodytes)*. *Child Development, 76(1)*, 291–306.

Leavens, D.A., Taglialatela, J.P., and Hopkins, W.D. (2014a). From grasping to grooming to gossip: Innovative use of chimpanzee signals in novel environments supports both vocal and gestural origins theories of language origins. In M. Pina and N. Gontier (Eds.), *The evolution of social communication in primates*. Switzerland: Springer International Publishing.

Lewis, J.W. (2006). Cortical networks related to human use of tools. *The Neuroscientist, 12(3)*, 211–231.

Liebal, K., Call, J., and Tomasello, C. (2004). Use of gesture sequences in chimpanzees. *American Journal of Primatology, 64*, 377–396.

Lieberman, P. (2007). The evolution of human speech: Its anatomical and neural bases. *Current Anthropology, 48(1)*, 39–66.

Loh, K.K., Petrides, M., Hopkins, W D., Procyk, E., and Amiez, C. (2016). Cognitive control of vocalizations in the primate ventrolateral-dorsomedial frontal (VLF-DMF) brain network. *Neuroscience and Biobehavioral Reviews*. doi:10.1016/j.neubiorev.2016.12.001

Lyn, H. (2012). Apes and the evolution of language: Taking stock of 40 years of research. In J. Vonk and T. Shackelford (Eds.), *Oxford Handbook of Comparative Evolutionary Psychology* Oxford: Oxford University Press.

Lyn, H. (in press). The question of capacity: Why enculturated and trained animals have much to tell us about the evolution of language. *Psychonomic Bulletin and Review*.

Makuuchi, M. (2005). Is Broca's area crucial for imitation? *Cerebral Cortex, 15*, 563–570.

Meador, K. J., Loring, D. W., Lee, K., Hughes, M., Lee, G., Nichols, M., and Heilman, K. M. (1999). Cerebral lateralization: Relationship of language and ideomotor apraxia. *Neurology, 53(9)*, 2028–2031.

Meguerditchian, A., Gardner, M.J., Schapiro, S.J., and Hopkins, W.D. (2012). The sound of one-hand clapping: handedness and perisylvian neural correlates of a communicative gesture in chimpanzees. *Proc Biol Sci, 279(1735)*, 1959–1966. doi:10.1098/rspb.2011.2485

Meguerditchian, A., and Vauclair, J. (2006). Baboons communicate with their right hand. *Behavioural Brain Research, 171*, 170–174.

Miller, E.K. (2000). The prefrontal cortex and cognitive control *Nature Reviews Neuroscience, 1, 59–65*.

Nielsen, M., Collier-Baker, E., Davis, J.M., and Suddendorf, T. (2005). Imitation recognition in a captive chimpanzee (*Pan troglodytes*). *Animal Cognition, 8(1)*, 31–36.

Nishitani, N., Schurmann, M., Amunts, K., and Hari, R. (2005). Broca's region: From action to language. *Physiology, 20*, 60–67.

Passingham, R. E. (1981). Broca's area and the origins of human vocal skill. *Philosophical Transactions of the Royal Society of London, B 292*, 167–175.

Petrides, M., Cadoret, G., and Mackey, S. (2005). Orofacial somatomotor responses in the macaque monkey homologue of Broca's area. *Nature, 435(7046)*, 1235–1238.

Petrides, M., and Pandya, D.N. (1999). Dorsolateral prefrontal cortex comparative cytoarchitectonic analysis in the human and the macaque brain and corticocortical connection patterns. *European Journal of Neuroscience, 11(3)*, 1011–1036.

Petrides, M., and Pandya, D.N. (2009). Distinct parietal and temporal pathways to the homologues of Broca's area in monkeys *PLoS Biology, 7(8)*, e1000170.

Phillips, K.A., Schaeffer, J., Barrett, E., and Hopkins, W.D. (2013). Performance asymmetries in tool use are associated with corpus callosum integrity in chimpanzees (*Pan troglodytes*): A diffusion tensor imaging study. *Behavioral Neuroscience, 127(1)*, 106–113.

Pika, S., Liebal, K., and Tomasello, M. (2003). Gestural communication in young gorillas (*Gorilla gorilla*): gestural repertoire, learning, and use. *American Journal of Primatology, 60(3)*, 95–111. doi:10.1002/ajp.10097

Pika, S., and Mitani, J.C. (2006). Referential gestural communication in wild chimpanzees (*Pan troglodytes*). *Current Biology, 16(6)*, R191–R192.

Pollick, A.S., and de Waal, F.M.B. (2007). Ape gestures and language evolution. *Proceedings of the National Academy of Sciences, 104(19)*, 8184–8189.

Pope, S.M., Russell, J.R., and Hopkins, W.D. (2015). The association between imitation recognition and socio-communicative competencies in chimpanzees (*Pan troglodytes*). *Frontiers in Psychology, 6*, Article 188. doi: 10.3389/fpsyg.2015.00188

Premack, D. (2004). Is language the key to human intelligence? *Science, 303*, 318–320.

Prieur, J., Pika, S., Barbu, S., and Blois-Heulin, C. (2016a). Gorillas are right-handed for their most frequent intraspecific gestures. *Animal Behaviour, 118*, 165–170.

Prieur, J., Pika, S., Barbu, S., and Blois-Heulin, C. (2016b). A multifactorial investigation of captvie chimpanzees' intraspecific gestural laterality. *Animal Behaviour, 116*, 31–43.

Rasmussen, T., and Milner, B. (1977). The role of early left-brain injury in determining lateralization of cerebral speech function. *Annals of the New York Academy of Sciences, 299*, 355–369.

Roby-Brami, A., Hermsdorfer, J., Roy, A.C., and Jacobs, S. (2012). A neuropsycholigical perspective on the link between language and praxis in modern humans. *Philosophical Transactions of the Royal Society B: Biological Sciences, 367*, 144–160.

Romanski, L.M., Averback, B., and Diltz, M. (2004). Neural representations of vocalizations in the primate ventrolateral prefrontal cortex. *Journal of Neurophysiology, 93*, 734–747.

Sanz, C.M., Morgan, D.B., and Hopkins, W.D. (2016). Lateralization and performance asymmetries in the temite fishing of wild chimpanzees in the Goualougo Triangle, Republic of Congo. *American Journal of Primatology, 78*, 1190–1200.

Schel, A.M., Townsend, S.W., Machanda, Z., Zuberbühler, K., and Slocombe, K.E. (2013). Chimpanzee alarm call production meets key criteria for intentionality. *PLoS One, 8(10)*, e76674.

Schenker, N.M., Hopkins, W.D., Spocter, M.A., Garrison, A.R., Stimpson, C.D., Erwin, J.M., Hof, P.R., and Sherwood, C.C. (2010). Broca's area homologue in chimpanzees (Pan troglodytes): probabilistic mapping, asymmetry, and comparison to humans. *Cerebral Cortex, 20*, 730–742.

Schenker, N.M., Sherwood, C.C., Hof, P.R., and Semendeferi, K. (2007). Microstructural asymmetries of the cerebral cortex in humans and other mammals. In W.D. Hopkins (Ed.), *Evolution of Hemispheric Specialization in Primates*: Academic Press.

Seyfarth, R.M., and Cheney, D.L. (2010). Production, usage, and comprehension in animal vocalizations. *Brain and Language, 115*, 92–100.

Sherwood, C.C., Broadfield, D.C., Holloway, R.L., Gannon, P.J., and Hof, P.R. (2003). Variability of Broca's area homologue in great apes: Implication for language evolution. *The Anatomical Record, 217A*, 276–285.

Slocombe, K.E., and Zuberbühler, K. (2005). Functional referential communication in a chimpanzee. *Current Biology, 15(19)*, 1779–1784.

Slocombe, K.E., and Zuberbühler, K. (2007). Chimpanzees modify recruitment screams as a function of audience composition. *Proceedings of the National Academy of Sciences, 104(43)*, 17228–17233.

Spocter, M.A., Hopkins, W.D., Barks, S.K., Bianchi, S., Hehmeyer, A.E., Anderson, S.M., Stimpson, C.D., Fobbs, A.J., Hof, P.R., and Sherwood, C.C. (2012). Neuropil distribution in the cerebral cortex differs between humans and chimpanzees. *Journal of Comparative Neurology, 520*, 2917–2929.

Stout, D., and Chaminade, T. (2012). Stone tools, language and the brain in human evolution. *Philosophical Transactions of the Royal Society B: Biological Sciences, 367*, 75–87.

Taglialatela, J.P., Cantalupo, C., and Hopkins, W.D. (2006). Gesture handedness predicts asymmetry in the chimpanzee inferior frontal gyrus. *Neuroreport, 17(9)*, 923–927.

Taglialatela, J.P., Reamer, L., Schapiro, S.J., and Hopkins, W.D. (2012). Social learning of a communicative signal in captive chimpanzees. *Biology Letters, 8*, 498–501. doi:10.1098/rsbl.2012.0113.

Taglialatela, J.P., Russell, J.L., Schaeffer, J.A., and Hopkins, W.D. (2008). Communicative signaling activates "Broca's" homologue in chimpanzees. *Current Biology, 18*, 343–348.

Taglialatela, J.P., Russell, J.L., Schaeffer, J.A., and Hopkins, W.D. (2011). Chimpanzee vocal signaling points to a multimodal origin of human language. *PLoS One, e18852*.

Tomaiuolo, F., MacDonald, J.D., Caramanos, Z., Posner, G., Chiavaras, M., Evans, A.C., and Petrides, M. (1999). Morphology, morphometry and probability mapping of the pars opercularis of the inferior frontal gyrus: an *in vivo* MRI analysis. *European Journal of Neuroscience, 11*, 3033–3046.

Uylings, H., Jacobsen, A., Zilles, K., and Amunts, K. (2006). Left-right asymmetry in volume and number of neurons in adult Broca's area. *Cortex, 42(4)*, 652–658.

Vargha-Khadem, F., Gadian, D.G., Copp, A., and Mishkin, M. (2005). FOXP2 and the neuroanatomy of speech and language. *Nature Reviews Neuroscience, 6(2)*, 131–138. doi:10.1038/nrn1605.

Volter, C.J., Rossano, F., and Call, J. (in press). Social manipulation in nonhuman primates: Cognitive and motivational determinants. *Neu-*

roscience and Biobehavioral Reviews. https://doi.org/10.1016/j.neubiorev.2016.09.008.

Zilles, K., and Amunts, K. (2010). Centenary of Brodmann's map-conception and fate. *Nature Reviews Neuroscience, 11*, 139–145.

Table 1: Descriptive Data on Area 44 and Area 45 in Humans and Chimpanzees

	Male		Females	
Measure	Left	Right	Left	Right
Volume				
Humans				
Area 44	5024 (1661)	3281 (1162)	2909 (652)	2036 (634)
Area 45	3155 (653)	4137 (1238)	2859 (341)	2218 (225)
Chimpanzees				
Area 44	625 (115)	544 (63)	577 (34)	752 (70)
Area 45	519 (76)	601 (77)	483 (88)	666 (96)
Neuron Number				
Humans				
Area 44	95.4 (19.1)	64.0 (13.2)	67.2 (12.3)	47.8 (13.4)
Area 45	68.2 (17.7)	79.0 (15.9)	74.6 (10.0)	54.8 (6.9)
Chimpanzees				
Area 44	9.98 (1.29)	11.40 (1.27)	9.88 (2.30)	8.55 (1.99)
Area 45	8.55 (1.59)	9.93 (0.91)	8.93 (1.47)	9.35 (0.73)
Neuropil Space				
Human				
Area 45	.278 (.020)	.267 (.016)	.316 (.020)	.328 (.016)
Chimpanzee				
Area 45	.215 (.022)	.179 (.017)	.228 (.019)	.226 (.015)

Volume measures are in mm^3. Neurons number ($x\ 10^6$)

Table 2: Descriptive Data on Sulci Used as Boundaries in Defining the Pars opercularis in Chimpanzees

	Frontal Orbital		Inferior Frontal		Precentral Inferior	
	Left	Right	Left	Right	Left	Right
Surface Area	**548.17**	**531.45**	966.07	933.46	*890.30*	*932.22*
	(6.92)	**(6.67)**	(16.91)	(18.92)	*(16.81)*	*(16.49)*
Mean Depth	**8.24**	**7.98**	**8.75**	**8.57**	8.79	8.87
	(.062)	**(.062)**	**(.079)**	**(.091)**	(.065)	(.065)
Gray Matter	3.499	3.521	3.373	3.426	**3.380**	**3.331**
Thickness	(.029)	(.031)	(.038)	(.032)	**(.024)**	**(.025)**

N = 271. Surface area measures are in mm^2. Mean depth and gray matter thickness measures are in mm. **Bolded** values indicate significant leftward asymmetries, while *italicized* values indicate rightward asymmetries.

Camille Coye [1], Simon Townsend [2,3] and Alban Lemasson [1]
[1]Laboratoire EthoS, Université´ de Rennes 1, Paimpont, France
[2]Language and learning group, Department of Psychology, University of Warwick, Coventry, United Kingdom,
[3]Department of comparative linguistics, University of Zurich, Zurich, Switzerland.

From Animal Communication to Linguistics and Back: Insight from Combinatorial Abilities in Monkeys and Birds

Abstract: For several decades, ethologists and comparative psychologists have been using a linguistic terminology to discuss complex communicative abilities in animals, with a particular focus on sound combinatorial rules. One historical example is the possible syntactic ability of songbirds. More recently, context-dependent call combinations have been described in nonhuman primates. This time, the detailed observational and experimental data gathered in this area has even drawn the attention of linguists and has given rise to studies highlighting the relevance of linguistic tools for the study of nonhuman primate communication systems. However, the parallels that can be drawn between humans', birds' and nonhuman primates' verbal/vocal combinations still remain the topic of intense debate possibly because mismatches between the terminologies used have confounding effects. The question is: can we go beyond the traditional dichotomy between phonological and lexical syntax to characterize the diversity of sound combinatorial rules found in animals? Here, we will adopt a two-step approach in order to discuss: (1) what forms sound combination takes in animals, based on structural and functional criteria and when it may or not be appropriate to use linguistic terms; (2) why sound combination may have evolved in some species more than others. We will notably illustrate our arguments with recent findings in some cooperative breeding birds and guenons, where cases of meaningful sound compositionality have been recently described.

Keywords: sounds combinations, verbal combinations, animal communication

1. Introduction

The communicative abilities of animals have received considerable research interest over the years. One likely contributing factor is that animal communication constitutes a particularly fruitful substrate for comparative analyses with human language. Language, in its full blown form is a product of a variety of communicative and socio-cognitive capacities[1]. This complexity, together with its relatively short window of emergence[2,3], has led to the suggestion that, as with other complex biological phenomena, language might have evolved from pre-existing capacities and structures initially serving other functions[3–5]. One way to shed light on the potential evolutionary path leading to the emergence of language is to decompose it into several core features and to explore their presence and role in non-human animals from various taxa[6,7]. This approach, using animals from distinct and phylogenetically distant taxa is particularly relevant in helping disentangle the relative influence of various factors involved in the evolution of complex communicative abilities culminating in human language.

One relevant historical example is that of capacities for vocal learning, which have been described in some birds (e.g. parrots, starlings and mockingbirds) and mammalian species (e.g. some bats and marine mammals)[8,9] in which individuals are able to acquire new vocalisations. Several studies provide convincing evidence for the importance of social learning and auditory feedback for call and song acquisition[10–13] which are central to language acquisition in humans. In addition, some studies even reported the cultural transmission of vocal dialects in several bird and mammal species[14–16]. Given, the distribution of vocal learning species in distant taxa these capacities are assumed to result from convergent evolution[17,18]. Moreover, the study of the neural substrates and genetics of species that possess vocal learning capacities, compared with humans and with non-learning species (e.g. gull, dove, suboscines) has generated hypotheses for the emergence of vocal learning[18–21]. For instance, comparative studies of the structure and expression patterns of the now famous FoxP2 gene, have shed light on its potential role in the emergence of vocal learning in the human lineage: using data from comparative studies and clinical studies in human, some authors proposed that ancient neural functions of *FoxP2* have been co-opted to subserve aspects of vocal communication, and notably vocal learning, in several species including humans[22–25]. This example demonstrates the utility of broad comparative studies to clarify particular aspects of the evolution of communication. However, the study of FOXP2 in isolation is not sufficient and other factors are likely involved and have to be explored regarding vocal learning in animals[22]. For instance, some so-called 'non-learners' have recently been shown to display significant abilities in vocal plasticity (e.g. elephant[26], goat[27], marmoset[28], guenons [29–31], gibbons[32,33]) questioning the relevance of the traditional learner/non-learner dichotomy.

Several other capacities involved in language have been studied and described to various extents in non-human species, including many non-human primates. For instance, various species display the capacity to produce intentional signals[34–36] or semantic-like signals (also termed referential signals i.e. that refer to an external object of the word)[35,37–43]. Also, vocal exchanges in some species are strictly organised and display "conversation-like" properties (i.e. based on call overlap avoidance and turn-taking between exchanging partners[44,45]). Furthermore, the capacity to combine sounds into complex structures has long been a topic of much contention[46]. Indeed, combinations of vocal units have been extensively reported in animals from various taxa, including numerous bird species (e.g. winter wren (*Troglodytes troglodytes*)[47], Bengalese finches (*Loncura striata*)[48], mockingbirds[49], European starlings (*Sturnus vulgaris*)[50], several species of chickadees[51,52], and blue-throated Hummingbirds (*Lampornis clemenciae*)[53]) as well as mammals such as rock hyraxes (*Procavia capensis*)[54], several species of bats (i.e. mustached bats (*Pteronotus parnellii*)[55]; free-tailed bats (*Tadarida brasiliensis*)[56,57]; sac-winged bats (*Saccopteryx bilineata*)[58]), whales (humpback whales (*Megaptera novaeangliae*)[59,60]; Killer whales (*Orcinus orca*)[61,62]; pilot whales (*Globicephala sp.*)[63]; sperm whales (*Physeter microcephalus*)[64]) and non-human primates (e.g cotton-top tamarins (*Saguinus Oedipus*)[65]; gorillas (*Gorilla sp.*)[66]; red-bellied titi monkeys (*Callicebus moloch*)[67]). Sound combinations occur in wide diversity of contexts, such as alarm contexts[68–70], socio-positive interactions[71–73], mate attraction or territorial defence[54,74–76] and can take various forms. Some animals, for example, can merge acoustic units (i.e. basic element consisting of a continuous mark on a sonogram, also termed notes in birds) into combined calls (e.g. consisting of several units merged linearly with little to no silence between them, also termed motifs in birds). Furthermore, calls (simple and/or combined ones) can then also be combined into higher-order call sequences (i.e. series of calls uttered in sequence and separated by a silent interval always shorter than silent gaps between sequences, also termed songs in birds)[56,71,77].

Combinatoriality is central to language, and drawing parallels between sound combinations in animals and the combinatorial systems of language is tempting. Language relies on the combination of sounds (phonemes, see Table 1 for a definition) into larger units (morphemes and words) which are themselves combined into larger utterances (sentences)[78]. However, useful comparisons are hard to achieve partly because, contrary to most ethological definitions given for sound combinations, linguistic definitions often heavily rely on functional aspects and include element's meaning or grammatical function as a way to characterise them[78–80].

Language's generativity (i.e. capacity to generate an infinite number of ends using finite means) is a product of dual articulation[81,82], which allows combination at two distinct layers: phonology and morphosyntax[83] (Table 1). Phonology corresponds to the combination of meaningless sounds (*i.e.* phonemes) into meaningful elements (*i.e.* morphemes and monomorphemic words). Simply put, a phoneme is a sound which, when added, deleted or used to replace another sound in a word, creates a phonemic contrast changing the meaning of the word. For example, in English the sounds /k/ and /b/ are phonemes as they differentiate the words 'cat' and 'bat'. Two words that differ only by one phoneme are termed 'minimal pair'. Morphosyntax (Table 1), corresponds to the second layer of combination, in which meaningful elements (morphemes and words) are combined into larger structures whose meaning depends on the elements composing them and their order. Some words consist of only one morpheme (*i.e.* monomorphemic words such as 'happy') but morphemes can also be combined together into polymorphemic words. For example, the word "happy" can be combined with the suffix "ness" to create the polymorphemic word "happiness" or with the prefix "un" to create 'unhappy'. Finally, in phrases, morphemes are combined according to grammatical rules[84]. These rules are a key contributor to language's generativity: with a finite number of rules it is possible to generate, using a finite number of elements, an infinite number of structures among which rules distinguish well-formed (or grammatical) syntactic structures from ill-formed (or non-grammatical) syntactic structures[78,85].

Table 1: Main linguistic concepts and definitions to be used in this chapter

Phoneme	Smallest meaning differentiating sounds in a language[78] i.e. meaningless sounds that allow to differentiate between two words.
Morpheme	A minimal unit of meaning or grammatical function[78].
Phonology	Combinatorial layer of language in which phonemes are combined to form morphemes and words[78,83].
Morphosyntax	Combinatorial layer of language in which meaningful elements are combined into larger structures whose meaning depends on the elements composing them and their order. Morphosyntax includes both morphology, where morphemes can be combined into more complex structures (*i.e.* polymorphemic words), and syntax, where mono- and polymorphemic words are combined into sentences[85,86].
Dual articulation	Characteristic of language whereby speech can be analysed at two complementary levels: phonology and morphosyntax. Duality of patterning has been characterised as a design-feature of language which is partly responsible for language's virtually infinite generativity[81,87].
Scalar implicatures	Linguistic concept related to pragmatic inference. The core idea is that the utterance of a sentence S implicates the falsity of stronger alternatives (i.e. more informative ones) as for any stronger alternative S' to S, a cooperative speaker would have used S' rather than S if s/he believed S' to be true[88–90]. *e.g.* the sentence S "some of my trees are oaks" implies that not all my tree are oaks as, if all were, I would have used the more informative S' sentence "all of my trees are oaks".

Peter Marler[91], proposed to differentiate between animal combinatorial structures depending on their organisation and the likely meaning of their components. To this end, he used terms borrowed from linguistics and distinguished two main types of organization: phonological syntax (or phonocoding) and lexical syntax (or lexicoding). He defined phonological syntax as the concatenation of sounds without independent information content and which are not used singularly, or meaningful sounds that lose their original content when combined. He defined lexical syntax as the level at which meaningful elements are combined. Whilst Marler borrowed terms from linguistics, several important differences with human phonology and morphosyntax remain. First, the concept of 'meaning differentiation" (i.e. change in meaning of the whole resulting from change in one of its meaningless elements) is absent from Marler's definition of phonological syntax. In addition, contrarily to phonology, Marler's phonological syntax allows the combination of meaningful units. Finally, the definition he proposed for lexical syntax – though closer to morphosyntax than phonological syntax is to phonology- differs strikingly from its 'human' counterpart as the importance of elements' order on combined call meaning is overlooked while the order of words and morphemes are of central importance in most human utterances.

Recently, the study of combinatorial systems in animals has received renewed interest from researchers from various fields including ethology, linguistics and psychology[46,86,88]. A multitude of studies have illuminated new perspectives and subsequently given rise to the development of interdisciplinary work[88,92,93] as well as prompting controlled experiments investigating combinatorial structures found in the natural communication of animals and their relevance to receivers[69,70,94–96]. Within this framework, some authors have questioned the relevance of the definitions proposed by Marler notably because the joint use of the terms 'phonological' and 'syntax', which correspond to very distinct linguistic concepts, is misleading[86]. We concur and forward that further reflection is now necessary in order to develop an accurate terminology to characterise the structure and functional aspects of sound combination in animals. In addition to providing the groundwork facilitating understanding of animal combinatorial systems, this will best serve comparative analyses with language.

This chapter therefore aims to review comparative work on sound combinations in humans and animals, with two intended outcomes:

- **to propose a basis for future interdisciplinary work aiming to develop a more appropriate terminology, and shed light on some potentially fruitful prospects for future studies of sound combination.**
- **To elucidate evolutionarily relevant factors likely to have influenced the development of combinatorial communication systems.**

We will firstly turn our attention to the variety of combinatorial structures found in primates and birds. We will (1) examine possible bases to define rudimentary parallels with sound combination in language and (2) review recent empirical studies providing convincing evidence for combinatorial capacities parallel to language in animals. In the third part of this chapter we will focus on the recent advances brought by the use of formal linguistic analyses on animal communication systems. Finally, in line with the comparative rationale adopted at the beginning of this chapter we will build on the taxonomic diversity of examples described as a way to formulate potential hypotheses regarding why combinatorial systems emerge.

2. Combinatorial systems: diversity and terminology

This section reviews various animal combinatorial systems and evaluates the terminology used to characterize them. Given the key role played by meaning in combinatorial systems, particularly when comparing combinatorial structures with those in human language, providing a clarification is important. We will use the term meaning in a form approaching Gricean natural meaning[97,98], i.e. as the significance/information that receivers derive from a signal and its context (because of its regular association with a given event, individual or object), without assuming emitter's intention to inform others[99,100].

2.1 Parallels with phonology?

Many bird species rely on the combination of apparently meaningless units into larger structures. However, studies describing such systems often lack information on the contextual correlates of the combinatorial variants emitted[49,101]. Furthermore, experiments testing a potential intrinsic meaning of single units (or changes in meaning accompanying changes in the type or order of units combined) are often missing[48,102,103]. Some experiments in songbirds have shown that receivers' reaction can be influenced by unit diversity[104], fine acoustic structure[105], or by the simultaneous modification of several frequency and temporal parameters [106]cited by[104], suggesting that information (about caller's quality, or identity) is conveyed. Nevertheless, as frequently suggested by behavioural observations, such modifications, as well as changes in unit's type or order, do not seem to alter the main function or "semantic content" of the sequence (i.e. mostly social bonding, mate attraction and/or territorial defence in the case of songbirds[47,49,74,86,104]). As a result, any parallel with the phonological layer of language is limited at best and subsequently it has been argued that such systems may be better described in terms of "phonetic patterning", that relates to the physical properties of sounds but does not characterise sounds as meaning-differentiating[86].

Previous work in non-human primates, particularly gibbons has suggested possible additional parallels with phonological organisation seen in language[107]. An observational study on white-handed gibbons (*Hylobates lar*), for example, indicated that their communication, like that of songbirds, relies on the combination of apparently meaningless units into sequences. However as far as we know, in contrast to songbirds, gibbons give two types of sequences that are associated with strikingly distinct contexts and functions: one is produced routinely in the morning, while the other functions to signal the presence of a predator[107]. In both contexts, these sequences are given in duets during which two partners produce song in a coordinated way but the organisation of sequences differs between morning duets and predatory ones. More precisely, morning and predatory duets differ in three ways: (1) in the proportion of one type of note (the "hoo" note, with on average 100 vs 10 "hoo" notes introducing predatory induced songs and morning duets respectively), (2) in the order of motifs involved (female-specific calls is given later and answered slower by her male partner in predatory contexts) and (3) in the presence of two note types as (i.e. 'learning-wa' notes are globally absent from predator-induced songs while 'sharp wow' notes are absent from morning duets). Moreover, natural observations indicate that wild individuals react differently to the distinct sequences suggesting that the structuring of the signal encodes information. Further experimental work is now required to clarify how. Playback experiments comparing receivers' reaction to natural sequences and artificial stimuli in which the order, proportion, and type of notes given are manipulated will be particularly necessary to identify what receivers use to discriminate between sequence types. In addition, further clarifications about whether notes' acoustic structure varies between contexts and whether they possess an intrinsic meaning (notably 'learning wa' and 'sharp wow' notes) could be obtained using acoustic analyses and playbacks. Such information would help determine the nature of the system (i.e. showing parallels with phonology, morphology or neither) and may also shed light on the possible cognitive processes underlying communication in this species.

Thus, although previous studies on birds and primates reviewed above match some of the criteria used to define phonological combination in language, none of them did so fully, primarily because the demonstration of changes in message according to sequence organisation (i.e. meaning-differentiation) was lacking or because the intrinsic meaning of notes was unclear. **We propose that convincing evidence for parallels with phonology in animals would require: (1) a combination involving units that are not associated to any particular behavioural context (hence from which receivers could not individually extract specific information about the environment, or caller's behaviour). (2) that the combination (or addition) of given "meaningless" units in a given order creates a signal which can be reliably associated with one (or several) external events or indeed a caller's behaviour(s)[96] and critically (3) that changes in unit order or composition triggers changes in signals' content. Finally, to parallel in a rudimentary way the productivity of language, we would also expect such systems to involve the reuse of units across distinct types of utterances.**

2.2 Parallels with morphology?

The second layer of language, morphosyntax, relies on the combination of meaningful sounds into larger structures whose meaning depends on their components and organisation. Several studies have described vocalisations composed of apparently meaningful calls but here, again, the parallel with the morphosyntactic organisation of language is not always clear. A series of studies investigating gorilla communication has described a potential combinatorial system in a great ape species[66,108]. Both mountain and Western gorillas possess a graded repertoire composed of five main types of close calls. Each type of call can be given alone or combined with every other close call unit in non-random ways. The authors analysed the contextual correlates of emission for three types of units and their most common combinations: atonal grunts (A1), short tonal grunts (T2) and grumbles (T4) as well as A1-T4 and T2-T4 combinations. The results show that, while T4 is given more in foraging contexts, in particular when there is no individual less than 5 meters from the emitter, A1 and T2 are associated with resting contexts, notably when other individuals are around the emitter (i.e. <5m) and do not differ in their context of emission. A1-T4 and T2-T4 combinations are given in the same context as A1 and T2 calls but, in contrast to single calls, combinations are associated strongly with vocal exchanges. These results suggest that in this system T4 units, which may serve as a « localisation » call due to their longer duration, can be added to A1 or T2 units (whose « normal » context of emission is thus respected) during vocal exchanges. However, whether combinations triggers changes in the information content of the calls remains unknown, because receiver's reaction to single and combined units have not been tested and, more importantly, because the contextual correlates of the vast majority of combinations given by gorillas (more than 150 different types[66]) have not been investigated yet. In addition, the role of repetition and call order in combined vocalisations that seem to vary greatly remains poorly understood[108].

A series of studies on the alarm call system of male putty-nosed monkeys also revealed an intriguing system which relies on the combination of calls that appear to carry meaning[109–113]. Indeed, male putty-nosed monkeys use two distinct loud calls « Pyow » and « Hack ». A first series of studies using natural observations, playbacks and predator presentation experiments suggested that sequences of « Pyows » were regularly given to leopards, while sequences of « Hacks » as well as transitional Hack series (i.e. several Hacks followed by several Pyows) were common responses to crowned-hawk eagles. Interestingly, Pyow-Hack sequences (i.e. 1–4 Pyows followed by 1–4 Hacks) reliably trigger movement (both natural sequences and sequences artificially composed of calls given in other contexts). The relationship between the apparent meaning of Pyow-Hack sequences and their components has raised questions and four main interpretations have been proposed[86,112]: (1) a phonological interpretation in which Pyow and Hack would work as « phonemes » i.e. allowing differentiation of meaning of single units (Pyow, Hack) and of their joint use (Pyow-Hack sequence)[86], (2) an idiomatic interpretation in which the original –compositional- meaning of Pyow-Hack sequence was blurred, similarly to human idioms (e.g. it's raining dogs and cats)[86,112,114] and two more « semantic » interpretations: (3) one in which Pyow and Hack would respectively carry the meaning « move on the ground » and « move in the air » while Pyow-Hack sequence would carry a combined general meaning « we move, let's go » as putty-nosed monkeys occupy various strata at a time and can travel on the ground as well as in the canopy[86] and (4) an interpretation based on weak meanings from Pyow (i.e. underspecified, general alarm) and Hack (i.e. non-ground movement or high arousal depending on the analysis) and inferences based on the pragmatic principles of competition and influence of contextual cues[114]. Now, further investigation of the possible mental representations triggered by conspecific calls as well as putty-nosed monkeys' capacities to handle and understand combinatorial structures more generally are necessary to determine which of these interpretation (or others) is most plausible.

The examples reviewed above show that combining meaningful calls into larger structures (either combined calls or call sequences) is not sufficient to offer a robust parallel with the morphosyntactic organisation of language. **In particular, we argue that to be considered as a rudimentary parallel with morphosyntax, a system would obviously need to (1) involve the combination of vocal units, from which receivers can individually extract information, into a larger structure. It would also need that the information content changes depending on and reflects (2) the units merged together and their respective content and (3) rules for unit combination (*i.e.* systematic order of combination and consistent alteration of the information conveyed by signal).**

3. Focus on promising examples: the cases of babblers and guenons

3.1 Parallels with phonology

To date and to our knowledge, only one study documenting note combinations in chestnut-crowned babblers has provided convincing evidence for a parallel with the phonological layer of language. In this study, Engesser et al.[96] combined natural observations, acoustic analyses and playback of natural and artificially recombined sounds in chestnut-crowned babblers. These cooperatively breeding birds living in arid areas of South-Western Australia possess a vocal repertoire of discrete calls, most of which are composed of apparently meaningless notes. Critically, some notes are reused across call types, such as the 'A' and 'B' notes that can be combined together into an 'AB' structure during flight (i.e. flight call) and a 'BAB' structure during nestling provisioning (i.e. prompt call) (Figure 1). Acoustic analyses showed no difference in notes' structure between call types.

Figure 1: Spectrogram of double-element flight call (i.e. F1 F2) and triple-element prompt call (i.e. P1 P2 P3) of adult chestnut-crowned babblers. Figure reproduced from Engesser et al. (2015) [94].

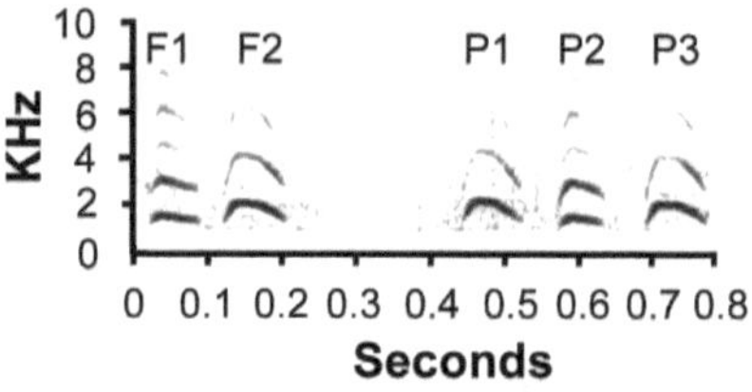

Receivers' reaction did not differ between natural and artificial stimuli (i.e. artificial flight calls created by deleting the first 'B' unit of a prompt call and artificial prompt calls created by adding a 'B" unit to a flight call) within a call type. In addition, the broadcast of single 'B' units and artificial 'CAB' stimuli ('C' being a call element naturally given in combination with other notes by chestnut babblers) triggered surprised reactions that differed from those obtained by the broadcast of flight or prompt calls. These additional testing conditions thus ruled out a possible 'priming effect' of a 'B' element as well as responses being driven by superstructure effects[96]. **Thus, the flight call/ prompt call complex in CCBs seems to match the three key criteria needed to draw a parallel with phonology i.e. (1) a combinations of 'meaningless' elements into (2) a structure meaningful to receivers and (3) which meaning changes if elements change order and presence.** Indeed, the authors argue this example represents a rudimentary form of phonemic contrasts. Given several other calls in the repertoire of these birds form call pairs (*i.e.* two calls given in distinct contexts and that differ only by one note)[115], future studies will be needed to determine whether they also make use of a similar combinatorial mechanism. Tackling this question would be of particular importance as this would also help shed light on the productivity of the system (i.e. the extent to which notes are reused across the repertoire to create various types of utterances).

3.2 Parallels with morphosyntax?

The combination of meaningful calls, via a mechanism resembling morphosyntax can involve the merging of sounds into combined calls as well as the combination of sounds (separated by silent gaps) into call sequences. In light of this, we therefore propose to differentiate between combined calls and call sequences. Such a distinction is advantageous for several reasons. Firstly, because this distinction falls in line with the traditional distinction forged between words and phrases (and between calls and call sequences in animals)[80,84]. Secondly, it may serve to facilitate analyses, notably during preliminary phases of investigation. Finally, it may also be more realistic, as different underlying capacities, such as working memory requirements[87,116,117] may be required to perform and interpret combinations of meaningful elements at these two levels in animals.

In addition, the combination of morphemes can involve two types of elements: bounded morphemes that are always used in conjunction with others (e.g. suffixes), and free morphemes that constitute monomorphemic words when used alone[78,85]. An analogous form of the combination of these two types of morphemes in animals would correspond to the merging of one individual call (that can be used alone) with a vocal unit that is never given by itself and the merging of two individual calls[86,94,95] respectively. Interestingly, evidence for both types of call combination (i.e. using individual calls or calls that are never used alone) have been recently reported in the literature of one cooperatively breeding bird (southern pied babbler, *Turdoides bicolor*) and of two species of guenons: Campbell's and Diana monkeys (see Table 2 for a summary).

Southern pied babblers

A recent study highlighted a combination mechanism in the alarm calls of the southern pied babbler, a cooperatively breeding bird living in the arid areas of South-Africa[118]. These birds emit an alert call with a broadband structure to low-urgency threats and a tonal, repetitive recruitment call in non-alarm contexts to attract group members to a new location (*e.g.* roosting or foraging). Critically, pied babblers sequentially combine alert and recruitment calls when encountering and mobbing terrestrial predators[118] (Figure 2, Table 2).

Figure 2: Spectrogram of a mobbing sequence composed of one alert and seven recruitment calls. Figure reproduced from Engesser et al. (2016)[129].

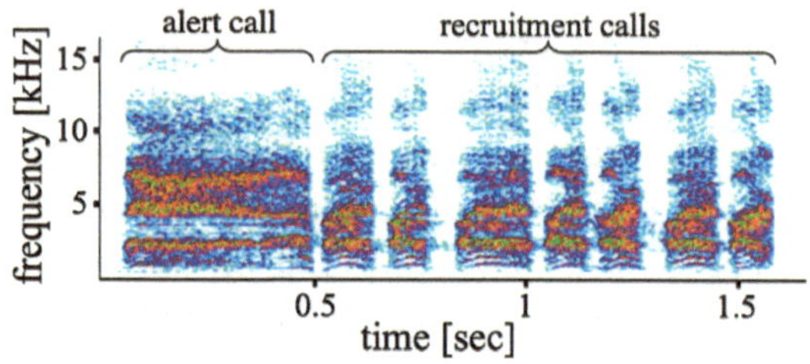

Using a playback experiment, the authors tested the combinatorial structure of the mobbing sequence and its relevance to receivers by comparing wild, but habituated, pied babblers' reaction to the broadcast of natural and artificially created stimuli (i.e. mobbing sequences created by combining alert and recruitment calls and single-call stimuli extracted from natural mobbing sequences). Subjects' reaction to natural and artificial stimuli did not differ, which demonstrated that mobbing sequences consisted in the linear merging of alarm and recruitment calls and thus confirmed their combinatorial nature. Also, the distinct reactions given to the three call types presented (i.e. alert call, recruitment call and mobbing sequence) demonstrated the relevance of these calls to receivers. **This study thus satisfies the three criteria proposed for parallels with linguistic morphosyntax in a non-human animal i.e. individually meaningful calls, combined into a meaningful structure whose meaning reflects that of the elements involved.** Interestingly, receivers' reaction to mobbing sequences exceeded the sum of reactions to their components (*i.e.* higher attentiveness and quicker approach). This suggests that, in this case, the combination of two elements did not simply lead to an addition of their meanings but potentially gave rise to a 'richer' meaning (*i.e.* 'mobbing a predator'), that is related to, yet goes beyond, the meaning of its parts[84,118].

Diana monkeys

Other studies, focussing on the communication of two cercopithecids, Diana and Campbell's monkeys, revealed meaningful combinatorial systems that could offer rudimentary parallels with the morphosyntactic organisation of language. These arboreal primates live sympatrically in the dense primary forests of West Africa[119,120] and their communication, which relies almost exclusively on sex-specific vocal signals, has been studied intensively over the past decades (e.g.[29,30,40,121–128]). In Diana monkeys, females possess four main types of social calls: H, L, R and A. The first three calls are associated with distinct contextual valences for the caller (very positive social context, neutral to mildly positive context and socio-negative or mildly dangerous context respectively). The last call (A) is given in a broad range of contexts and strongly signals caller's identity[29,121]. Each of these calls can be given alone or in combination according to the following pattern: a contextual unit (i.e. H, L or R) merged with an arched unit (i.e. A) (Figure 3).

Figure 3: combined calls of female Diana monkeys. (a) HA call (socio-positive contexts), (b) LA (neutral to positive contexts) and (c) RA calls (negative contexts and mild danger).

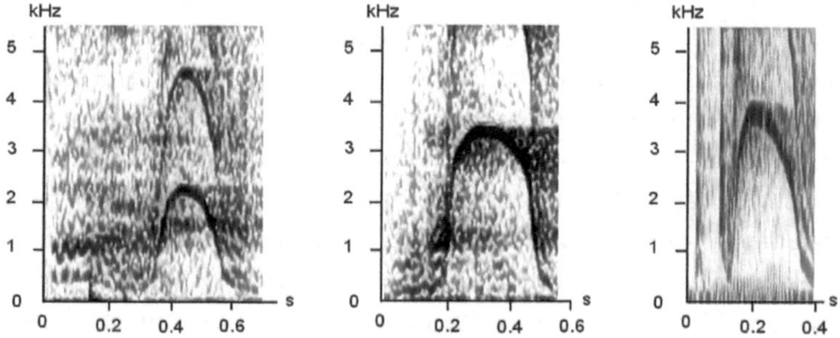

To verify the combinatorial structure (i.e. whether apparently combined calls consisted in the merging of individual calls) and test the relevance to receivers of distinct combined calls, Coye et al.[95] conducted a playback experiment on females from a wild habituated group of Diana monkeys. In particular, to determine the relevance to receivers of the contextual unit, they compared subjects' reaction to the stimuli created by merging L or R units (i.e. relating to distinct contexts) with an A call from a group member (*i.e.* LA_{Group} and RA_{Group} stimuli). To determine whether A calls allowed receivers to identify the caller, they compared subject's reaction to stimuli created by merging an R unit with either A calls from group members or A calls from females in a neighbouring group (*i.e.* RA_{Group} and $RA_{Neighbour}$ stimuli). The change of one unit systematically triggered predictable changes in receivers' reaction. **The results strongly suggest that the contact call system of female Diana monkeys relies on a combinatorial operation through which two independent calls are combined into a larger structure whose information content reflects its components.** Hence, female Diana monkey's contact call system matches the three criteria we proposed and may offer a parallel with morphosyntax.[95]

Campbell's monkeys

Similarly to Diana monkeys, adult Campbell's monkeys possess a sex-specific vocal repertoire as females' communication relies mostly on social calls while males give mainly alarm calls[116–117]. A notable example from female Campbell's monkeys involves the merging of a low-pitched trill (resembling Diana monkey's L call), which can also be used alone and varies with caller's emotional state[129], with an arched unit that strongly signals a caller's identity and social affiliation[122,130] (i.e. resembling Diana monkey's A call). However, contrarily to Diana monkeys, the second unit combined (i.e. identity-rich arch) by Campbell's monkeys is never used alone, suggesting a mechanism more akin to suffixation. Playbacks verifying the combinatorial structure of complex calls remain to be performed.

Intriguingly, male Campbell's monkeys also use a combinatorial system resembling suffixation in their alarm calls (Figure 4, Table 2)[131,132]. More precisely, in Campbell's monkeys, males possess two urgent alarm calls, Krak and Hok. While the first generally signals the presence of an urgent ground danger (i.e. classically a leopard in the Taï National park), the latter signals urgent aerial dangers (i.e. classically an eagle)[133]. These calls can also be combined with a unique 'oo' unit to create Krak-oo and Hok-oo calls (Figure 4). While the 'oo' unit is never used alone, its addition to Krak or Hok calls seems to reduce the danger signalled given Krak-oo and Hok-oo calls signal a general disturbance (e.g. a duiker) and an aerial danger of lesser urgency (*e.g.* a fight in an associated group of red colobus)[40].

Figure 4: Sonograms of the calls of adult male Campbell's monkeys. (a) Krak (urgent ground danger), (b) Krak-oo (non-urgent general disturbance), (c) Hok (urgent aerial danger) and (d) Hok-oo calls (non-urgent aerial danger). Figure reproduced from Coye et al., 2015[92]. On sonograms (b) and (d), the black arrow signals the position of the 'oo' unit.

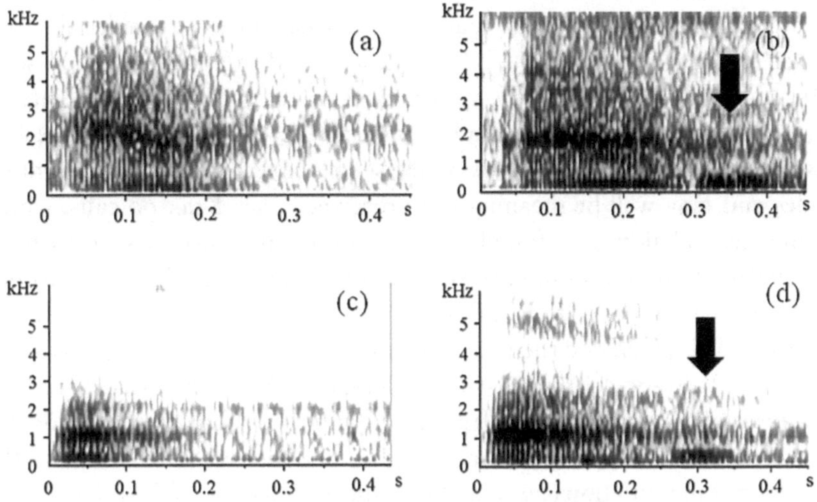

Natural observations were complemented by a playback experiment aiming to verify the combinatorial nature of Krak/Krak-oo calls in this alarm system[94]. The authors analysed the reaction of Diana monkeys (which react to the distinct alarm calls of Campbell's monkeys with their own referential alarm calls[134]) to broadcasted natural and artificially recombined Krak and Krak-oo calls created by deleting the 'oo' part of a Krak-oo or by adding an 'oo' part to a Krak call respectively. Subjects' reaction to Krak and Krak-oo calls reflected their distinct levels of urgency, regardless of their origin (i.e. natural or artificially created). Statistical analysis suggested that, although subtle changes in acoustic structure of the Krak part were perceived by receivers (possibly as a result of caller's emotional state at the time of calling), the presence or absence of a suffix was the main factor leading subject's reaction. **Thus, in addition to confirming the combinatorial nature of Krak-oo calls (i.e. which result from the linear merging of a Krak call with an 'oo' unit), this experiment demonstrated that changes in call structure triggered predictable changes in receiver's reactions and**

confirmed the biological relevance of the addition of an 'oo' unit to decrease the urgency of Krak calls[94]. Here again, we can thus conclude that the three criteria proposed to define rudimentary parallels with morphosyntax in animals are met.

The long-term study of Campbell's monkeys' communication also revealed another type of combination in the alarm calls of males that are given in long sequences, the organisation of which appears to vary with the context[40]. **Although this system does not possess the complexity of syntactic structures occurring in language, the type of calls involved as well as the position of some call types in the sequence seem to obey non-random rules and may well be meaningful to receivers**[40,133]. Krak-oo calls, which signal general alerts, are found in most (but not all) alarm sequences given by males[124]. In addition, several regularities have been described, as not only sequence composition varies according to the context but also to the order of calls given and their rhythm of emission. Distinct call types can be added to Krak-oo sequences, such as Krak and Hok calls (that appear at the beginning of a sequence), depending on the type of danger, in particular the type of predator (leopard and eagle) encountered[40]. In addition, the urgency of the situation (e.g. visual vs auditory detection of the predator) influences the speed of call delivery of Krak-oo calls in the sequence[135] while the speed of call delivery for Hok calls (when an eagle is detected) relates to a male's willingness to attack the predator[135]. Boom calls (i.e. another call type) are always given in pairs and trigger group gathering and movement when produced singly. However, when Booms are followed by other calls, they signal non-predatory events and the calls following them vary with the context. For example, a Krak-oo sequence follows Booms when a large branch or tree is falling down[40,136]. The insertion of Hok-oo calls, systematically between Boom and Krak-oo series, to these "tree-falling" sequences (i.e. Booms-Krak-oos) occurs during inter-group encounters with neighbours (*i.e.* Booms-Hok-oos-Krak-oos)[40]. A first playback study has investigated the 'non-predatory' modification of a sequences' message through the addition of Boom calls by comparing receivers' reaction to natural predator-deterring sequences preceded or not by Boom calls[137]. Now, several studies will be required to experimentally verify the other interesting patterns of organisation derived from observational data.

Table 2: Summary of the main characteristics of the combined calls in the species under the focus of this chapter.

Species	Structure of combined vocalizations	Meaningful elements?	Link between the meaning of units & combined structures	Mechanism for meaning differentiation	References
Chestnut-crowned babbler (*Pomatostomus ruficeps*)	Calls composed of several notes	No	No	Rudimentary form of 'phonemic contrast'	[94, 155, 157]
Southern pied babbler (*Turdoides bicolor*)	Combination of individual calls into call sequence	Yes	Yes	Rudimentary form of 'morphosyntax': combination of single calls	[129, 153–154]
Diana monkey (*Cercopithecus diana*)	Combination of individual calls into larger calls	Yes	Yes	Rudimentary form of morphosyntax: combination of single calls	[29, 93, 113, 115]
Campbell's monkey (*Cercopithecus campbelli*)	Combination of alarm calls with a call unit never used alone. Combination of calls in sequences	Yes	Yes	Rudimentary form of morphosyntax: combination of calls with a 'proto-suffix'	[40, 92, 118]

The experimental results presented in this section demonstrate that some animal species combine meaningful structures in non-random ways to create richer signals (i.e. conveying more complex information) or to diversify the messages conveyed with only a limited number of distinct calls (Table 2). In each example described in this section, the three criteria we proposed to classify calls as phonological or morphosyntactic structures were met. Additional testing will clearly be necessary to further our understanding of the relevant mechanisms underlying such combinatoriality and subsequent changes in meaning. Notably, it will be necessary to replicate the recombination experiments on other calls of Campbell's monkeys (i.e. males Hok/Hok-oo calls) and Diana monkeys (e.g. HA calls) to determine the pervasiveness of sound combinations in these species. Experiments manipulating the order of units in combined calls and call sequences will also be required to fully determine how such changes alter the information extracted by receivers.

4. Formal linguistic analysis of combinatorial systems:

Whilst ethologists have relied on linguistics as a source of inspiration for years, more recently linguists have also begun to systematically compare and contrast animal and human communication systems applying methods from formal linguistics (i.e. posing rules to define formally a 'lexicon', a 'syntax' and 'semantics' for a given system)[88,92]. Among other primates, the vocal systems of Campbell's and Diana monkeys have been subjected to such analyses in studies by Schlenker and colleagues. The authors reanalysed existing data on these guenons providing complementary investigations to the ethological approach[88,92,138].

A first study on Campbell's monkeys, focussed on the possible semantic content of Krak, Hok and their 'suffixed' versions. It compared models built using methodologies from the field of formal semantics to shed light on the possible meanings of these calls and on the mechanism by which the addition of an 'oo' unit alters the meaning of the call 'stems' (i.e. Krak and Hok). Authors specifically focused on the distinct calling patterns of males from two populations of Campbell's monkeys in Ivory Coast (Taï National Park) and Sierra Leone (Tiwai island)[88]. Crowned-hawk eagles are present in both areas and leopards still being present in Taï, but absent from Tiwai for as long as thirty years[139]. Importantly, while Hok functions to signal the presence of an eagle in both populations, Krak is used primarily to signal the presence of a leopard in Taï but it has the distribution of a general alarm call on Tiwai (*i.e.* given to a broad range of disturbances including falling trees and eagles)[40,124,140]. **To determine which 'semantic' explanation best captured the patterns observed, the authors systematically tested the predictions of two models against the data.** The first model posits the hypothesis that in both populations Krak and Hok calls have the same "innate" meanings (*i.e.* Krak: general disturbance; Hok: aerial predator) and that the addition of an 'oo' unit decreases the urgency of the innate meaning of both calls (*i.e.* Krak-oo: general and less urgent disturbance; Hok-oo less urgent aerial disturbance). Finally, this model hypothesises that, while Krak-oo is derived from the innate meaning of Krak in both areas (i.e. the Krak 'root' of Krak-oos kept its original meaning), the 'lexical entry' for Krak in Taï has changed to 'leopard-related disturbance'. The second model proposes an alternative hypothesis to explain the pattern described: the innate meaning of Krak and Hok calls is the same in both populations (*i.e.* Krak: general disturbance; Hok: aerial predator) and it holds in both unsuffixed and suffixed calls. But while both Krak-oo (*i.e.* non-urgent danger) and Hok (*i.e.* aerial predator) are specific, Krak has a rather broad meaning (*i.e.* general alarm call). The second model thus proposes that the competition between more specific calls and Krak calls may lead to the strengthening of the meaning of Krak in a mechanism akin to scalar implicatures[88,92] (see Table 1 for a definition). Specifically, when a male gives Krak calls, a receiver might infer that there is a non-weak and non-aerial disturbance as the call given is not Krak-oo nor Hok. Hence, the meaning of Krak calls can be strengthened from 'general urgent disturbance' into 'dangerous non-aerial predator'. In Taï the presence of leopards led to the strengthening of the meaning of Krak calls as 'dangerous non-aerial predators' but not in Tiwai where the absence of ground predators prevented it. From this, the authors concluded that the second model was more parsimonious and more likely to describe the associated 'meanings' of calls in the call system of Campbell's monkeys than the first one[88].

A second study led by Schlenker and collaborators proposed to analyse the communication of female Diana monkeys, using both a statistical analysis of transition probabilities between units and call types and a formal semantic analysis of utterances based on their context of emission[93]. Again, the authors proposed two alternative competing hypotheses to describe the system observed. The first hypothesis proposed that combined calls consisted of two simple calls given in close succession (*i.e.* maximized adjacency hypothesis). Contrarily, the second hypothesis proposed that combined calls (*i.e.* HA, LA and RA calls) resulted from the combination of two units that were subsequently used as one call (*i.e.* combined calls hypothesis). To determine which hypothesis was the more likely, the authors developed two corresponding models (*e.g.* putative 'rules' of call use describing the observed patterns) and compared them. **This work showed that treating 'combined' calls as sequences of simple calls given in close succession failed to account for their distribution in sequences.** The most parsimonious model was obtained under the 'combined call hypothesis' (*i.e.* 'combined calls result from the combination of vocal units and are used as one call) as the alternative hypothesis (i.e. maximized adjacency hypothesis) would need to be supplemented by phonological complexity in order to account for the data with respect to maximal sequence length and call repetition[93].

Other recent articles by the same authors offer analyses of the calling systems of additional species using similar methods (e.g. black-fronted titi monkeys and putty-nosed monkeys)[92,114]. **The results obtained converged with field observations and these articles are key in not only generating testable hypotheses but also confirming the relevance of using linguistic methodologies to analyse combinatorial systems in non-human animals.** We argue these studies bring key additional support to our findings while adopting different, yet complimentary methodological approaches. Indeed, although previous studies had also described non-random patterns of transitions between elements comprising vocal sequences produced by animals (*e.g.* marine mammals[141], bats[57], birds[48,53,142]), they failed to take into account the meaning and relevance to receivers of sequence organisation and composition. For instance, Kershenbaum and collaborators[80] analysed the vocal sequences produced by animals from several taxa (*i.e.* killer and pilot whales, rock hyraxes, Bengalese finches, Carolina chickadee, free-tailed bats and orangutans) using various transition models of increasing complexity to determine which one matched best the transition between elements in the sequences recorded. Such studies are very informative regarding the possible evolution of sequence complexity in animals and may participate in bridging the gap between human language and animal communication[80]. However in language, combination is relevant only because it is meaningful[84]. The work reviewed in this chapter highlights the need to include a more systematic analysis of animal sequence structure, meaning and composition and its relevance to receivers, in particular, when one aims to undertake a comparative approach with human language. We believe that the current progress on animal combinatorial abilities, together with future developments in complementary methodological approaches and appropriate terminology, will pave the way to a more comprehensive understanding of the evolution of sound combination in animals. In the final section we will discuss the evolutionary insights such comparative data can provide on the drivers of the emergence of combinatorial abilities.

5. Evolutionary relevant insights from animal combinatorial systems?

The topic of language origins is frequently accompanied by heated debates over the analogous (i.e. convergent evolution) or homologous (i.e. inherited from a common ancestor) nature of some parallel features of language described in non-human primates, including combinatorial abilities (e.g. [3,6,143–148]). However, we forward that this is not the most pressing question, because we can learn a lot from the study of animal communication regardless of its shared or distinct evolutionary history with language. Indeed, if language is a unique communication system, it is also clearly the product of a gradual evolutionary process and, in this regard, it does not differ from other animal communication systems. Thus, in our opinion, a **more important question to tackle would be – what pressures drove the evolution of combinatorial abilities?**

Social complexity

Social life is often viewed as a major driver of communicative complexity and this hypothesis has been supported by empirical studies highlighting a positive relationship between indexes of social complexity and signal diversity for both social and alarm calls[47,149–152]. The description and testing of combinatorial systems in animals suggests that sound combination may allow the diversification of a species' repertoire using a limited number of signals. Interestingly, two studies comparing closely related species of non-human primates and *herpestidaes* reported a correlation between the complexity of a species' social life and the presence, diversity and frequency of use of combinatorial structures[153,154]. In line with this, each species in which meaningful sound combinations were described has also been reported to reside in a complex and strongly bonded social group[30,155–158]. These observations support the idea that **increased needs for complex communication**

resulting from social complexity might have played an important role in the emergence of combinatorial capacities in animals.

Phonatory limits

Another possible factor leading to the emergence of combinatorial capacities results from the phonatory limits that some species face, notably in non-human primates. Work from computational modelling provides relevant additional insight here. Nowak and colleagues[159] modelled scenarios for the emergence and propagation of certain language features in a population, such as arbitrary signals, sound combinations and grammatical rules. Nowak et al. proposed that combinatorics would emerge after a communication system reaches a threshold number of signals above which the addition of new signals (because they would be likely to resemble existing ones) ultimately increases the error risk due to mis-comprehension. In this case, **the combination of sounds would allow a continued increase in a language's fitness (through addition of new signals) without increasing the risk of ambiguous information transfer.** This rationale relies on the hypothesis that a species is capable of increasing its repertoire via the acoustic diversification of signals in the first place. We propose that it is also valid in species with limited capacities of vocal production but that in this case, the first limit to signal diversification might be the species' lack of vocal plasticity rather than the breadth of the existing repertoire. This hypothesis is supported by the fact that all the species in which sound combination has been shown to play a meaning-differentiating role display limited capacities for vocal production[118,121,130,156,160]. Further studies investigating the presence of meaningful call combinations in species characterised by various levels of social complexity as well as distinct capacities for vocal learning are key to testing these hypotheses with more extensive empirical data.

Habitat and constraints on communication

Finally, habitat has often been proposed as a factor influencing species' communication. In particular, it has been proposed that dense habitats, which impose constraints on sound propagation and visual access to others, may favour the emergence of discrete communication signals (i.e. as opposed to communication systems of graded signals, whose acoustic structures form a continuum without distinct boundaries between call types[161–164]). In dense habitats, discrete signals would allow more robust communication and prevent ambiguities resulting from poor visual access to others[65,163,165]. Sound combination may therefore benefit animals through the production of more efficient communication signals. This is, for instance, the case in female Diana monkeys, whose combined calls linearly convey information about a caller's emotional state and identity. Here, females concatenate signals sequentially which might have already evolved to ensure maximal efficiency of information transfer (e.g. calls with more salient identity cues or with an improved acoustic adaptation to propagation constraints). This organisation allows information to be temporally segregated creating richer signals without increasing ambiguity due to the accumulation of information. Interestingly, a combinatorial system resembling that of Diana monkeys has been described in the graded contact calls of desert-living banded mongoose. More precisely, banded mongooses use a contact call composed of two segments, given in three distinct contexts: when the caller is digging, searching and moving[166]. The first segment relates strongly to a caller's identity and remains identical in the three contexts. A playback experiment confirmed that between-caller variations in the identity segment were relevant to receivers[167]. The second segment has a graded structure and varies with caller's activity: when the caller is digging the segment is absent (or very short), it's duration increases when the caller is searching, and reaches its maximal duration (together with more pronounced harmonics) when the caller is moving. Thus, here again, the use of combinatoriality seems to increase the information content of calls while maintaining a low level of ambiguity.

Importantly, banded mongoose live in an open habitat but lack visual access to conspecifics because their foraging strategy constrains them to face the ground most of the time[168,169]. Although literature traditionally pitted species with graded and discrete repertoires against each other, multiple concerns with the relevance of this dichotomy have been raised, notably because of evidence for subtle gradation in the communication of 'discrete' species as well as evidence for categorical perception of graded signals by receivers[162,170–173]. Taken together, these observations suggest that, **more than habitat-based propagation constraints, the lack of visual access to**

others may be important in influencing the emergence of short-distance combined social call structures that convey complementary information about a caller's identity, activity and localisation. Finally, this hypothesis can be aligned with the theoretical work of Nowak and collaborators[159]. Indeed, if combination has emerged to limit the risk of ambiguous communication, the inability to disambiguate the context of calling or caller's identity using visual cues (e.g. due to habitat constraints or foraging strategy) is a possible additional factor triggering its emergence. To investigate the potential relative impacts of habitat density and actual visual access to others on the development of combinatorial capacities, we would need to extend the comparison to other species whose visual access to others contradicts the predictions that could be proposed by simply looking at their habitat density (e.g. other species than banded mongoose with poor visual access in spite of an open habitat).

The hypotheses proposed above shed further light on the factors involved in the evolution of language and other communication systems and have largely resulted from data provided by only a few species where receivers have been experimentally documented to process and use combinatorial structures. Various additional examples also exist that have not yet been completely described and are likely to fit the definitions we have previously used (i.e. in section 1). In particular, several systems in which apparently meaningful calls are combined into larger structures whose context of emission reflects that of its parts have been described in wedged-caped capuchins[174], cotton-top tamarins[65], female Campbell's monkeys[122], red-bellied titi monkeys[67], black-fronted titi monkeys[68,175], red-capped mangabeys[72], bonobos[176] and chimpanzees[177] as well as in non-primate species such as Japanese great tits[70], banded mongooses[154,166] and meerkats[178]. **The diversity of species in which meaningful sound combinations have been documented (e.g. birds and mammals, including primates and *herpestidae*) suggests that it may be an evolutionary solution to deal with communicative demands** and the comparative study of these species is central to test any hypotheses regarding the potential drivers promoting the emergence of combinatoriality[46,87].

6. Conclusion: Towards a more comprehensive approach of combinatorial abilities

The study of animal combinatorial abilities appears to be a promising research area, with a number of avenues open to exploration. In this chapter, we restrict discussion to examples exhibiting more or less marked parallels with language. However, a large number of other combinatorial systems in animals remain to be investigated, among which some have been described but remain only partially understood (e.g. gorillas, putty-nosed monkeys, gibbons, rock hyraxes, mustached bats[54,55,66,92,107–109,114]). In addition, some animal combinatorial systems may differ strikingly from language in their organisation and underlying mechanisms facilitating information transfer. For instance, in some systems the diversity of units (e.g. in some songbirds[104]) or the proportion of various units (e.g. in bonobos[176,179]) seem to play a role in meaning generation.

Joint efforts from linguists, psychologists and ethologists is clearly necessary to provide a unified and more relevant framework. **One possible first step would be to develop a terminology suitable to describe the vast diversity of combinatorial systems found through the animal kingdom.** Indeed, whilst some rare examples developed in this chapter can be captured by pre-set definitions, it seems clear that a number of sound combination systems will not. However, even (if not especially) in those cases, the use of strict definitions is essential. This is important firstly to provide a clearer view of the diversity and complexity of combinatorial organisations in the animal kingdom. Secondly, and perhaps more pertinently, because the study of varied examples relating to potentially meaningful, non-random and contextually flexible combination patterns may be an important step to further understand the biological relevance of vocal combination in animals and its evolution(s).

The rationale adopted to build the two definitions we propose for parallels with phonology and morphosyntax (section 1) could be generalised to develop a more suitable terminology for alternative systems of sound combination. **Notably, the definitions we proposed involve three components: (1) whether the vocal units combined possess an intrinsic meaning, (2) whether (and how) the meaning of their combination reflects the meaning of individual elements (if they possess one) and (3) which rules (if any)**

best describe the mechanism for meaning-differentiation between combined utterances with distinct functions. In addition, the third component of the definitions would also allow us to capture systems that differ strongly from language, such as those relying on the proportion of one call type, or on the diversity of units involved. As such, this three-component structure could be further expanded to characterise the variety of animal combination systems described while offering a systematic basis for interspecific comparison.

In addition, it may be useful that authors specialised in various taxa (e.g. ornithologists, primatologists, marine biologists) and disciplines (e.g. ethologists, philosophers, linguists) readdress questions pertaining to the nature of meaning as this central question may be approached from various directions (e.g. can regional 'dialectal' variations of a song also be considered as changes in meaning?). In any case, future studies focussing on diverse combinatorial systems, including systems that differ strongly from language, are likely to be fruitful. Indeed, **understanding the organisation and evolution of systems that differ strongly from ours will, if anything, bring insights into the various evolutionary paths that the human lineage did not follow and may be a relevant source of information to identify important turning points in our "history".**

This chapter focussed mostly on studies that relied on the simultaneous use of natural observations of calling contexts and experiments. These two complementary approaches are essential to investigating the combinatorial structure (i.e. transferability of units in combination) of complex utterances and the relevance to receivers of changes in meaning as a result of changes in the combinatorial structure. We argue that this is a key first question to tackle in order to provide a comprehensive description of animal communication systems. However, beyond the combination of phonemes into words and words into sentences, language relies on a set of rules that allow interlocutors to produce and understand completely novel utterances[84,87,159]. **The cognitive mechanisms underlying our processing of rules and our capacity to generalise them are at the very base of language generativity**[81,87]. Now that some animal communication systems have been characterised in terms of their basic structure and meaning-differentiating mechanisms, the next step should be dedicated to clarifying the cognitive mechanisms underlying their combinatoriality. In particular, it will be important to determine whether animals perceive combined utterances as a mosaic of elements whose message can be inferred from the elements and their relationships (i.e. as compound signals) or as unique elements whose combinatorial nature is only structural[159]. For instance, can Campbell's monkeys learn the meaning of Krak, Krak-oo, Hok and Hok-oo calls independently or do they learn the meaning of Krak, Hok and the alteration of meaning associated with the presence of an oo unit? And would they be capable to generalise that "rule"? Previous studies have shown that some non-human primates possess particularly sophisticated social cognition skills involving a hierarchically structured representational knowledge of social relationships, governed by rules and involving causal inference – a likely result of their complex social life[147,180–182]. In addition, studies based on experimental tasks suggest that some animals possess, to some extent, capacities to parse combinatorial and sequential artificial structures or rules[183]. For instance non-human primates of several species have been shown to learn sequential lists of items[184,185], and to compute probabilities of occurrence, and dependencies between syllables or letters[186–189]. Cotton-top tamarins have the capacity to acquire general 'rules' of structuring such as simple pFSA grammar[190] and starlings, as well as language-trained animals (apes, dolphins and parrots), even acquired more complex rules such as recursive structures[191] or generative languages[192–196]. In line with this, a particularly interesting perspective would be to compare the experimental performances in laboratory settings (e.g. assessing a species' capacity to acquire and handle various artificial grammars) of species using combinatorial structures to various extents and determine whether their performances at parsing artificially constructed structures correlates with their natural tendency to use sound combination during communication. Finally, studies clarifying the ontogeny and acquisition mechanisms of vocal repertoires involving combinatorial structures would be important to complete our knowledge, especially in species that seem to rely on meaningless notes, as the arbitrariness of combinations should be based on a mandatory learning phase.

To conclude, we can say that there now exists a growing array of species that rely on sound combinations and as such provides an intriguing starting point for investigations into the evolution and emergence of these abilities. When reviewing combinatorial structures across species, however, it becomes clear that questions associated with meaning and information conveyed by animal signals are central and should be taken into account in the development of an appropriate terminology to describe sound combinations in animals. Finally we actively encourage interdisciplinary research uniting linguists, ethologists, psychologists and anthropologists, to build a unified framework and to further explore the links between human language and animal communication.

References

[1] Pinker, S. and Poeppel, D. (2000.) Words and rules: the ingredients of language. *Nature*, *403*, 361–361.

[2] Aiello, L.C. and Dunbar, R.I. (1993). Neocortex size, group size, and the evolution of language. *Current Anthropology*, *34*, 184–93.

[3] Pinker, S. and Bloom, P. (1990). Natural language and natural selection. *Behavioral and Brain Sciences*, *13*, 707–727.

[4] Enard, W., Przeworski, M., Fisher, S.E., Lai, C.S.L., Wiebe, V., Kitano, T. et al. (2002). Molecular evolution of FOXP2, a gene involved in speech and language. *Nature*, *418*, 869–72.

[5] Zuberbühler, K. (2003). Referential Signaling in non-human primates: cognitive precursors and limitations for the evolution of language. *Advances in the Study of Behavior*, Academic Press. p. 265–307.

[6] Lemasson, A. (2011). What can forest guenons "tell" us about the origin of language. In A. Vilain, J.L. Schwartz, C. Abry, and J. Vauclair (Eds.), *Primate communication and human language: vocalisation, gestures, imitation and deixis in humans and non-humans (pp.* 39–70). Amsterdam: John Benjamins.

[7] Zuberbühler, K. (2015). Linguistic capacity of non-human animals. *Wiley Interdisciplinary Reviews: Cognitive Science*, 6, 313–21.

[8] Janik, V.M. and Slater, P.J. (1997). Vocal learning in mammals. *Advances in the Study of Behaviour*, *26*, 59–100.

[9] Snowdon, C.T. and Hausberger, M. (1997). Social influences on vocal development. Cambridge University Press.

[10] Konishi, M. (1965). The role of auditory feedback in the control of vocalization in the White-crowned Sparrow. *Zeitschrift Für Tierpsychologie*, *22*, 770–83.

[11] Marler, P. (1970). A comparative approach to vocal learning: song development in White-crowned Sparrows. *Journal of Comparative and Physiological Psychology*, *71*, 1.

[12] Price, P.H. (1979). Developmental determinants of structure in zebra finch song. *Journal of Comparative and Physiological Psychology*, *93*, 260.

[13] Thorpe, W.H. (1958). The learning of song patterns by birds, with especial reference to the song of the chaffinch Fringilla coelebs. *Ibis*, *100*, 535–70.

[14] Deecke, V.B., Ford, J.K. and Spong, P. (2000). Dialect change in resident killer whales: implications for vocal learning and cultural transmission. *Animal Behaviour*, *60*, 629–38.

[15] Henry, L., Barbu, S., Lemasson, A. and Hausberger, M. (2015). Dialects in animals: Evidence, development and potential functions. *Animal Behavior and Cognition*, 2, 132–55.

[16] Jenkins, P.F. (1978). Cultural transmission of song patterns and dialect development in a free-living bird population. *Animal Behaviour*, *26*, 50–78.

[17] Gahr, M. (2000). Neural song control system of hummingbirds: comparison to swifts, vocal learning (songbirds) and nonlearning (suboscines) passerines, and vocal learning (budgerigars) and nonlearning (dove, owl, gull, quail, chicken) nonpasserines. *Journal of Comparative Neurology*, *426*, 182–96.

[18] Nottebohm, F. (1972). The origins of vocal learning. *American Naturalist*, 116–40.

[19] Jarvis, E.D. (2004). Learned birdsong and the neurobiology of human language. *Annals of the New York Academy of Sciences*, *1016*, 749–77.

[20] Okanoya, K. (2002). Sexual display as a syntactical vehicle: the evolution of syntax in birdsong and human language through sexual selection. *The Transition to Language*, 2, 46–63.

[21] Webb, D.M. and Zhang, J. (2005). FoxP2 in song-learning birds and vocal-learning mammals. *Journal of Heredity*, *96*, 212–6.

[22] Fisher, S.E. and Marcus, G.F. (2006). The eloquent ape: genes, brains and the evolution of language. *Nature Reviews Genetics*, 7, 9–20.

[23] Haesler, S., Wada, K., Nshdejan, A., Morrisey, E.E., Lints, T., Jarvis, E.D. et al. (2004). FoxP2 expression in avian vocal learners and non-learners. *The Journal of Neuroscience*, *24*, 3164–75.

[24] Scharff, C. and White, S.A. (2004). Genetic components of vocal learning. *Annals of the New York Academy of Sciences*, *1016*, 325–47.

[25] Teramitsu, I., Kudo, L.C., London, S.E., Geschwind, D.H. and White, S.A. (2004). Parallel FoxP1 and FoxP2 expression in songbird and human brain predicts functional interaction. *The Journal of Neuroscience*, *24*, 3152–63.

[26] Poole, J.H., Tyack, P.L., Stoeger-Horwath, A.S. and Watwood, S. (2005). Animal behaviour: elephants are capable of vocal learning. *Nature*, *434*, 455–6.

[27] Briefer, E.F. and McElligott, A.G. (2012). Social effects on vocal ontogeny in an ungulate, the goat, Capra hircus. *Animal Behaviour*, *83*, 991–1000.

[28] Snowdon, C.T. and Elowson, A.M. (1999) Pygmy marmosets modify call structure when paired. *Ethology*, *105*, 893–908.

[29] Candiotti, A., Zuberbühler, K. and Lemasson, A. (2012). Convergence and divergence in Diana monkey vocalizations. *Biology Letters*, *8*, 382–5.

[30] Lemasson, A. and Hausberger, M. (2004). Patterns of Vocal Sharing and Social Dynamics in a Captive Group of Campbell's Monkeys (Cercopithecus campbelli campbelli). *Journal of Comparative Psychology*, *118*, 347–59.

[31] Lemasson, A., Ouattara, K., Petit, E.J. and Zuberbühler, K. (2011). Social learning of vocal structure in a nonhuman primate? *BMC Evolutionary Biology*, *11*, 362.

[32] Koda, H., Lemasson, A., Oyakawa, C., Pamungkas, J. and Masataka, N. (2013). Possible role of mother-daughter vocal interactions on the development of species-specific song in gibbons. *PLoS One*, *8*, e71432.

[33] Dooley, H. and Judge, D. (2007). Vocal responses of captive gibbon groups to a mate change in a pair of white-cheeked gibbons (Nomascus leucogenys). *Folia Primatologica*, *78*, 228–39.

[34] Cartmill, E.A. and Byrne, R.W. (2010). Semantics of primate gestures: intentional meanings of orangutan gestures. *Animal Cognition*, *13*, 793–804.

[35] Gaunet, F. and Deputte, B.L. (2011). Functionally referential and intentional communication in the domestic dog: effects of spatial and social contexts. *Animal Cognition*, *14*, 849–60.

[36] Schel, A.M., Townsend, S.W., Machanda, Z., Zuberbühler, K. and Slocombe, K.E. (2013). Chimpanzee alarm call production meets key criteria for intentionality. *PLoS One*, *8*, e76674.

[37] Caesar, C. and Zuberbühler, K. (2012). Referential alarm calling behaviour in New World primates. *Current Zoology 58*, 680–697.

[38] Cäsar, C., Byrne, R.W., Hoppitt, W., Young, R.J. and Zuberbühler, K. (2012). Evidence for semantic communication in titi monkey alarm calls. *Animal Behaviour*, *84*, 405–11.

[39] Di Bitetti, M.S. (2003). Food-associated calls of tufted Capuchin monkeys (Cebus apella nigritus) are functionally referential signals. *Behaviour*, *140*, 565–92.

[40] Ouattara, K., Lemasson, A. and Zuberbühler, K. (2009). Campbell's monkeys concatenate vocalizations into context-specific call sequences. *Proceedings of the National Academy of Sciences*, *106*, 22026–31.

[41] Seyfarth, R.M., Cheney, D.L. and Marler, P. (1980). Vervet monkey alarm calls: semantic communication in a free-ranging primate. *Animal Behaviour*, *28*, 1070–94.

[42] Zuberbühler, K. (2000). Causal cognition in a non-human primate: field playback experiments with Diana monkeys. *Cognition*, *76*, 195–207.

[43] Zuberbühler, K., Cheney, D.L. and Seyfarth, R.M. (1999). Conceptual semantics in a nonhuman primate. *Journal of Comparative Psychology*, *113*, 33–42.

[44] Chow, C.P., Mitchell, J.F. and Miller, C.T. (2015). Vocal turn-taking in a non-human primate is learned during ontogeny. *Proceedings of the Royal Society of London B: Biological Sciences*, *282*, 20150069.

[45] Lemasson, A., Glas, L., Barbu, S., Lacroix, A., Guilloux, M., Remeuf, K. et al. (2011). Youngsters do not pay attention to conversational rules: is this so for nonhuman primates? *Scientific Reports*, *1*, 22.

[46] Zuberbühler, K. and Lemasson, A. (2014). Primate communication: meaning from strings of calls. In F. Lowenthal and L. Lefebvre (Eds.), *Language and Recursion (*pp. 115–125). New York: Springer.

[47] Kroodsma, D.E. (1977). Correlates of song organization among North American wrens. *American Naturalist*, *111*, 995–1008.

[48] Honda, E. and Okanoya, K. (1999). Acoustical and syntactical comparisons between songs of the white-backed munia (Lonchura striata) and its domesticated strain, the Bengalese finch (Lonchura striata var. domestica). *Zoological Science*, *16*, 319–26.

[49] Gammon, D.E. and Altizer, C.E. (2011). Northern mockingbirds produce syntactical patterns of vocal mimicry that reflect taxonomy of imitated species. *Journal of Field Ornithology*, *82*, 158–64.

[50] Hausberger, M. (1990). Organization of whistled song sequences in the European starling. *Bird Behavior*, *9*, 81–7.

[51] Hailman, J.P. and Ficken, M.S. (1986). Combinatorial animal communication with computable syntax: chick-a-dee calling qualifies as 'language'by structural linguistics. *Animal Behaviour*, *34*, 1899–901.

[52] Lucas, J.R. and Freeberg, T.M. (2007). Information and the chick-a-dee call: communicating with a complex vocal system. In K.A. Otter (Ed.) *Ecology and Behavior of Chickadees and Titmice* (pp. 199–213). *Oxford: Oxford University Press.*

[53] Sigler Ficken, M., Rusch, K.M., Taylor, S.J. and Powers, D.R. (2000). Blue-throated Hummingbird song: A pinnacle of nonoscine vocalizations. *The Auk*, *117*, 120–8.

[54] Kershenbaum, A., Ilany, A., Blaustein, L. and Geffen, E. (2012). Syntactic structure and geographical dialects in the songs of male rock hyraxes. *Proceedings of the Royal Society of London B: Biological Sciences*, *279*, 2974–81.

[55] Kanwal, J.S., Matsumura, S., Ohlemiller, K. and Suga, N. (1994). Analysis of acoustic elements and syntax in communication sounds emitted by mustached bats. *The Journal of the Acoustical Society of America*, *96*, 1229–54.

[56] Bohn, K.M., Schmidt-French, B., Ma, S.T. and Pollak, G.D. (2008). Syllable acoustics, temporal patterns, and call composition vary with behavioral context in Mexican free-tailed bats. *The Journal of the Acoustical Society of America*, *124*, 1838–48.

[57] Bohn, K.M., Schmidt-French, B., Schwartz, C., Smotherman, M. and Pollak, G.D. (2009). Versatility and stereotypy of free-tailed bat songs. *PLoS One*, **4**, e6746.

[58] Behr, O. and von Helversen, O. (2004). Bat serenades—complex courtship songs of the sac-winged bat (Saccopteryx bilineata). *Behavioral Ecology and Sociobiology*, **56**, 106–15.

[59] Au, W.W., Pack, A.A., Lammers, M.O., Herman, L.M., Deakos, M.H. and Andrews, K. (2006). Acoustic properties of humpback whale songs. *The Journal of the Acoustical Society of America*, **120**, 1103–10.

[60] Payne, R.S. and McVay, S. (1971). Songs of humpback whales. *Science*, **173**, 585–97.

[61] Riesch, R., Ford, J.K.B. and Thomsen, F. (2008). Whistle sequences in wild killer whales (Orcinus orca). *The Journal of the Acoustical Society of America*, **124**, 1822–9.

[62] Riesch, R., Ford, J.K. and Thomsen, F. (2006). Stability and group specificity of stereotyped whistles in resident killer whales, Orcinus orca, off British Columbia. *Animal Behaviour*, **71**, 79–91.

[63] Tyack, P. (1998). Acoustic communication under the sea. *Animal Acoustic Communication*, Springer. p. 163–220.

[64] Weilgart, L. and Whitehead, H. (1997). Group-specific dialects and geographical variation in coda repertoire in South Pacific sperm whales. *Behavioral Ecology and Sociobiology*, **40**, 277–85.

[65] Cleveland, J. and Snowdon, C.T. (1982). The complex vocal repertoire of the adult cotton-top tamarin (Saguinus oedipus oedipus) 1). *Zeitschrift Für Tierpsychologie*, **58**, 231–70.

[66] Hedwig, D., Hammerschmidt, K., Mundry, R., Robbins, M.M. and Boesch, C. (2014). Acoustic structure and variation in mountain and western gorilla close calls: a syntactic approach. *Behaviour*, *151*, 1091–120.

[67] Robinson, J.G. (1979) An analysis of the organization of vocal communication in the titi monkey Callicebus moloch. *Zeitschrift Für Tierpsychologie*, *49*, 381–405.

[68] Cäsar, C., Byrne, R.W., Young, R.J. and Zuberbühler, K. (2012). The alarm call system of wild black-fronted titi monkeys, Callicebus nigrifrons. *Behavioral Ecology and Sociobiology*, *66*, 653–67.

[69] Collier, K., Radford, A.N., Bickel, B., Manser, M.B. and Townsend, S.W. (2016). Dwarf mongooses combine meaningful alarm calls. In S.G. Roberts, C. Cuskley, L. McCrohon, L. Barceló-Coblijn, O. Fehér, and T. Verhoef (Eds.), *The Evolution of Language: Proceedings of the 11th International Conference (EVOLANG11)*. http://evolang.org/neworleans/papers/114.html

[70] Suzuki, T.N., Wheatcroft, D. and Griesser, M. (2016). Experimental evidence for compositional syntax in bird calls. *Nature Communications*, 7, 10986.

[71] Berwick, R.C., Okanoya, K., Beckers, G.J. and Bolhuis, J.J. (2011). Songs to syntax: the linguistics of birdsong. *Trends in Cognitive Sciences*, *15*, 113–21.

[72] Bouchet, H., Pellier, A.-S., Blois-Heulin, C. and Lemasson, A. (2010). Sex differences in the vocal repertoire of adult red-capped mangabeys (Cercocebus torquatus): a multi-level acoustic analysis. *American Journal of Primatology*, 72, 360–375.

[73] Kondo, N. and Watanabe, S. (2009). Contact calls: Information and social function. *Japanese Psychological Research*, *51*, 197–208.

[74] Dahlin, C.R. and Wright, T.F. (2012). Does syntax contribute to the function of duets in a parrot, Amazona auropalliata? *Animal Cognition*, *15*, 647–56.

[75] Holy, T.E. and Guo, Z. (2005). Ultrasonic songs of male mice. *PLoS Biol*, *3*, e386.

[76] Robinson, J.G. (1979). Vocal regulation of use of space by groups of titi monkeys Callicebus moloch. *Behavioral Ecology and Sociobiology*, *5*, 1–15.

[77] ten Cate, C. and Okanoya, K. (2012). Revisiting the syntactic abilities of non-human animals: natural vocalizations and artificial grammar learning. *Philosophical Transactions of the Royal Society of London B: Biological Sciences*, *367*, 1984–94.

[78] Yule, G. (2014). *The study of language*. Cambridge, USA: Cambridge University Press.

[79] Hurford, J.R. (2002). The roles of expression and representation in language evolution. In A. Wray (Ed.) *The Transition to Language* (pp. 311–34). Oxford (UK): Oxford University Press.

[80] Kershenbaum, A., Bowles, A.E., Freeberg, T.M., Jin, D.Z., Lameira, A.R. and Bohn, K. (2014.) Animal vocal sequences: not the Markov chains we thought they were. *Proceedings of the Royal Society of London B: Biological Sciences*, *281*, 20141370.

[81] Hockett, C. (1960). The origin of speech. *Scientific American*, *203*, 88–111.

[82] Hurford, J.R. (2008). The evolution of human communication and language. In P. d'Etore and D.P. Hughes (Eds.), *Sociobiology of Communication: An Interdisciplinary Perspective* (pp. 249–264). Oxford: Oxford University Press.

[83] de Boer, B., Sandler, W. and Kirby, S. (2012). New perspectives on duality of patterning: Introduction to the special issue. *Language and Cognition*, *4*, 251–9.

[84] Hurford, J.R. (2011). *The Origins of grammar: language in the light of evolution II*. Oxford: Oxford University Press.

[85] Tellier, I. (2008). *Introduction au TALN et à l'ingénierie linguistique*. Université de Lille 3.

[86] Collier, K., Bickel, B., Schaik, C.P. van, Manser, M.B. and Townsend, S.W. (2014). Language evolution: syntax before phonology? *Proceedings of the Royal Society B: Biological Sciences*, *281*, 20140263.

[87] Hauser, M.D., Chomsky, N. and Fitch, W.T. (2002). The faculty of language: what is it, who has it, and how did it evolve? *Science*, *298*, 1569–79.

[88] Schlenker, P., Chemla, E., Arnold, K., Lemasson, A., Ouattara, K., Keenan, S. et al. (2014). Monkey semantics: two 'dialects' of Campbell's monkey alarm calls. *Linguistics and Philosophy*, *37*, 439–501.

[89] Chemla, E. and Singh, R. (2014). Remarks on the experimental turn in the study of scalar implicature, Part I. *Language and Linguistics Compass*, *8*, 373–86.

[90] Carston, R. (1998). Information, relevance and scalar implicature. In R. Carston and S. Uchida (Eds.), *Relevance Theory: Applications and Implications*, (pp. 179–236). Amsterdam: John Benjamins.

[91] Marler, P. (1977). The structure of animal communication sounds. In T.H. Bullock (Ed.) *Recognition of complex acoustic signals* (pp. 17–35). Berlin: Springer.

[92] Schlenker, P., Chemla, E., Schel, A.M., Fuller, J., Gautier, J.-P., Kuhn, J. et al. (2016). Formal monkey linguistics. *Theoretical Linguistics*, *42*, 1–90.

[93] Veselinovic, D., Candiotti, A. and Lemasson, A. (Under revision). Female Diana monkeus (Cercopithecus diana) have complex calls.

[94] Coye, C., Ouattara, K., Zuberbühler, K. and Lemasson, A. (2015). Suffixation influences receivers' behaviour in non-human primates. *Proceedings of the Royal Society of London B: Biological Sciences*, *282*, 20150265.

[95] Coye, C., Zuberbühler, K. and Lemasson, A. (2016). Morphologically structured vocalizations in female Diana monkeys. *Animal Behaviour*, *115*, 97–105.

[96] Engesser, S., Crane, J.M., Savage, J.L., Russell, A.F. and Townsend, S.W. (2015). Experimental Evidence for Phonemic Contrasts in a Nonhuman Vocal System. *PLoS Biol*, *13*, e1002171.

[97] Grice, H.P. (1957). Meaning. *The Philosophical Review*, *66*, 377–88.

[98] Scott-Phillips, T.C. (2015). Meaning in animal and human communication. *Animal Cognition*, *18*, 801–5.

[99] Font, E. and Carazo, P. (2010). Animals in translation: why there is meaning (but probably no message) in animal communication. *Animal Behaviour*, *80*, e1–6.

[100] Smith, W.J. (1997). The behavior of communicating, after twenty years. In D.H. Owings, M.D.Beecher, and N.S. Thompson (Eds.), *Communication* (pp. 7–53). New York: Plenum.

[101] Beer, C. (1976). Some complexities in the communication behavior of gulls. *Annals of the New York Academy of Sciences*, *280*, 413–32.

[102] Leger, D.W. (2005). First documentation of combinatorial song syntax in a suboscine passerine species. *The Condor*, *107*, 765–74.

[103] Trillo, P. and Vehrencamp, S. (2005). Song types and their structural features are associated with specific contexts in the banded wren. *Animal Behaviour*, *70*, 921–35.

[104] Bremond, J. (1968). Recherches sur la semantique et les elements vecteurs d'information dans les signaux acoustiques du rouge-gorge (Erithacus rubecula L.). La Terre et la vie, 2, 109–220.

[105] Holland, J., Dabelsteen, T. and Paris, A.L. (2000). Coding in the song of the wren: importance of rhythmicity, syntax and element structure. *Animal Behaviour*, *60*, 463–70.

[106] Thielcke, G. (1962). Versuche mit Klangattrappen zur Klärung der Verwandtschaft der BaumläuferCerthia familiaris L., C. brachydactyla Brehm undC. americana Bonaparte. *Journal Für Ornithologie*, *103*, 266–71.

[107] Clarke, E., Reichard, U.H. and Zuberbühler, K. (2006). The Syntax and Meaning of Wild Gibbon Songs. *PLoS ONE*, *1*, e73.

[108] Hedwig, D., Mundry, R., Robbins, M.M. and Boesch, C. (2015). Contextual correlates of syntactic variation in mountain and western gorilla close-distance vocalizations: Indications for lexical or phonological syntax? *Animal Cognition*, *18*, 423–35.

[109] Arnold, K., Pohlner, Y. and Zuberbühler, K. (2011). Not words but meanings? alarm calling behaviour in a forest guenon. In V. Sommer, C. Ross and L. Barrett (Eds.), *Primates of Gashaka* (pp. 437–68). New York: Springer.

[110] Arnold, K. and Zuberbühler, K. (2006). The alarm-calling system of adult male putty-nosed monkeys, Cercopithecus nictitans martini. *Animal Behaviour*, *72*, 643–53.

[111] Arnold, K. and Zuberbühler, K. (2008). Meaningful call combinations in a non-human primate. *Current Biology*, *18*, R202–3.

[112] Arnold, K. and Zuberbühler, K. (2012). Call combinations in monkeys: Compositional or idiomatic expressions? *Brain and Language*, *120*, 303–9.

[113] Arnold, K. and Zuberbühler, K. (2013). Female Putty-Nosed Monkeys Use Experimentally Altered Contextual Information to Disambiguate the Cause of Male Alarm Calls. *PLoS ONE*, *8*, e65660.

[114] Schlenker, P., Chemla, E., Arnold, K. and Zuberbühler, K. (2016). Pyow-hack revisited: Two analyses of putty-nosed monkey alarm calls. *Lingua*, *171*, 1–23.

[115] Crane, J.M., Savage, J.L. and Russell, A.F. (2016). Diversity and function of vocalisations in the cooperatively breeding Chestnut-crowned Babbler. *Emu*, *116*, 241–253.

[116] Carey, S. (2001). Cognitive foundations of arithmetic: Evolution and ontogenisis. *Mind & Language*, *16*, 37–55.

[117] Todt, D. and Hultsch, H. (1998). How songbirds deal with large amounts of serial information: retrieval rules suggest a hierarchical song memory. *Biological Cybernetics*, *79*, 487–500.

[118] Engesser, S., Ridley, A.R. and Townsend, S.W. (2016). Meaningful call combinations and compositional processing in the southern pied babbler. *Proceedings of the National Academy of Sciences*, *113*, 5976–5981.

[119] Buzzard, P.J. (2006). Ecological Partitioning of Cercopithecus campbelli, C. petaurista, and C. diana in the Taï Forest. *International Journal of Primatology*, *27*, 529–58.

[120] McGraw, W.S., Zuberbühler, K. and Noë, R. (2007). *Monkeys of the tai forest: an African primate community*. Cambridge University Press.

[121] Candiotti, A., Zuberbühler, K. and Lemasson, A. (2012). Context-related call combinations in female Diana monkeys. *Animal Cognition*, *15*, 327–39.

[122] Lemasson, A. and Hausberger, M. (2011). Acoustic variability and social significance of calls in female Campbell's monkeys (Cercopithecus campbelli campbelli). *The Journal of the Acoustical Society of America*, *129*, 3341–52.

[123] Lemasson, A., Gandon, E. and Hausberger, M. (2010). Attention to elders' voice in non-human primates. *Biology Letters*, *6*, 325–328.

[124] Ouattara, K., Lemasson, A. and Zuberbühler, K. (2009). Campbell's Monkeys Use Affixation to Alter Call Meaning. *PLoS ONE*, *4*, e7808.

[125] Ouattara, K., Zuberbühler, K., N'goran, E.K., Gombert, J.-E. and Lemasson, A. (2009). The alarm call system of female Campbell's monkeys. *Animal Behaviour*, *78*, 35–44.

[126] Uster, D. and Zuberbühler, K. (2001). The functional significance of Diana monkey clear' calls. *Behaviour*, *138*, 741–56.

[127] Zuberbühler, K. (2000). Referential labelling in Diana monkeys. *Animal Behaviour*, *59*, 917–27.

[128] Zuberbühler, K., Noë, R. and Seyfarth, R.M. (1997). Diana monkey long-distance calls: messages for conspecifics and predators. *Animal Behaviour*, *53*, 589–604.

[129] Lemasson, A., Remeuf, K., Rossard, A. and Zimmermann, E. (2012). Cross-Taxa similarities in affect-induced changes of vocal behavior and voice in arboreal monkeys. *PLoS ONE*, *7*, e45106.

[130] Lemasson, A., Gautier, J.-P. and Hausberger, M. (2003). Vocal similarities and social bonds in Campbell's monkey (Cercopithecus campbelli). *Comptes Rendus Biologies*, *326*, 1185–93.

[131] Furrer, R.D. and Manser, M.B. (2009). The Evolution of urgency-based and functionally referential alarm calls in ground-dwelling species. *The American Naturalist*, *173*, 400–10.

[132] Macedonia, J.M. and Evans, C.S. (1993). Essay on contemporary issues in ethology: variation among mammalian alarm call systems and the problem of meaning in animal signals. *Ethology*, *93*, 177–197.

[133] Zuberbühler, K. (2001). Predator-specific alarm calls in Campbell's monkeys, Cercopithecus campbelli. *Behavioral Ecology and Sociobiology*, *50*, 414–22.

[134] Zuberbühler, K. (2000). Interspecies semantic communication in two forest primates. *Proceedings of the Royal Society of London Series B: Biological Sciences*, *267*, 713–8.

[135] Lemasson, A., Ouattara, K., Bouchet, H. and Zuberbühler, K. (2010). Speed of call delivery is related to context and caller identity in Campbell's monkey males. *Naturwissenschaften*, *97*, 1023–7.

[136] Ouattara, K. (2009). *Communication vocale chez la mone De Campbell sauvage (Cercopithecus campbelli campbelli) au parc national de Taï- Côte d'Ivoire : flexibilité acoustique et proto-syntaxe* [Internet]. Université Rennes 2, Université de Cocody-Abidjan.

[137] Zuberbühler, K. (2002). A syntactic rule in forest monkey communication. *Animal Behaviour*, *63*, 293–9.

[138] Kuhn, J., Keenan, S., Arnold, K. and Lemasson, A. (2014). On the /-oo/ 'suffix' of Campbell's monkeys (C. Campbelli). [Internet].

[139] Stephan, C. and Zuberbühler, K. (2008). Predation increases acoustic complexity in primate alarm calls. *Biology Letters*, *4*, 641–4.

[140] Arnold, K., Keenan, S., Lemasson, A. and Zuberbühler, K. (2013). *Population differences in wild Campbell's monkeys alarm call use.* University of St Andrews.

[141] Saulitis, E.L., Matkin, C.O. and Fay, F.H. (2005). Vocal repertoire and acoustic behavior of the isolated AT1 killer whale subpopulation in southern Alaska. *Canadian Journal of Zoology*, *83*, 1015–29.

[142] Dobson, C.W. and Lemon, R.E. (1979). Markov sequences in songs of American thrushes. *Behaviour*, *68*, 86–105.

[143] Bickerton, D. and Szathmáry, E. (2009). *Biological foundations and origin of syntax*. Mit Press.

[144] Bouchet, H., Coye, C. and Lemasson, A. (2016). Le langage est-il le propre de l'homme ? Apport des études sur les primates non humains. *Tétralogiques*, *21*, 87–133

[145] Chomsky, N. (1981). Knowledge of language: Its elements and origins. *Philosophical Transactions of the Royal Society of London B: Biological Sciences*, *295*, 223–34.

[146] Fitch, W.T. (2005). The evolution of language: a comparative review. *Biology and Philosophy*, *20*, 193–203.

[147] Seyfarth, R.M., Cheney, D.L. and Bergman, T.J. (2005). Primate social cognition and the origins of language. *Trends in Cognitive Sciences*, *9*, 264–6.

[148] Vilain, A., Abry, C., Schwartz, J. and Vauclair, J. (2011). *Primate communication and human language: vocalisation, gestures, imitation and deixis in humans and non-humans*. Amsterdam: John Benjamins Publishing Company.

[149] Blumstein, D.P. (2003). Social complexity but not the acoustic environment is responsible for the evolution of complex alarm communication. In R. Ramousse, D. Allaine, and M. Le Berre (Eds.), *Adaptive Strategies and Diversity in Marmots (*pp 31–38). International Network on Marmots.

[150] Gustison, M.L., Roux, A. le and Bergman, T.J. (2012). Derived vocalizations of geladas (Theropithecus gelada) and the evolution of vocal complexity in primates. *Philosophical Transactions of the Royal Society of London B: Biological Sciences*, *367*, 1847–59.

[151] May-Collado, L.J., Agnarsson, I. and Wartzok, D. (2007). Phylogenetic review of tonal sound production in whales in relation to sociality. *BMC Evolutionary Biology*, 7, 1.

[152] McComb, K. and Semple, S. (2005). Coevolution of vocal communication and sociality in primates. *Biology Letters*, *1*, 381–5.

[153] Bouchet, H., Blois-Heulin, C. and Lemasson, A. (2013). Social complexity parallels vocal complexity: a comparison of three non-human primate species. *Frontiers in Psychology*, *4*, 390.

[154] Manser, M.B., Jansen, D.A., Graw, B., Hollén, L.I., Bousquet, C.A., Furrer, R.D. et al. (2014). Vocal complexity in meerkats and other mongoose species. *Advances in the Study of Behavior*, *46*, 281.

[155] Buzzard, P. and Eckardt, W. (2007). The social systems of the guenons. *Cambridge Studies in Biological and Evolutionary Anthropology*, *1*, 51–71.

[156] Golabek, K.A. (2010). *Vocal communication and the facilitation of social behaviour in the southern pied babbler (Turdoides bicolor)*. University of Bristol – http://ethos.bl.uk/OrderDetails.do?uin=uk.bl.ethos.535476.

[157] Radford, A.N. and Ridley, A.R. (2008). Close calling regulates spacing between foraging competitors in the group-living pied babbler. *Animal Behaviour*, *75*, 519–27.

[158] Russell, A., Portelli, D., Russell, D. and Barclay, H. (2010). Breeding ecology of the chestnut-crowned babbler: a cooperative breeder in the desert. *Emu*, *110*, 324–31.

[159] Nowak, M.A., Komarova, N.L. and Niyogi, P. (2001). Evolution of universal grammar. *Science*, *291*, 114–8.

[160] Crane, J.M., Savage, J.L. and Russell, A. (2016). Diversity and function of vocalisations in the cooperatively breeding chestnut-crowned babbler. *Emu*, *116*, 241–253.

[161] Brown, C.H. and Waser, P.M. (1988). Environmental influences on the structure of primate vocalizations. In D. Todt, P. Goedeking, and D. Symmes (Eds.), *Primate Vocal Communication* (pp. 51–66). Berlin: Springer.

[162] Keenan, S., Lemasson, A. and Zuberbühler, K. (2013). Graded or discrete? A quantitative analysis of Campbell's monkey alarm calls. *Animal Behaviour*, *85*, 109–18.

[163] Marten, K. and Marler, P. (1977). Sound transmission and its significance for animal vocalization. *Behavioral Ecology and Sociobiology*, *2*, 271–90.

[164] Waser, P.M. and Brown, C.H. (1986). Habitat acoustics and primate communication. *American Journal of Primatology*, *10*, 135–54.

[165] Marler, P. (1975). On the origin of speech from animal sounds. Inj. F. Kavanagh and J. E. Cutting (Eds.), *The Role of Speech in Language* (pp. 11–37). Cambridge, USA: M.I.T. Press.

[166] Jansen, D.A., Cant, M.A. and Manser, M.B. (2012). Segmental concatenation of individual signatures and context cues in banded mongoose (Mungos mungo) close calls. *BMC Biology*, *10*, 97.

[167] Müller, C.A. and Manser, M.B. (2008). Mutual recognition of pups and providers in the cooperatively breeding banded mongoose. *Animal Behaviour*, *75*, 1683–92.

[168] Jansen, D.A. (2013). *Vocal communication in banded mongoose.* Doctoral dissertation, University of Zürich.

[169] Rood, J.P. (1975). Population dynamics and food habits of the banded mongoose. *African Journal of Ecology*, *13*, 89–111.

[170] Fedurek, P. and Slocombe, K.E. (2011). Primate vocal communication: a useful tool for understanding human speech and language evolution? *Human Biology*, 83, 153–73.

[171] Fischer, J. (1998). Barbary macaques categorize shrill barks into two call types. *Animal Behaviour*, *55*, 799–807.

[172] Fischer, J., Cheney, D.L. and Seyfarth, R.M. (2000). Development of infant baboons' responses to graded bark variants. *Proceedings of the Royal Society of London B: Biological Sciences*, *267*, 2317–21.

[173] May, B., Moody, D.B. and Stebbins, W.C. (1989). Categorical perception of conspecific communication sounds by Japanese macaques, Macacafuscata. *The Journal of the Acoustical Society of America*, *85*, 837–47.

[174] Robinson, J.G. (1984). Syntactic structures in the vocalizations of wedge-capped capuchin monkeys, Cebus olivaceus. *Behaviour*, *90*, 46–78.

[175] Cäsar, C., Byrne, R.W., Hoppitt, W., Young, R.J. and Zuberbühler, K. (2012). Evidence for semantic communication in titi monkey alarm calls. *Animal Behaviour*, *84*, 405–11.

[176] Clay, Z. and Zuberbühler, K. (2009). Food-associated calling sequences in bonobos. *Animal Behaviour*, *77*, 1387–96.

[177] Crockford, C. and Boesch, C. (2005). Call combinations in wild chimpanzees. *Behaviour*, *142*, 397–421.

[178] Collier, K., Townsend, S.W. and Manser, M.B. (Accepted). Call concatenation in wild meerkats. *Animal Behaviour*.

[179] Clay, Z. and Zuberbühler, K. (2011). Bonobos extract meaning from call sequences. *PLoS ONE*, *6*, e18786.

[180] Bergman, T.J., Beehner, J.C., Cheney, D.L. and Seyfarth, R.M. (2003). Hierarchical classification by rank and kinship in baboons. *Science*, *302*, 1234–6.

[181] Cheney, D.L. and Seyfarth, R.M. (1999). Recognition of other individuals' social relationships by female baboons. *Animal Behaviour*, *58*, 67–75.

[182] Cheney, D.L., Seyfarth, R.M. and Silk, J.B. (1995). The responses of female baboons (Papio cynocephalus ursinus) to anomalous social interactions: evidence for causal reasoning? *Journal of Comparative Psychology*, *109*, 134.

[183] Endress, A.D., Cahill, D., Block, S., Watumull, J. and Hauser, M.D. (2009). Evidence of an evolutionary precursor to human language affixation in a non-human primate. *Biology Letters*, *5*, 749–51.

[184] Terrace, H.S., Son, L.K. and Brannon, E.M. (2003). Serial expertise of rhesus macaques. *Psychological Science*, *14*, 66–73.

[185] Wright, A.A., Santiago, H.C., Sands, S.F., Kendrick, D.F. and Cook, R.G. (1985). Memory processing of serial lists by pigeons, monkeys, and people. *Science*, *229*, 287–9.

[186] Grainger, J., Dufau, S., Montant, M., Ziegler, J.C. and Fagot, J. (2012). Orthographic processing in baboons (Papio papio). *Science*, *336*, 245–8.

[187] Hauser, M.D., Newport, E.L. and Aslin, R.N. (2001). Segmentation of the speech stream in a non-human primate: statistical learning in cotton-top tamarins. *Cognition*, *78*, B53–64.

[188] Ziegler, J.C., Hannagan, T., Dufau, S., Montant, M., Fagot, J. and Grainger, J. (2013). Transposed-letter effects reveal orthographic processing in baboons. *Psychological Science*, *24*, 1609–11.

[189] Newport, E.L., Hauser, M.D., Spaepen, G. and Aslin, R.N. (2004). Learning at a distance II. Statistical learning of non-adjacent dependencies in a non-human primate. *Cognitive Psychology*, *49*, 85–117.

[190] Fitch, W.T. and Hauser, M.D. (2004). Computational constraints on syntactic processing in a nonhuman primate. *Science*, *303*, 377–80.

[191] Gentner, T.Q., Fenn, K.M., Margoliash, D. and Nusbaum, H.C. (2006). Recursive syntactic pattern learning by songbirds. *Nature*, *440*, 1204–7.

[192] Savage-Rumbaugh, S., Rumbaugh, D.M. and McDonald, K. (1986). Language learning in two species of apes. *Neuroscience & Biobehavioral Reviews*, *9*, 653–65.

[193] Sevcik, R.A. and Savage-Rumbaugh, E.S. (1994). Language comprehension and use by great apes. *Language & Communication*, *14*, 37–58.

[194] Herman, L.M., Richards, D.G. and Wolz, J.P. (1984). Comprehension of sentences by bottlenosed dolphins. *Cognition*, **16**, 129–219.

[195] Pepperberg, I.M. (2009). *The Alex studies: cognitive and communicative abilities of grey parrots*. Harvard University Press.

[196] Pepperberg, I.M. (1981). Functional vocalizations by an African Grey parrot (Psittacus erithacus). *Zeitschrift Für Tierpsychologie*, *55*, 139–60.

Klaus Zuberbühler
Institute of Biology, University of Neuchatel, Switzerland School of Psychology and Neuroscience, University of St Andrews, Scotland (UK)

Primate Roots of Speech and Language

Abstract: Human language is largely a vocal behaviour that has evolved from a more ancient primate communication system. Although vocalizations are also the main way by which nonhuman primates communicate and interact socially it has been difficult to demonstrate direct transitions from non-linguistic primate vocal communication to human language. Nevertheless, several continuities are apparent. First, primates produce and perceive sounds by specialized anatomical and neural structures also present in humans. Compared to humans, however, nonhuman primates are severely limited in the amount of control they have over vocal production, which restricts their ability for phonology, syntax, and vocal learning. But language is also a cognitive capacity and here there is good evidence that primates understand others' calls as given by specific individuals to specific events or social interactions. In great apes, moreover, callers can take the past history with their audience into account, by suppressing, exaggerating and socially directing their calls in strategic ways. Yet, there is no clear evidence that primates, apart from humans, perceive others as governed by complex mental states, such as shared knowledge or false beliefs, during acts of communication. Also, primates do not seem to be motivated to convey knowledge relevant to their audience and there is no clear indication that they use vocal behaviour for the purpose of social bonding. The current hypothesis is that these differences in cognition and motivation have prevented the evolution of flexible, combinatorial vocal communication in nonhuman primates.

Keywords: nonhuman primate communication, intentionality in communication, referential communication

1. Introduction

Considerable debate surrounds the question of how and why human language has evolved from non-linguistic forms of primate communication. From an evolutionary perspective, it is implausible that language has emerged without any relevant precursors, so the debate is largely on the nature of continuities and discontinuities between nonhuman primate and human communication (Fitch and Zuberbühler, 2013). Most likely, language is a mosaic of older components that have emerged during different evolutionary time periods, some of which perhaps only recently during hominine evolution. The goal of this chapter is to identify key components of the language faculty, particularly vocal production and the cognitive mechanisms underlying language, and to explore how they might have evolved from earlier forms. This is done within a comparative approach and special focus on great ape natural communication. Throughout the chapter, a distinction is made between communication, speech and language. Communication is defined as the exchange of thoughts, messages, or information, as by speech, signals, writing or behaviour (n.d. 2011b). Speech is defined as an the act of expressing or describing thoughts, feelings, or perceptions by articulation of words (n.d. 2011a) and language as communication of thoughts and feelings through a system of arbitrary signals, such as voice sounds, gestures or written symbols (n.d. 2011c).

2. Vocal production[1]

The basic mechanism for sound production in mammals is described by the source-filter theory. Both animal vocalisations and human speech sounds are produced by an apparatus that consists of two independent mechanisms, the source (larynx) and the filter (supra-laryngeal vocal tract (Fant, 1960). Compared to other primates and most other mammals, however, human vocal communication is highly unusual. Humans not only possess a species-specific repertoire of non-linguistic vocalisations, but they also have the capacity to actively control the apparatus in order to produce sustained airflow, which then generates stable vibration of the larynx, the fundamental frequency and acoustic source of speech (Herbst, 2016). In addition, humans have fine motor control of various anatomical structure

1 Based on previously published material by A. R. Lameira, I. Maddieson, and K. Zuberbühler, 'Primate Feedstock for the Evolution of Consonants', *Trends in Cognitive Sciences,* 18/2 (Feb 2014a), 60–62. and W.T. Fitch and K. Zuberbühler, 'Primate Precursors to Human Language: Beyond Discontinuity', in Eckart Altenmuller, Sabine Schmidt, and Elke Zimmerman (Eds.), *Evolution of Emotional Communication: From Sounds in Nonhuman Mammals to Speech and Music in Man* (Oxford: Oxford University Press, 2013).

involved in speech production, so-called articulators, the result of which are several hundred perceptually distinct phonemes, the building blocks of the 7,000 or so currently spoken languages. Each language only uses a small fraction of this species-specific repertoire and speakers loose the ability to discriminate the different sound contrasts during ontogeny (Crystal 1997). The variability across languages is enormous. Some languages use nine vowels and ten consonants (Andoke; Colombia) while others only three vowels and 22 consonants (Diyari; Australia; Lameira et al., 2014). Vowels and consonants are the product of complex vocal tract configurations by the tongue, lips and jaw, while the sound producing activity of the larynx remains relatively constant. Some consonants are produced as co-articulations with vowels, while others are produced without vocal-fold vibration (e.g. voiceless stops /p/, /t/, /k/). Recent work has shown that the human vocal apparatus is not fundamentally different from those of non-human primates, suggesting that there are no anatomical reasons that would prevent other species from producing speech. The vocal production apparatus of higher primates, in other words, is speech-ready (Boë et al., 2017; Fitch et al., 2016).

If the main evolutionary transition towards speech has been at the level of larynx control and not, as claimed for decades, due to differences in vocal tract anatomy (Fitch and Zuberbühler, 2013; Lieberman, 2012), then the main question is how and why humans have evolved such an unprecedented level of vocal control. One line of research addresses this kind of human uniqueness at the neural level, insofar as the muscles that operate the larynx appear to be governed by projections from the motor cortex to the brainstem nuclei that steer laryngeal muscles (Jurgens, 2002). One hypothesis is that this direct cortical pathway may be the neural cause of the fine motor control that humans have over their laryngeal musculature to produce the fundamental acoustic source for speech and for singing. Whether or not the duet songs in primates, such as gibbons or indris, are governed by human-like laryngeal control is currently unknown (Filippi, 2016; Gamba et al., 2016; Geissmann, 2002). In humans, laryngeal control is responsible for the rhythmic and intonational aspect of language, i.e. speech prosody (Hirano et al., 1969; Ohala, 1990). Yet, the physiological mechanisms underlying prosodic features are still poorly understood (Erickson, 1995; Finnegan et al., 2000; Lecuit and Demolin, 1998). Some progress has been made by modelling the relationships between fundamental frequency and subglottal pressure and with other techniques, such as laryngeal electromyography (e.g. Riede et al., 2011; Riede, 2011; Zhang and Ghazanfar, 2016). Further research in this field is likely to contribute to explain the human uniqueness of speech.

3. Vocal learning

Within the primates, humans are the only species that can vocally imitate, a behaviour that starts early in ontogeny. One hypothesis for why non-human primates are prevented from vocal learning is in terms of differences in laryngeal control. If an individual cannot control sound production at will, then it cannot vocally imitate arbitrary sound pattern, a basic requirement for speech acquisition. Non-human primate communication, instead, is limited to biologically hardwired vocal repertoires, a collection of signals given in context-related and often age- and sex-specific ways (Zuberbühler, 2016b). Typically, different call types have different biological functions, such as 'greeting' calls to facilitate social interactions or 'alarm calls' to avoid predation. Other examples are food calls, copulation calls, movement calls, long-distance calls, or lost calls. For some call types, phylogenetic relatedness and acoustic similarities are linked, with more closely related species producing more similar calls, a finding that includes human vocalisations (Davila Ross et al., 2009; Kersken et al., 2017).

Some call structures appear to have evolved to be psychologically effective on receivers, either by having physiological effects or by triggering relevant psychological processes in recipients, such as increased attention or facilitated learning (Owren and Rendall, 2001). However, all this is not to say that primate sound production is completely inflexible. Various studies on monkeys have shown that social variables can influence the acoustic structure of some call types. For instance, in Campbell's monkeys, the strength of a social bond between two individuals significantly correlates with the acoustic similarity of their contact calls, independent of genetic relatedness (Lemasson et al., 2011). Evidence of this type is relatively common, but is usually in terms of acoustic variations within an existing call type rather than the emergence of new vocal structures, which is very different from the ease by which human infants produce, combine and learn complex sound utterances. Also relevant is research on captive primates, which has shown that subjects can be trained to produce some calls on command and to modify the acoustic structure of their calls (Fitch and Zuberbühler, 2013).

Given the seemingly unbridgeable gap in motor control between human and nonhuman primate vocal behaviour, how did speech evolve? A relevant finding here is that motor control of the supra-laryngeal vocal tract may be less of a constraint, at least for great apes, compared to control of the larynx. Great apes possess call repertoires that comprise both voiced and voiceless calls, with some voiceless calls being subject to social learning (Lameira et al. 2014). In wild orang-utans, for example, there are population differences in calls produced during nest building with some populations producing 'raspberries' and others 'smacks', in the absence of genetic or habitat differences. Captive orang-utans can imitate human voiceless whistling sounds, a clear demonstration of their ability to control key speech articulators and airflow. In chimpanzees, an interesting case study is 'Viki', who underwent intensive language training by her human caretakers and so learned to shape the vocal tract to imitate few English words ('mama', 'papa', 'cup'), although unable to activate the larynx as a sound source (Hayes, 1951).

A reasonable conclusion from this literature is that the common ancestor of humans and great apes already possessed the ability to control their supra-laryngeal vocal tract, while vocal fold control emerged later and perhaps only after the phylogenetic split from the other great apes (de Boer et al., 2015).

4. Concepts and categories[2]

Human language is not just a vocal behaviour but the product of a cognitive architecture that is in constant interaction with social partners. The main tools for these social interactions are mental representations of the world, organised in humans as concepts (defined as mental representations

2 Based on material published by K. Zuberbühler, 'Social Concepts and Communication in Nonhuman Primates', in Mark A. Bee and Cory T. Miller (Eds.), *Psychological Mechanisms in Animal Communication* (New York: Springer, 2016a).

that refer to categories, i.e. sets of entities), with corresponding symbolic representations in a speaker's linguistic vocabulary. Nonhuman primates do not have linguistic lexicons but they may nevertheless have mental concepts akin to ours. So how do primates and other animals organise their worlds internally? What are natural concepts in primate minds and how are they linked to vocal communication? In the following, evidence for different types of social concepts is reviewed, including dominance, kinship, friendship, or group.

4.1 Dominance

A common challenge in social groups is that some individuals exert social power over others, which can lead to conflicts. Possessing an abstract notion of social rank, thus, is likely to be adaptive to navigate in such systems and there is some evidence for this in nonhuman primates. In chacma baboons, Bergman et al. (2003) showed that subjects responded strongly to vocal interactions between two group members, provided this suggested a rank reversal. The effect was even more pronounced if the rank change occurred between individuals that belonged to different matrilines, which suggested a larger upheaval within the group. The conclusion was that these primates comprehend the invisible hierarchical social structure that is characterised by dominance ranks. The study has recently been partly replicated in vervet monkeys, with similar results (Borgeaud et al., 2013), suggesting that primates in general represent social rank in complex ways.

But how do primates learn to understand their own and others' social rank in a group? One possibility is a cognitive operation termed transitive inference by which subjects observe social interactions between others to compute an invisible dominance matrix (e.g. Tromp et al., 2015). Another mechanism is more proactive, by directly challenging other group members that are presumed to be close in social rank. Male bonobos, for example, actively provoke other group members by approaching them with acoustically distinct 'contest hoots' combined with aggressive gestures, for the sole purpose of provoking a social reaction, usually agonistic chase. The social targets are chosen very carefully, by selecting group members of adjacent rank, a behaviour that also seems to function as a way to advertise the caller's own social position in the group to bystanders (Genty et al., 2014).

4.2 Kinship and friendship

Primate social life is not just about social rank but also about social relationships. In an early study long-tailed macaques were trained to differentiate mother-offspring pairs from other types of social dyads. In subsequent transfer trials, subjects were able to identify untrained kin dyads, suggesting a concept of kinship (Dasser, 1988). Other social concepts in primates refer to what has loosely been called 'friendship' (measured as social bonds, e.g. Silk et al. 2010), which in primates can be maintained over long time periods, sometimes between the sexes and independent of kin relations. The ability to form social bonds with nonrelatives may thus be one of the most important features of primate sociality, which has also been linked with the evolution of large brains (Dunbar and Shultz, 2007). According to this 'social brain' hypothesis, forming, maintaining and monitoring others' social bonds is computationally very demanding, suggesting that large brains evolved in response to the cognitive demands of sociality (Dunbar, 1993). However, this focus has come under increasing scrutiny, due to energetic constraints in maintaining brain tissue (Isler and Van Schaik, 2014).

Although social bonds are apparent from patterns of social interactions, there is not much direct evidence that primates represent others' social relations as distinct concepts (e.g. friend vs. foe). In an early field experiment, Kummer et al. (1974) investigated the mechanisms of bond formation between male and female hamadryas baboons. In one condition, an observer male was allowed to watch how another male from the same troop interacted with a new female before being admitted to the pair. The striking finding was that the observer male respected the new pair bond, even if he was dominant over the rival. This respect of "ownership," however, was not generalizable and did not transfer to situations were two males competed over access to food. Using playback experiments, Wittig et al. (2014) recently showed that, even hours after a natural aggressive interaction had occurred, chimpanzees were still strongly affected if they heard the aggressive "waa" barks of a "friend" of the former aggressor, which was not the case if they heard the "waa" barks of other group members. Chimpanzees, in other words, are able to recruit memories of past social interactions from different sources to make inferences about current interactions.

4.3 Group

Another potential social concept concerns group membership. Almost all higher primates form groups that are defined by individualised memberships, but do they also categorise others in terms of in- and out-group members? In one field experiment, chimpanzees reacted to the calls of out-group members more strongly than to own group members (Herbinger et al., 2009), while males appear to assess the vocal behaviour of neighbouring males to numerically assess their current part size (Wilson et al., 2001). Again, these results indicate that primates can distinguish in- from out-group members but this is not direct evidence of a corresponding concept that could generalise to novel situations.

In sum, there is a large literature suggesting that non-human primates appear to structure the real world along mental concepts akin to humans, such as rank, kinship, friendship and so on. In most cases, however, the existence of these concepts is only vaguely inferred by patterns of behaviour. Hence, although behaviour patterns are indicative that primates categorise others in terms of friends, rivals, relatives, there is no real direct evidence that this is based on corresponding mental concepts. Related to this, there is no good evidence that primates possess vocal labels for any of these putative concepts, as it is the case for human language.

5. Intentionality[3]

5.1 Audience effects

There is little doubt that nonhuman primates, and probably many other animals, can extract useful information from signals, but this in itself is perhaps not so interesting. As Tomasello (2008, p. 19) puts it: "…the monkey has simply learned that one thing predicts another, or even causes another, in the same basic way as many other phenomena in their daily lives." But

3 Based on material published by K. Zuberbühler and J.C. Gomez, 'Communication, Primate Intentional ', in C Power (Ed.), *International Encyclopedia of Anthropology* (Wiley, in press).

when humans extract meaning from speech signals, this usually goes beyond the utterance's (literal) linguistic meaning. Instead, semantic inferences are made from shared experiences with the signaller, the prior interaction history and, crucially, the speaker's intended meaning. Human communication, thus, operates not just on what is signalled, but also on what the signaller intends to communicate (Grice, 1969). Recipients, in turn, assume that signallers intend to say something that is relevant for them, which requires higher levels of intentionality than simple goal directedness.

The ability to ascribe intentions is based on the cognitive ability to attribute mental states to others, which then opens the possibly most important question in comparative research – do nonhuman primates base their communication on the mental states of others? Do they perceive others as independent minds with own intentions, beliefs, and knowledge? There is some evidence that primates can go beyond simple signal-response arithmetic and consider social factors when producing and responding to each other's signals. Chimpanzees, for example, time the delivery of some of their calls to a partner's attention, and inhibit call production in the presence of some audiences (Hostetter et al., 2001). For chimpanzee greeting signals, the pant grunts, vocal production is often inhibited by the presence of the top-ranking bystanders, demonstrating that call production is audience dependent (Laporte and Zuberbühler, 2010). Similarly, victims of aggression can alter the acoustic structure of screams, not only depending on the severity of aggression experienced, but also in relation to the composition of the nearby audience. Victims of severe attacks tend to exaggerate the aggression experienced, but only if at least one high-ranking listener is in the audience who could interfere (Slocombe and Zuberbühler, 2007). Other evidence is females who are more likely to give copulation calls if high-ranking males are in the vicinity and less likely if other females are nearby (Townsend et al., 2008). Finally, when finding food chimpanzees produce acoustically distinct rough grunts which attract other group members to the site, but again these calls are especially common if subjects travel with social allies or high-ranking group members (Slocombe et al., 2010; Schel et al., 2013a). In monkeys, audience-aware vocal communication has been found in predation contexts (Wich and de Vries, 2006; Papworth et al., 2008) and during social interactions (Semple et al., 2009) and it is very well documented in great ape gestural communication (e.g. Cartmill and Byrne, 2007).

In sum, nonhuman primate signalling can be affected by the presence of others and the presumed social consequences of signal production, most likely because they perceive others as intentional beings. Whether or not the primates perceive their audiences as possessing more complex mental states, such as knowledge or beliefs, is not addressed by these studies.

5.2 Mental state attribution

Most theories of animal communication grant primate signals an imperative function, whereas human communication, from an early age, appears to have additional declarative, informative or interrogative functions. This is manifested by the fact that, from a young age, human infants point to objects, not because they want them, but with the sole aim of sharing attention or information with others. Apes, in contrast, typically use gestures with the only purpose of requesting or directing others to objects or activities. For example, if apes indicate hidden tools or food to a human, this act is performed with the final aim of obtaining food, not just to share a target of common attention (Roberts et al., 2014).

A key component in any theory of language evolution thus has to do with how aware signallers are of the mental states of their recipients during acts of communication. In humans, communication is the product of intention, both in the sense of goal-directedness and aboutness (Brentano, 1874). Do primates perceive others as governed by mental states during acts of communication? If so, how complex are these mental states? Are they able and motivated to convey knowledge relevant to their audience?

An early approach to study the problem was to raise great apes in an environment entirely structured by humans in order to document the development of their communication skills. One famous case study was a chimpanzee called 'Sarah', who acquired over 100 arbitrary signs-meaning linkages, which enabled her to generate sentence-like strings of up to eight units long (Premack, 1970). Similarly, two chimpanzees, 'Sherman' and 'Austin', learned to use geometric symbols (lexigrams) to make requests to one another (Savage-Rumbaugh et al., 1980), while 'Kanzi', a male bonobo learned to master a lexigram system of over 300 arbitrary symbols (Savage-Rumbaugh and Lewin, 1994). A related approach was to teach apes American Sign Language, which resulted in repertoires of over 100 signals of the major grammatical classes. Despite these findings, the apes did not make much use of their acquired skills during natural social interactions and there was no good evidence that subjects demonstrated a deep understanding of the symbolic nature and basic grammatical rules of their acquired language systems (Terrace et al., 1979).

5.3 Levels of intentionality

Intentionality derives from Latin *intendere*, i.e. being directed towards a goal or thing (Jacob, 2014). However, people's psychological states are not always about such entities, but sometimes also about other individuals' psychological states, which makes it necessary to distinguish between different levels of intentionality (Jacob, 2014). Whether or not animals (or young human infants) can also attribute psychological states with intentionality to others has thus been the topic of much empirical research, usually under the notion of 'theory of mind' (Premack and Woodruff, 1978; Tomasello, 2014). For example, apes use a rich repertoire of gestures when interacting with social partners, with evidence of audience awareness, social directedness, and communicative persistence (Call and Tomasello, 2007), similar to what has been found in pre-linguistic human infants. For vocal signals, a relevant finding emerged from free-ranging vervet monkeys where it was shown that playbacks of acoustically distinct alarm calls given to eagles, leopards, and pythons were enough to make others respond in adaptive ways, even if no actual predator was present. For instance, after hearing a snake alarm, monkeys responded by bipedally scanning the surrounding area, as if trying to locate the putative snake (Cheney and Seyfarth, 1992). A philosophical approach to the problem of differing levels of intentionality has been proposed by Dennett (1983), and called the 'intentional stance'. This framework has produced landmark progress in assessing the question of intentionality in ethological data and relating it to intentionality in human communication (Table 1).

Table 1: Dennett's (1983) recursive grades of intentionality in animal communication

Intention	Content
0 order	A recognizes x
1st order	A wants B to do x
2nd order	A wants B to recognize x
3rd order	A wants B to recognize that A wants B to do x
4th order	A wants B to recognize that A wants B to recognize x
5th order	....

For the vervet monkeys, the proposal then was that, when a monkey produces an eagle alarm call in response to which other group members run into cover, this may be the result of different 'grades' of intentionality. The null hypothesis, or 0-order intentionality, attributes no intentionality at all but callers merely react automatically to the perceptual recognition of each type of predator. They are simply "...prone to three flavours of anxiety or arousal: leopard anxiety, eagle anxiety, and snake anxiety". Each anxiety has an evolved link to one call type, and listeners may form associations between external events and call types that allow them to react appropriately. Signallers and recipients benefit in their own ways, but are not mentally connected, in a sort of "by-product semanticity".

First-order intentionality is different in that the monkey produces alarm calls with the goal to influence the behaviour of others (Table 1). Although there was no direct evidence for 1st order intentionality in the original studies with vervet monkeys (Seyfarth et al., 1980), other work has suggested that this is within the cognitive capacities of non-human primates. For example, female Diana monkeys continue to produce alarm calls, until the group's single adult male also produces the same alarm calls, i.e. the alarm call type that corresponds to the predator the females have perceived, as if they are soliciting the male's acknowledgment of the situation (Stephan and Zuberbühler, 2016). These results indicate that primate alarm calls are not just automatic and direct responses to an external disturbance, but that calling is additionally governed by the social interaction history between signallers and recipients, and may require some level of intentionality in Dennett's scale. More specifically, first-order intentionality requires that signals are produced to cause a specific and deliberate effect in an audience. For this to be the case, the signal needs to be socially addressed, the recipient's response should be monitored and, if the desired outcome is not achieved, signalling should be modified accordingly (Bruner, 1981).

There is good evidence that primates deploy some of their signals in such ways, particularly in the realm of gestural communication. For example, bonobos produce beckoning gestures with persistence and signal elaboration to persuade a sexual partner to follow to a distant location (Genty and Zuberbühler, 2014). This type of goal-directed intentionality is common in ape gestural communication but also visible in facial signals, such as chimpanzee lip-smacking during grooming, where it appears to cause longer and more reciprocal grooming bouts, probably by expressing benign intention (Fedurek et al., 2015).

While primate gestural communication appears to qualify as the result of first-order intentionality this is less clear for vocal signals. Vocal signals do not require visual contact between partners so it is not always clear whether or whom the caller wishes to address. Also, as many calls are given in relatively rigid context-specific ways, it is less plausible that they are active attempts to alter a recipient's behaviour (let alone mental states), compared to being mere vocal tags of specific external events. At the same time, several studies have shown that primates can use vocalisations in goal-directed ways, specifically to influence a recipient in some way or another. In one key study, chimpanzees were more likely to produce alarm calls to snakes if their audience was unaware of the danger, suggesting some active assessments of others' mental states (Crockford et al., 2012). Similarly, chimpanzees prior to travel sometimes give specific "travel hoos" in connection with complex departure behaviour that includes audience checking and other signs of goal-directedness (Gruber and Zuberbühler, 2013).

A more complex sense of intentional communication is displayed if both signallers and recipients appear to take into account each other's mental states when producing and responding to signals. In this situation, a signaller not only 'wants' a recipient to do something specific, but she also wants to be understood. For primate communication, there is no good evidence for such second order intentionality, a controversy directly linked to whether primates can attribute mental states to others, that is, whether they have a 'theory of mind' (Zuberbühler and Gomez, in press). In captivity, chimpanzees have demonstrated mental state attribution, first in terms of judgments about what others can or cannot see (Hare et al., 2000) and, more recently, what they can and cannot know (Krupenye et al., 2016), but these capacities have not manifested themselves in their communication behaviour, with a few exceptions. As mentioned before, in more natural situations when encountering snakes, chimpanzees direct their alarm calls to arriving group members, not to the snake, and call more if the arriving individuals are socially close to them and ignorant about the danger (Schel et al. 2013b; Crockford et al., 2012). Equally relevant are studies on ape gestures, for instance bonobos are more likely to repeat their gestures to a reluctant familiar than unfamiliar keeper, whereas to an unfamiliar keeper they were more likely to change gestures, as if taking into account the differences in the keepers' knowledge rather than just perceptions (Genty et al., 2015).

In sum, some of the primate gestural literature is in line with the hypothesis that 2nd order intentional states can guide primate communication, although it is usually possible to propose simpler explanatory models based on behaviour-reading and associative learning mechanisms, as a string of associations between behaviour cues and event outcomes.

5.4 Ostension

Zuberbühler and Gomez (in press) argue that there is another sense of intentional communication, a 'Mitteilungsbedürfnis', which is different because it refers to others' mental states in a particular way (Fitch, 2010). Here, information transfer is not just by linguistic coding and decoding but also by producing ostensive indicators of meaning, which recipients use to infer that the sender is trying to convey a message (Grice, 1957). Humans often use finger pointing to this effect, something that nonhuman primates do not do. However, in theory almost any behaviour can serve as an ostensive signal, including gaze fixation, which is more common in primates (Zuberbühler, 2008). Ostension, in its simplest form, may be little more than 'addressing' behaviours, e.g., calling or claiming attention, something that has been repeatedly described in great apes.

5.5 Shared intentionality

A recent theory of human evolution suggests that humans are special in their ability to share goals and intentions when participating in collaborative activities (Tomasello et al., 2005). According to this line of argument, subjects require not only powerful forms of mindreading but also a profound motivation to share their mental states with others to enter some kind of shared cognitive representation of joint intentions. There is some evidence that primates can engage in joint activities, which require that the intentions match (Pika and Zuberbühler, 2008), but whether this is equivalent to a sense of sharing cannot be decided. According to this theory, the implications of being able to share intentions are enormous, enabling subjects to create linguistic conventions, social norms, social institutions and many other types of human activities (Heesen et al., in press).

5.6 Conclusions

There is compelling evidence that primates use both vocal and gestural signals in a goal-directed, first-order intentionality sense. Some further evidence suggests that there might also be second-order intentionality, but it remains to be determined in what sense primates attribute mental states to others and whether they are able to communicate about this. Primates show intentions to communicate in an ostensive sensc, if one accepts evidence of addressing and seeking others' attention as a demonstration of an ostensive function. Finally, there is no good evidence of shared intentionality. Although great apes understand the basics of intentional action, and may use communicative signals to affect others' intentions, they do not appear to experience as sense of 'shared' intentionality.

6. Referential communication[4]

6.1 Aboutness

One philosophical definition of intentionality is in the sense of aboutness—as being about or directed at a particular object (Brentano 1874). In this

4 Based on material published by ibid. and K. Zuberbühler and C. Neumann, 'Referential Communication in Nonhuman Animals', in J. Call (Ed.), *Apa Hand-*

view, a signal is intentional if it is emitted or understood as being about an object (Zuberbühler and Gomez, in press). In the vervet monkey example, recipients may interpret an alarm call as indicating, not just that there is a predator, but that a sender has found a predator and that the alarm call is about this predator. In humans, aboutness is in that humans can think about non-existing things or situations, including reflecting on others' false beliefs. There is no good evidence that non-human animals are able to engage in such mental operations. Nevertheless, some findings are relevant and suggest, at least, some precursor abilities. For example, in playback experiments, free-ranging baboons recognised when a call was directed at themselves as opposed to at other individuals (Engh et al., 2006). In chimpanzees, victims of aggression retreated from playback of aggressive barks given by an ally of the former opponent, but ignored the same barks if given by other group members, even hours after the conflict, again suggesting that they understood that the barks were about them (Wittig et al., 2014). In sum, when attending to vocal signals, baboons, chimpanzees, and probably other primates take into account the target of others' attention, but mostly if this is about themselves.

6.2 Functional reference

The vervet monkey experiments discussed earlier, together with more recent studies, amount to a considerable literature on what is often called 'functionally referential' communication. Empirical evidence is usually in the form of structurally unique signals given to identifiable events that are external to the signaller. 'Functional reference' has become one of the most influential models of animal communication, mainly because it is conceptually simple and makes predictions that are easily quantifiable. According to the original definition (Macedonia and Evans, 1993), a functionally referential signal should (a) exhibit a degree of stimulus specificity and (b) be sufficient to allow receivers to select appropriate responses, even in the absence of the eliciting stimulus and other normally available cues. Vervet monkey alarm calling was said to fulfil both criteria and has thus become

book of Comparative Psychology (Vol. 1. Basic Concepts, Methods, Neural Substrate, and Behavior: APA, 2017).

the textbook example of functionally referential signalling in non-human animals (Macedonia and Evans, 1993).

More recent fieldwork has shown that the main carriers of meaning are not always the individual calls, but sometimes sequences or combinations of calls (Arnold and Zuberbühler, 2006; Arnold and Zuberbühler, 2008). Here, the evidence is not restricted to primates, but also includes studies on birds (Engesser et al., 2015, 2016). In addition, in some communication systems distinct call types can appear as call variants or compound calls, which are given to subtle changes in external events. For example, male Campbell's monkeys produce acoustically distinct alarm calls to crowned eagles ('hok') and leopards ('krak'), but add an acoustically distinct 'oo' suffix if the danger imposed by these two predators is not imminent (i.e., 'hok-oo', 'krak-oo', respectively) (Ouattara et al., 2009b), an acoustic variation that is perceived and discriminated by recipients (Coye et al., 2015). Moreover, the different call types can be concatenated into long call sequences, which are given in context-specific ways to predation events and various non-predatory disturbances (Ouattara et al., 2009a), some of which are perceived and discriminated by recipients (Zuberbühler, 2002).

These findings have led to an important theoretical discussion, namely whether animal signals are really equivalent to human referential communication or just a by-product of evolved behaviour that grants the signaller with a fitness payoff. One problem is what exactly qualifies as an external event; the 'stimulus specificity' criterion. Detecting a predator or a novel food source are relatively obvious examples, but what about calls given when encountering another group member or calls given during mating? Chimpanzees, as mentioned, sometimes produce acoustically distinct vocalisations when encountering higher-ranking group members, a reliable indicator of the type of social interaction the caller is engaged in. Although these calls refer to relatively specific social events, it is not clear whether they should qualify as functionally referential, mainly because the event (stimulus) is not strictly 'external' to the caller.

Another problem with 'functional reference' is that animals often produce alarm calls to a range of events that, in human terms, cannot be subsumed in a common category. For example, monkeys may give the same alarm call type to a range of events, which are represented by different mental concepts in humans. Alarm calls to terrestrial disturbances are often particularly broad. Interestingly, in response to 'terrestrial alarms' other monkeys often first look in the direction of the caller, most likely to obtain additional behavioural information, such as gaze direction. Also, monkeys often take the more general context into account, when responding to alarm calls. For example, in response to playbacks of 'terrestrial alarms', putty-nosed monkeys spend less time looking at the caller if the alarms were preceded by additional information that disambiguates the cause of the call, such as the loud thundering sound of a falling tree (Arnold and Zuberbühler, 2013).

The deeper problem is that there are virtually no empirical studies on whether or how wild animals organise their worlds in mental concepts, as discussed in a previous section. Until there are data to systematically describe how wild animals categorise and represent the real world mentally, the production specificity criterion of functional reference will remain a contentious notion. For example, it is possible that animal vocalisations are not the direct product of specific mental representations, triggered by specific external events, but are mere indicators of behavioural intentions. Diana monkey eagle alarm calls, for instance, may not have evolved to warn others about eagle presence, but to indicate that the caller is about to engage with a perched eagle (Stephan and Zuberbühler, 2016). Again, future research should address this possibility with targeted field experiments.

6.3 Psychological reference

The term functional reference has also been created specifically to avoid discussions about underlying intention, whether animals deliver signals in intentional ways. The fact that most animal signals are part of a relatively hard-wired species-specific repertoire is not necessarily a problem for questions about referential signalling. One of the most powerful referential signals in humans is pointing, a non-linguistic, species-specific, innate signal used to direct a recipient towards a shared experience. Human signallers actively seek to point out relevance to their recipients, in order to share common ground. Although vervet monkey alarm calls inform recipients about specific external events, it is unclear whether these signals are produced with the intention to do so, as explained earlier. Human-like referential communication goes beyond mere audience awareness and goal-directedness, as discussed before, and usually includes some assessment of others' mental states – their perceptions, intentions, knowledge or beliefs. Whether animals can take their audiences into account at this level is controversial. The evidence for referencing as a psychological act is currently best for wild chimpanzees, particularly from a series field experiments with detecting snake models. As mentioned, alert call production was strongly determined by the presence of friends but also whether recipients were ignorant or aware of the snake (Crockford et al. 2012, 2015; Schel et al., 2013b). Although more targeted experiments are needed, results tentatively suggest that chimpanzees can take the knowledge state of a receiver into account, as opposed to mere perceptions, when referring others to a relevant external event.

7. Communication as social bonding[5]

The biological success of our species is partly grounded in a major evolutionary transition in mental capacities from self-serving, competitive to group-oriented, cooperative social relationships. Compared to other primates, humans are much more collaborative, prosocial, and amenable to social norms, which has far-reaching implications at almost every level of human activity, including language (Tomasello, 2014). Human social interactions are highly structured joint activities during which partners are in tune with each other's intentions -- the human 'interaction engine' (Levinson, 2006). Linguistic discourse is a specific type of joint activity, which is characterized by rapid exchanges of syntactic units, while its semantic content is often about others' social behaviour, or gossip, and this exchange of social information appears to have a strong bonding effect on the interlocutors (Dunbar, 1993).

In non-human primates, social bonding is mainly achieved by manual grooming. In wild chimpanzees, for example, urinary oxytocin levels are higher after grooming with bond partners than other group members (Crockford et al., 2013) and there is also good evidence that grooming has a direct

---

5 Based on material published by K. Zuberbühler and P. Fedurek, 'Vocal Grooming', in Todd K. Shackelford and Viviana A. Weekes-Shackelford (Eds.), *Encyclopedia of Evolutionary Psychological Science* (Berlin: Springer, in press).

impact on an individual's fitness. In primates, grooming increases the probability that partners will help each other during fights, whereas individuals with strong social networks have higher life-time reproductive success than less socially integrated individuals (Silk et al., 2010). Across species, there is a positive relation between social grooming time and group size, suggesting that living in large groups leads to relatively more time demands for manual grooming, most likely due to an exponentially increased number of social relationships. This has led to the hypothesis that, during human evolution, manual grooming has been gradually replaced by vocal grooming as the main mechanism for social bonding, which has paved the way for the evolution of language (Dunbar, 1993). Vocal grooming, in other words, has become something like the missing link in this theory of language evolution.

How did the transition from manual to vocal grooming take place? One candidate for vocal grooming in primates is chorusing. In chimpanzees, males of the same group chorus with their pant hoots, an acoustically unique long-distance vocalization. There is evidence that these joint vocal displays facilitate tolerance at feeding sites and predict support in agonistic interactions on a short-term basis, while manual grooming seems to be a better predictor of long-term social bonds (Fedurek et al., 2013). Another candidate is vocal convergence, usually seen between affiliated individuals that adjust the acoustic structure of their calls during vocal interactions (Candiotti et al., 2012). In humans, vocal convergence is a well-documented phenomenon, described as 'speech accommodation', although similar effects are also seen in gestures, suggesting that people try to either emphasise or minimise the social difference to their interlocutors, reflecting their desire or refusal to strengthen a social bond (Clark and Schaefer, 1989). Another candidate that might have paved the way for a transition from manual to vocal grooming is call exchanges, as found in many primates, and often between closely affiliated individuals. Call exchange networks tend to be socially more rigid than grooming networks, insofar as socially important individuals are more likely to elicit vocal responses than others. Related to call exchanges is duetting, a vocal behaviour seen in gibbons and other monogamous primates. Duetting has a presumed bonding function (Geissmann, 2002; Filippi, 2016), but the behaviour is relatively inflexible in the sense that once a relationship is established it simply functions to broadcast this fact to potential sexual rivals. Vocal turn-taking is also a key feature of human linguistic discourse, which is characterized by highly structured, rapid exchanges of short syntactic units, around 1500 per day (Levinson, 2016). In humans, the length of turns can be flexible, as well as the number of speakers, but participants rigidly observe short time gaps of about 200ms between turns. This highly conserved pattern requires participants to predict when a partner's turn comes to an end, which is only possible if interlocutors can anticipate the forthcoming semantic content, while constructing their next utterance. This pattern is found across languages, suggesting that it is based on a biological predisposition, with possibly deep evolutionary roots (Levinson, 2016).

Recent empirical research has suggested another candidate behaviour for the transition from manual to vocal grooming. In chimpanzees, manual grooming is often accompanied by a specific acoustic signal, lip-smacking, especially between closely affiliated individuals. Lip-smacking plays a key role in coordinating grooming bouts by making them longer, more reciprocated and more intimate (Fedurek et al., 2015). Many primates lip-smack during affiliative social interactions (Bergman, 2013), but so far there is only evidence in chimpanzees for a direct link between this behaviour and manual grooming (Zuberbühler and Fedurek, 2017). Chimpanzee lip-smacking has thus become somewhat of the prime candidate as the missing link to human vocal interactions. While chimpanzee lip-smacking is semantically empty, it similarly seems to function in social bonding and is delivered in temporarily structured ways. One evolutionary scenario therefore is that lip-smacking may have served as a precursor for the evolution of articulatory control (Ghazanfar et al., 2012). As argued earlier, there is good evidence that great apes have substantial control over their supra-laryngeal vocal tract but relatively little active control over the larynx (Lameira et al., 2014). The transition from manual to vocal grooming, in other words, may have been linked with gaining active control over sound production in line with concurrent underlying changes in the cognitive architecture.

8. General conclusions

Human communication is strikingly different from any other known natural communication system. From an evolutionary perspective, this is particularly striking because, biologically, humans are primates whose communication system has evolved during a long and shared phylogenetic history. One way to investigate the roots of human language is with comparative studies of primate cognition, particularly the basic processes required for language production and perception.

Humans are clearly unique in the amount of control they have over sound production, and this has considerable effects on other important features of language, such as vocal imitation, phonemic repertoires and the ability to produce syntactic structures. Humans are the only primate species capable of controlling and socially learning their vocal output, which as a consequence becomes part of shared communicative conventions. Although non-human primates and most other animals do not have active control over their vocal output, some studies suggest that they can use parts of their signal repertoire in referential ways, to inform others about relevant events in their environment. Current evidence also suggests that the primate vocal tract is essentially speech-ready, and that great apes have good voluntary control over most speech articulators, apart from the larynx. Why and how humans have evolved the ability to control the larynx is an open question, but it has been suggested that this has evolved in the context of cooperative breeding (Zuberbühler, 2011).

Language is also a cognitive system based on conceptual thought and the ability to categorise the world in mental concepts. Here, the evidence for equivalent structures in nonhuman primates is not strong, although much of their behavioural patterns suggest similar representational organisation. A key feature of any language definition is that, during acts of communication, signallers draw their recipients' attention to what they consider relevant entities, both real or imagined. Humans routinely refer others to what they perceive as relevant aspects of the world, a psychological propensity that was probably one of the major drivers of language evolution. Linguistically, this is achieved with arbitrary acoustic conventions, but referring can also happen non-linguistically, for example with iconic gestures or pointing. In humans, this ability emerges early during development, mainly in the form of pointing that functions, similar to linguistic expressions, as a referential signal. An important evolutionary question therefore is whether this characteristically human way of communicating is also present in some animal signalling and, if so, whether the underlying cognitive processes are similar. While in humans referential communication appears to be the result of mental state attributions, in the sense that signallers communicate content relevant to their audience, only very few studies in animals have found evidence of this kind, all of them in wild chimpanzees communicating about food or danger. In human communication, in contrast, there seem to be no such contextual limitations, a possible consequence of our highly cooperative nature.

Acknowledgements

I am grateful to the European Research Council (PRILANG 283871) and to the Swiss National Science Foundation for past and on-going support.

References

Arnold, K., and Zuberbühler, K. (2006). Semantic combinations in primate calls. *Nature,* 441, 303–303.

Arnold, K., and Zuberbühler, K. (2008). Meaningful call combinations in a non-human primate. *Current Biology, 18*, R202-R203.

Arnold, K., and Zuberbühler, K. (2013). Female putty-nosed monkeys use experimentally altered contextual information to disambiguate the cause of male alarm calls. *Plos One, 8*, e65660.

Bergman, T. J. (2013). Speech-like vocalized lip-smacking in geladas. *Current Biology,* 23, R268-R269.

Bergman, T. J., Beehner, J. C., Cheney, D. L., and Seyfarth, R. M. (2003). Hierarchical classification by rank and kinship in baboons. *Science,* 302, 1234–1236.

Boë, L. J., Berthommier, F., Legou, T., Captier, G., Kemp, C., Sawallis, T. R., Becker, Y., Rey, A., and Fagot, J. (2017). Evidence of a vocalic proto-system in the baboon (Papio papio) suggests pre-hominin speech precursors. *PLoS One, 12*, e0169321.

Borgeaud, C., van de Waal, E., and Bshary, R. (2013). Third-party ranks knowledge in wild vervet monkeys (Chlorocebus aethiops pygerythrus). *PLoS One, 8*, 1–4.

Brentano, F. (1874). *Psychologie vom empirischen Standpunkt.* Leipzig: Duncker & Humblot.

Bruner, J. (1981). Intention in the structure of action and interaction. In L. Lipsitt (Ed.) *Advances in infancy research* (pp. 41–45). Norwood: Ablex.

Call, J., and Tomasello, M. (Eds.) (2007). *The gestural communication of apes and monkeys.* Mahwah, New Jersey, & London: Lawrence Erlbaum.

Candiotti, A., Zuberbühler, K., and Lemasson, A. (2012). Convergence and divergence in Diana monkey vocalizations. *Biology Letters, 8*, 382–385.

Cartmill, E. A., and Byrne, R. W. (2007). Orangutans modify their gestural signaling according to their audience's comprehension. *Current Biology, 17*, 1345–1348.

Cheney, D. L., and Seyfarth, R. M. (1992). How monkeys see the world. *Behavioral and Brain Sciences, 15*, 135–147

Clark, H. H., and Schaefer, E. F. (1989). Contributing to discourse. *Cognitive Science, 13*, 259–294.

Coye, C., Ouattara, K., Zuberbühler, K., and Lemasson, A. (2015). Suffixation influences receivers' behaviour in non-human primates. *Proceedings of the Royal Society B-Biological Sciences,* 282 (1807).

Crockford, C., Wittig, R. M., Langergraber, K., Ziegler, T. E., Zuberbühler, K., and Deschner, T. (2013). Urinary oxytocin and social bonding in related and unrelated wild chimpanzees. *Proceedings of the Royal Society B-Biological Sciences, 280*, 20122765.

Crockford, C., Wittig, R. M., Mundry, R., and Zuberbühler, K., (2012). Wild chimpanzees inform ignorant group members of danger. *Current Biology, 22*, 142–146.

Crockford, C., Wittig, R. M., and Zuberbühler, K. (2015). An intentional vocalization draws others' attention: A playback experiment with wild chimpanzees. *Animal Cognition, 18*, 581–591.

Crystal, D. (1997). *The Cambridge encyclopedia of language.* Cambridge: Cambridge University Press.

Dasser, V. (1988). A social concept in Java monkeys. *Animal Behaviour, 36*, 225–230.

Davila Ross, M., Owren, M. J., and Zimmermann, E. (2009). Reconstructing the evolution of laughter in great apes and humans. *Current Biology, 19*, 1106–1111.

de Boer, B., Wich, S. A., Hardus, M. E., Lameira, A. R. (2015). Acoustic models of orangutan hand-assisted alarm calls. *Journal of Experimental Biology, 218*, 907–914.

Dennett, D. C. (1983). Intentional systems in cognitive ethology: The "Panglossian paradigm" defended'. *Behavior and Brain Sciences, 6*, 343–355.

Dunbar, R. I. M. (1993). Co-evolution of neocortical size, group size and language in humans. *Behavioral and Brain Sciences, 16*, 681–735.

Dunbar, R. I. M., and Shultz, S. (2007). Evolution in the social brain. *Science, 317*, 1344–1347.

Engesser, S., Crane, J. M. S., Savage, J. L., Russell, A. F., and Townsend, S. W. (2015). Experimental evidence for phonemic contrasts in a nonhuman vocal system. *Plos Biology, 13*, e1002171.

Engesser, S., Ridley, A.R., and Townsend, S.W. (2016). Meaningful call combinations and compositional processing in the southern pied babbler. *Proceedings of the National Academy of Sciences, 113*, 5976–5981.

Engh, A. L., Hoffmeier, R. R., Cheney, D. L., and Seyfarth, R. M. (2006). Who, me? Can baboons infer the target of vocalizations? *Animal Behaviour, 71*, 381–387.

Erickson, D. (1995). The production of low tones in English intonation. *Journal of Phonetics, 231*, 179–188.

Fant, G. (1960). *Acoustic theory of speech production*. The Hague: Mouton.

Fedurek, P., Machanda, Z. P., Schel, A. M., and Slocombe, K. E. (2013). Pant hoot chorusing and social bonds in male chimpanzees. *Animal Behaviour, 86*, 189–196.

Fedurek, P., Slocombe, K. E., Hartel, J. A., and Zuberbühler, K. (2015). Chimpanzee lip-smacking facilitates cooperative behaviour. *Scientific Reports, 5*, 13460.

Filippi, P. (2016). Emotional and interactional prosody across animal communication systems: a comparative approach to the emergence of language', *Frontiers in Psychology, 7*, 1393.

Finnegan, E. M., Luschei, E. S., and Hoffman, H. T. (2000). Modulations in respiratory and laryngeal activity associated with changes in vocal intensity during speech. *Journal of Speech Language and Hearing Research, 43*, 934–950.

Fitch, W. T. (2010). *The evolution of language*. Cambridge: Cambridge University Press.

Fitch, W. T, de Boer, B., Mathur, N., and Ghazanfar, A. A. (2016). Monkey vocal tracts are speech-ready. *Science Advances, 2*, e1600723.

Fitch, W.T., and Zuberbühler, K. (2013). Primate precursors to human language: beyond discontinuity. In E. Altenmuller, S. Schmidt, and E. Zimmerman (Eds.), *Evolution of emotional communication: from sounds in nonhuman mammals to speech and music in man* (pp. 26–48). Oxford: Oxford University Press.

Gamba, M., Torti, V., Estienne, V., Randrianariso, R. M., Valente, D., Rovara, P., Bonadonna, G., Friard, O., and Giacoma, C. (2016). The indris have got rhythm! Timing and pitch variation of a primate song examined between sexes and age classes. *Frontiers in Neuroscience, 10*, 249.

Geissmann, T. (2002). Duet-splitting and the evolution of gibbon songs. *Biological Reviews, 77*, 57–76.

Genty, E., Clay, Z., Hobaiter, C., and Zuberbühler, K. (2014). Multi-modal use of a socially directed call in bonobos. *Plos One, 9*, e84738.

Genty, E., Neumann, C., and Zuberbühler, K. (2015). Bonobos modify communication signals according to recipient familiarity. *Scientific Reports, 5*, 16442.

Genty, E., and Zuberbühler, K. (2014). Spatial reference in a bonobo gesture. *Current Biology, 24*, 1601–1605.

Ghazanfar, A. A., Takahashi, D. Y., Mathur, N., and Fitch, W. T. (2012). Cineradiography of monkey lip-smacking reveals putative precursors of speech dynamics. *Current Biology, 22*, 1176–1182.

Grice, H. P. (1957). Meaning. *The Philosophical Review, 66*, 377–288.

Grice, H. P. (1969). Utterers' meaning and intentions. *The Philosophical Review,* 78, 147–177.

Gruber, T. and Zuberbühler, K. (2013). Vocal recruitment for joint travel in wild chimpanzees. *Plos One, 8*, e76073.

Hare, B., Call, J., Agnetta, B., and Tomasello, M. (2000). Chimpanzees know what conspecifics do and do not see. *Animal Behaviour, 59*, 771–785.

Hayes, C. (1951). *The Ape in Our House*. New York: Harper.

Heesen, R., Genty, E.; Rossano, F.; Zuberbuhler, K., and Bangerter, A. (in press). Social play as joint action: a framework to study the evolution of shared intentionality as an interactional achievement. *Learning & Behavior*.

Herbinger, I., Papworth, S., Boesch, C., and Zuberbühler, K. (2009). Vocal, gestural and locomotor responses of wild chimpanzees to familiar and unfamiliar intruders: a playback study. *Animal Behaviour, 78*, 1389–1396.

Herbst, C. T. (2016). Biophysics of vocal production in mammals. In R. A. Suthers, W. Fitch, R. Fay, and A. Popper (Eds.), *Vertebrate Sound Production and Acoustic Communication* (pp. 159–189). Springer Handbook of Auditory Research.

Hirano, M., Ohala, J., and Vennard, W. (1969). The function of laryngeal muscles in regulating fundamental frequency and intensity of phonation. *Journal of Speech Language and Hearing Research, 12*, 616–28.

Hostetter, A. B., Cantero, M., and Hopkins, W. D. (2001). Differential use of vocal and gestural communication by chimpanzees (Pan troglodytes) in response to the attentional status of a human (Homo sapiens). *Journal of Comparative Psychology, 115*, 337–343.

Isler, K., and Van Schaik, C. P. (2014). How humans evolved large brains: Comparative evidence. *Evolutionary Anthropology, 23, 65–75.*

Jacob, P. (2014). Intentionality. In E. N. Zalta (Ed.), *The Stanford Encyclopedia of Philosophy* (Winter 2014 edn.).

Jurgens, U. (2002). Neural pathways underlying vocal control. *Neuroscience and Biobehavioral Reviews, 26*, 235–258.

Kersken, V., Zuberbühler, K., and Gomez, J. C. (2017). Listeners can extract meaning from non-linguistic infant vocalisations cross-culturally. *Scientific Reports, 7*, 41016.

Krupenye, C., Kano, F., Hirata, S., Call, J., and Tomasello, M. (2016). Great apes anticipate that other individuals will act according to false beliefs. *Science, 354*, 110–14.

Kummer, H., Gotz, W., and Angst, W. (1974). Triadic differentiation: An inhibitory process protecting pair bonds in baboons. *Behaviour, 49*, 62–87.

Lameira, A. R., Maddieson, I., and Zuberbühler, K. (2014). Primate feedstock for the evolution of consonants. *Trends in Cognitive Sciences, 18*, 60–62.

Laporte, M. N. C., and Zuberbühler, K. (2010). Vocal greeting behaviour in wild chimpanzee females. *Animal Behaviour, 80*, 467–73.

Lecuit, V., and Demolin, D. (1998). The relationship between intensity and subglottal pressure with controlled pitch. In R. H. Mannell, and J. Robert-Ribes (Eds.), *Proceedings 5th International Conference on Spoken Language Processing (ICSLP)* (pp. 3079–3082). Sydney: Australian Speech Science and Technology Association, Incorporated (ASSTA).

Lemasson, A., Ouattara, K., Petit, E.J. and Zuberbühler, K. (2011). Social learning of vocal structure in a nonhuman primate? *BMC Evolutionary Biology, 11*, 362.

Levinson, S. C. (2006). On the human "interaction engine". In N. J. Enfield, and S. C. Levinson (Eds.), *Roots of human sociality: Culture, cognition and interaction* (pp. 39–69). Oxford: Berg.

Levinson, S. C. (2016). Turn-taking in human communication – origins and implications for language processing. *Trends in Cognitive Sciences, 20*, 6–14.

Lieberman, P. (2012). Vocal tract anatomy and the neural bases of talking. *Journal of Phonetics, 40*, 608–622.

Macedonia, J. M., and Evans, C.S. (1993). Variation among mammalian alarm call systems and the problem of meaning in animal signals. *Ethology, 93*, 177–197.

n.d. (2011a). 'Speech'. *American Heritage® Dictionary of the English Language* (5 edn.).

n.d. (2011b). 'Communication'. *American Heritage® Dictionary of the English Language* (5 edn.).

n.d. (2011c). 'Language'. *American Heritage® Dictionary of the English Language* (5 edn.).

Ohala, J. J. (1990). Respiratory activity in speech. In W. J. Hardcastle, and A. Marchal (Eds.), *Speech Production and Speech Modelling* (pp. 25–53). Dordrecht: Kluwer.

Ouattara, K., Lemasson, A., and Zuberbühler, K. (2009a). Wild Campbell's monkeys concatenate vocalizations into context-specific call sequences. *Proceedings of the National Academy of Sciences of the United States of America, 106*, 22026–22031.

Ouattara, K., Lemasson, A., and Zuberbühler, K. (2009b). Campbell's monkeys use affixation to alter call meaning. *PLoS One, 4*, e7808.

Owren, M. J., and Rendall, D. (2001). Sound on the rebound: bringing form and function back to the forefront in understanding nonhuman primate vocal signaling. *Evolutionary Anthropology, 10*, 58–71.

Papworth, S., Papworth, S., Böse, A. S., Barker, J., Schel, A. M., and Zuberbühler, K. (2008). Male blue monkeys alarm call in response to danger experienced by others. *Biology Letters, 4*, 472–475.

Pika, S., and Zuberbühler, K. (2008). Social games between bonobos and humans: Evidence for shared intentionality? *American Journal of Primatology, 70*, 207–210.

Premack, D. (1970). The education of Sarah: A chimp learns the language. *Psychology Today, 4*, 54–58.

Premack, D., and Woodruff, G. (1978). Does the chimpanzee have a theory of mind? *Behavioral and Brain Sciences, 1*, 515–526.

Riede, T. (2011). Subglottal pressure, tracheal airflow, and intrinsic laryngeal muscle activity during rat ultrasound vocalization. *Journal of Neurophysiology, 106*, 2580–2592.

Riede, T., Tokuda, I. T., and Farmer, C. G. (2011). Subglottal pressure and fundamental frequency control in contact calls of juvenile Alligator mississippiensis. *Journal of Experimental Biology, 214*, 3082–3095.

Roberts, A. I., Vick, S. J., Roberts, S. G. B., and Menzel, C. R. (2014). Chimpanzees modify intentional gestures to coordinate a search for hidden food. *Nature Communications, 5*, 3088.

Savage-Rumbaugh, E. S. and Lewin, R. (1994). *Kanzi. The ape at the brink of the human mind.* New-York: Wiley.

Savage-Rumbaugh, E. S., Rumbaugh, D. M., Smith, S. T., and Lawson, J. (1980). Reference: the linguistic essential. *Science, 210*, 922–925.

Schel, A. M., Machanda, Z., Townsend, S. W., Zuberbühler, K., and Slocombe, K. E. (2013a). Chimpanzee food calls are directed at specific individuals. *Animal Behaviour,* 86, 955–965.

Schel, A. M., Townsend, S. W., Machanda, Z., Zuberbühler, K., and Slocombe, K. E. (2013b). Chimpanzee alarm call production meets key criteria for intentionality. *PLoS One, 8*, e76674.

Semple, S., Gerald, M. S., and Suggs, D. N. (2009). Bystanders affect the outcome of mother-infant interactions in rhesus macaques. *Proceedings of the Royal Society B-Biological Sciences, 276*, 2257–2262.

Seyfarth, R. M., Cheney, D. L., and Marler, P. (1980). Monkey responses to three different alarm calls: Evidence of predator classification and semantic communication. *Science, 210*, 801–03.

Silk, Joan B., Beehner, J. C., Bergman, T. J., Crockford, C., Engh, A L., Moscovice, L. R., Wittig, R. M., Seyfarth, R. M., and Cheney, D. L. (2010). Strong and consistent social bonds enhance the longevity of female baboons. *Current Biology, 20*, 1359–61.

Slocombe, K. E., Kaller, T., Turman, L., Townsend, S. W., Papworth, S., Squibbs, P., and Zuberbühler, K. (2010). Production of food-associated calls in wild male chimpanzees is dependent on the composition of the audience. *Behavioral Ecology and Sociobiology, 64*, 1959–66.

Slocombe, K.E., and Zuberbühler, K. (2007). Chimpanzees modify recruitment screams as a function of audience composition. *Proceedings of the National Academy of Sciences of the United States of America, 104*, 17228–17233.

Stephan, C. and Zuberbühler, K. (2016). Persistent females and compliant males coordinate alarm calling in diana monkeys. *Current Biology, 26*, 2907–2912.

Terrace, H. S., Petitto, L. A., Sanders, R. J., and Bever, T. G. (1979). Can an ape create a sentence? *Science, 206*, 891–902.

Tomasello, M. (2008). *Origins of human communication*. Cambridge: MIT Press.

Tomasello, M. (2014). *A natural history of human thinking*. Cambridge, Massachusetts: Harvard University Press.

Tomasello, M., Carpenter, M., Call, J., Behne, T., and Moll, H. (2005). Understanding and sharing intentions: The origins of cultural cognition. *Behavioral and Brain Sciences, 28*, 675–735.

Townsend, S. W., Deschner, T., and Zuberbühler, K. (2008). Female chimpanzees use copulation calls flexibly to prevent social competition. *PLoS One, 3*, e2431.

Tromp, D., Meunier, H., and Roeder, J. J. (2015). Transitive inference in two lemur species (Eulemur macaco and Eulemur fulvus. *American Journal of Primatology, 77*, 338–345.

Wich, S. A., and de Vries, H. (2006). Male monkeys remember which group members have given alarm calls. *Proceedings of the Royal Society B-Biological Sciences, 273*, 735–740.

Wilson, M. L., Hauser, M. D., and Wrangham, R. W. (2001). Does participation in intergroup conflict depend on numerical assessment, range location, or rank for wild chimpanzees? *Animal Behaviour, 61*, 1203–1216.

Wittig, R. M., Wittig, R. M., Crockford, C., Langergraber, K. E., and Zuberbühler, K. (2014). Triadic social interactions operate across time: a field experiment with wild chimpanzees. *Proceedings of the Royal Society B-Biological Sciences, 281*, 20133155.

Zhang, Y. S., and Ghazanfar, A. A. (2016). Perinatally influenced autonomic system fluctuations drive infant vocal sequences. *Current Biology, 26*, 1249–60.

Zuberbühler, K. (2002). A syntactic rule in forest monkey communication. *Animal Behaviour, 63*, 293–99.

Zuberbühler, K. (2008). Gaze following. *Current Biology, 18*, R453-R455.

Zuberbühler, K. (2011). Cooperative breeding and the evolution of vocal flexibility. In M. Tallerman, and K. Gibson (Eds.), *Oxford handbook of language evolution* (pp. 71–81). Oxford: Oxford University Press.

Zuberbühler, K. (2016a). Social concepts and communication in nonhuman primates. In M. A. Bee, and C. T. Miller (Eds.), *Psychological Mechanisms in Animal Communication* (pp. 251–270). New York: Springer.

Zuberbühler, K. (2016b). Vocal communication. In Agustin Fuentes (Ed.), *International Encyclopedia of Primatology*. UK: Blackwell Publishing.

Zuberbühler, K. and Fedurek, P. (2017). Vocal grooming. In T. K. Shackelford, and V. Weekes-Shackelford (Eds.), *Encyclopedia of Evolutionary Psychological Science* (pp. 1–5). Berlin: Springer.

Zuberbühler, K., and Gomez, J. C. (in press). Communication, primate intentional. In C. Power (Ed.), *International Encyclopedia of Anthropology*. New-York: Wiley.

Zuberbühler, K., and Neumann, C. (2017). Referential communication in nonhuman animals. In J. Call (Ed.), *APA Handbook of Comparative Psychology* (Vol. 1. Basic Concepts, Methods, Neural Substrate, and Behavior, pp. 645–661). APA.

Katja Liebal
Department of Education and Psychology, Comparative Developmental Psychology, Freie Universität Berlin

What Gestures of Nonhuman Primates Can (and Cannot) Tell Us about Language Evolution

Abstract: There is a variety of different evolutionary scenarios hypothesizing how human language might have evolved. While some suggest that language evolved from scratch in humans only, others propose that precursors to human language were already present in our shared last common ancestor. Consequently, comparative researchers suggest that at least some of the abilities necessary for language to evolve are shared with other primates. However, which aspects of primate communication are studied to shed light on language evolution heavily depends on which communicative modality is studied. This chapter focuses on the gestural communication of nonhuman primates. The aim is contribute to the ongoing debate about how language might have evolved by evaluating findings from comparative research on the different building blocks of language, and by discussing how these data support a gestural origin of human language.

Keywords: nonhuman primate communication, gestural communication, language evolution

1. Introduction

Many theories of language evolution are based on comparative evidence of the communicative abilities of our closest relatives, the nonhuman primates. In searching for the roots of human language, researchers therefore aim at identifying potential precursors to language in other primate species. The aims of this chapter are, first, to give an overview of the building blocks of language, which are usually studied by researchers proposing a gestural origin of human language, such as intentional and flexible use of gestures, mechanisms of gesture acquisition, and the potential for rule-governed, meaningful combinations. Second, for each of these characteristics, the current evidence from gestural communication of apes and monkeys is evaluated to answer the question if these findings support a gestural origin of language. The chapter will close by suggesting some future directions for research into primate gestural communication necessary to inform theories of language evolution in more comprehensive ways.

2. Different approaches to language evolution

Although there is general agreement that language is unique to the human species, there is much debate about whether language is fundamentally different from other animals' communication systems. Thus, some scholars propose that language evolved from scratch in humans only and that we can therefore not gain any knowledge about language evolution by comparing human language to the communicative systems of other, even closely related species (Hauser et al., 2014). Animal communication is suggested to be fundamentally different, both with regard to its structural properties as well as cognitive foundations (Bickerton, 1992; Chomsky, 1966; Scott-Phillips, 2015) and comparisons should be limited – if conducted at all – to comparing the nonverbal "gesture-call systems" of humans to that of nonhuman primates (Burling, 1993).

Others argue that it is highly unlikely that a trait as complex as language evolved in such a short time in the human lineage only (Pinker and Bloom, 1990) and that language built on traits already present in our shared common ancestor, including neurobiological substrates (Arbib, 2005, 2016), anatomical structures such as the vocal apparatus (Fitch et al., 2016; Riede et al., 2005), or cognitive skills (Seyfarth et al., 2005). Therefore, proponents of this continuity approach suggest that a comparative approach, which investigates the communicative abilities of nonhuman primates, is useful to identify potential precursors to human language in our closest relatives (King, 1999; Tomasello, 2008; Zuberbühler, 2005).

It is important to highlight that the answer to the question whether proposed differences between human language and animal communication are a matter of degree or kind heavily depends on how language is defined. However, a universally accepted definition, which covers the many different facets of language, has not been achieved yet. Hauser and colleagues (2002) suggested that for studying how language might have evolved, "… it is unproductive to discuss language as an unanalyzed whole" (see also Fitch, 2005, p. 194). Rather, it is important to decompose language into its many different mechanisms necessary for language to emerge, and to differentiate between more general mechanisms involved in language processing, which are potentially shared with other species ("faculty of language in the broad sense") compared to those that are *specific to language* and *uniquely human* ("faculty of language in the narrow sense") (Hauser et al., 2002).

Importantly, which aspects of primate communication are studied to identify those abilities shared across primates, and to differentiate them from those unique to humans, heavily depends on whether researchers are interested in vocal, gestural, or facial communication. Thus, the overwhelming majority of existing research into primate communication uses a unimodal approach, to study one specific modality in isolation, while ignoring other modes of communication (Slocombe et al., 2011). Therefore, in searching for the origins of human language, a major debate centers on the question which communicative modality provided the starting point for language to emerge, with scholars defending a gestural, vocal, or orofacial origin. Proponents of each origin usually focus on one specific modality, and use their findings to argue why the corresponding other modalities are not suitable candidates to explain how language might have emerged (Slocombe et al., 2011).

The focus of this chapter is on primates' gestural communication, but it is important to note that depending on the type of communicative modality studied, different aspects of primate communication are investigated to shed light on the origins of human language. Therefore, in the following section, those aspects of primate gestural communication are introduced, which are usually studied to find evidence for a gestural origin of human language.

3. Which aspects of primate gestural communication are studied to find evidence for a gestural origin of human language?

The first of modern gestural theories, which heavily influenced other theories suggesting a gestural origin, was the Gestural Primacy Hypothesis postulated by Hewes (1973). In general, hypotheses proposing a gestural origin of human language assume that spoken language was preceded by a gestural stage and that our ancestors therefore initially communicated by using voluntarily produced manual gestures (Armstrong et al., 1995; Corballis, 2002; Hewes, 1973; Tomasello, 2008).

Since *intentional production* is key to human language, comparative researchers supporting gestural scenarios of language evolution focus on gestures of nonhuman primates, because primates use them intentionally in a purposeful, goal-directed way and are therefore able to voluntarily control their production. Gesture researcher usually argue that in contrast to gestures, the majority of vocalizations and facial expressions are involuntary expressions of internal emotional states, and that it therefore seems unlikely that language has emerged from these communicative modalities (Tomasello, 2008). A second heavily investigated aspect of primate gestural communication is the usage of gestures across different social contexts, representing a marker for the *flexibility* of this modality. Thus, gesture research is interested in those gestures, which are flexibly used across different contexts to achieve different social goals (Arbib et al., 2008; Call and Tomasello, 2007), in contrast to vocal research, which focuses mostly on context-specific vocalizations, such as food-related or predator-specific calls (Kalan et al., 2015; Murphy et al., 2013; Schel et al., 2013a; Zuberbühler, 2001). These different research foci result in different conclusions across modalities: because gesture researcher focus on the flexibility of gesture usage, they conclude that gestures have no inherent meaning, since the information they convey is defined by the context in which they are used (Call and Tomasello, 2007). On the other hand, the focus of vocal researchers on context-specific vocalizations leads to the conclusion that most vocalizations have specific meanings (Seyfarth et al., 1980; Slocombe and Zuberbühler, 2005; Zuberbühler, 2000), and that they can be combined into meaningful sequences (Ouattara et al., 2009; Zuberbühler, 2002). Therefore, unlike vocal researchers, gesture researchers are less interested in the questions if primate gestures have specific *meanings* and if they are *combined into meaningful sequences*. However, there is an increasing number of studies also addressing these issues (Cartmill and Byrne, 2010; Hobaiter and Byrne, 2014; Liebal et al., 2004a). Finally, gesture researchers study the mechanisms of gesture *acquisition* during ontogeny, to investigate when and how gestures emerge and whether gestural repertoires vary between individuals (Arbib, 2016; Liebal et al., submitted). It seems that primates can create new gestures and incorporate them into their repertoires (Goodall, 1986). They are therefore able to increase the number of gesture types they are using, resulting in more open gestural repertoires. Vocal or facial repertoires, on the other hand, seem to be closed and species-specific, since primates are not able to acquire new, additional signals (Owren et al., 1992).

In sum, to find evidence for a gestural origin of human language, gesture researchers focus on the intentional production and flexible usage of gestures, they investigate if gestures have specific meanings and are combined into meaningful sequences, and how they emerge in ontogeny. Before the remaining chapter evaluates the current evidence for each of these different features of primate gestural communication, the next two sections briefly describe how the term gesture is defined, and how the field of gesture research has developed over the last decades.

4. What is a gesture?

Primate gestures are considered voluntarily produced, purposeful behaviors, directed at specific individuals to influence their behavior (Benga, 2005). Criteria applied to identify such intentional use include that their communicative behavior depends on the presence of an audience, and that they are tailored to the recipient's behavior, like their attentional state or response to the other's communicative attempts (Leavens et al., 2005b). Regarding gesture modality, visual gestures are differentiated from tactile and auditory (or audible) gestures. Visual gestures are distant signals and include manual gestures, such as "extend arm", but also body postures, such as "present". Tactile gestures, such as "slap" or "gentle touch", involve the physical contact between two interacting partners, but are motorically ineffective. That means that if an individual wants another one to follow her, grabbing the other's arm and dragging him along would not count as a gesture, since this was not accompanied by waiting for the other's response to this behavior. However, briefly pulling the other's arm and then waiting for his response to follow is a potential gesture, because the recipient has the chance to respond – or not. Auditory gestures generate a sound, which can be produced with different body parts, such as clapping hands or slapping the chest or belly (Kalan and Rainey, 2009; Pika, Liebal, and Tomasello, 2003).

It is important to note that the definition of a gesture and its discrimination from other signal types is often confusing. Gestures include visual, tactile and auditory signals, and they may therefore overlap with the sensory modalities of other signal types. For example, like vocalizations, auditory gestures are acoustic signals, but they do not engage the vocal cords. Visual gestures are considered a different signal type than facial expressions (Call and Tomasello, 2007; Pollick and de Waal, 2007), although they both rely on the visual modality. This indicates that signal types are not necessarily classified based on their sensory channels, but that they are differentiated based on the traditional dichotomy between voluntarily produced gestures in comparison to more reflexive, involuntarily produced vocalizations and facial expressions (Liebal et al., 2013b; Tomasello, 2008).

5. How did the field of gesture research develop?

Initially, gestural communication was not considered a separate communicative modality, but gestures were mentioned as components of the general behavioral repertoire describing a given species. These very first studies focused primarily on primates in their natural habitats, such as baboons (Kummer, 1968), gibbons (Carpenter, 1940), orangutans (MacKinnon, 1974; Rijksen, 1978), gorillas (Schaller, 1963), chimpanzees (Goodall, 1986) and bonobos (Kuroda, 1980). The first studies on captive apes were conducted by van Hooff (1973), who investigated chimpanzees, and de Waal (1988), who compared chimpanzees and bonobos. These studies focused on establishing gestural repertoires and provided detailed descriptions of each gesture and the contexts of their use.

A more psychological approach, applying definitions derived from developmental psychology (Bates et al., 1979; Leavens et al., 2005b), was initiated by Michael Tomasello, Josep Call and their colleagues (Tomasello et al., 1994; Tomasello et al., 1997; Tomasello et al., 1989). Instead of focusing on phylogenetically ritualized signals and their adaptive function (Maynard-Smith and Harper, 2003; Smith, 1977), they were interested in proximate factors and therefore more cognitive aspects of primate communication. In their initial series of studies, they used observational methods to study different groups of captive chimpanzees (Tomasello et al., 1985). This work was later extended to other primates in captive settings, such as gorillas (Pika et al., 2003), bonobos (Pika et al., 2005), Sumatran orang-utans (Liebal et al., 2006), and one species of small apes (Liebal et al., 2004c).

Maestripieri conducted the very first systematic studies on the gestural repertoires of captive rhesus, pigtail and stump-tail macaques. Interestingly, these species differ in their social systems and rank relationships (Maestripieri, 1996a, 1996b, 1997; Maestripieri and Wallen, 1997). A major conclusion from these studies is that gestural communication varies as a function of the social system, with more despotic species using a more limited repertoire of gestures, mostly in the contexts of submission and assertion, while more egalitarian species seem to offer more opportunities for negotiating social relationships, evident in the high variety and variability of gestures in affiliative contexts (Maestripieri 1999, 2005).

The first systematic studies on wild populations were conducted by Emily Genty and Cat Hobaiter, together with Richard Byrne, who studied the gestural repertoires and their usage in gorillas (Genty et al., 2009) and chimpanzees (Hobaiter and Byrne, 2011a). In line with the results from captive research, they found that gorillas and chimpanzees used their gestures intentionally, in many different contexts, and that they adjust them to the recipient's behavior. However, they found a greater repertoire (> 100 gestures) than previously reported for captive gorillas (Pika et al., 2003), with limited individual variability and large overlap of individual repertoires, even across species. They therefore concluded that gestural repertoires are species-specific and might even be shared across species, indicating that they are most likely biologically determined (Genty et al., 2009; Hobaiter and Byrne, 2011a).

Experimental settings, which include the communication between a great ape or monkey with a human experimenter, are frequently used to study both the production as well as perception of gestural signals. The basic procedure across these studies is that primates are required to request food from the human, which is either visible or hidden, and which they cannot obtain themselves. Thus, primates need to use the human as a tool to get the food, since they are not able to reach the food by themselves. Much attention has therefore been paid to gesture types produced in this experimental setting. At least great apes readily learn to use pointing gestures, such as whole hand or index finger pointing to request food (Call and Tomasello, 1994; Leavens and Hopkins, 1999; Leavens et al., 1996). There is increasing evidence that different monkey species also use pointing gestures (Anderson et al., 2007; Kumashiro et al., 2002; Meguerditchian and Vauclair, 2006; Mitchell and Anderson, 1997), but these studies often involve special training, possibly indicating that monkeys' pointing behavior might be less flexible than that of great apes.

Experimental settings have also been widely used to study primates' understanding of human gestures, specifically pointing gestures, for example, by applying the object-choice paradigm (for a review, see Miklósi and Sopròni, 2006). After hiding a food item under one of two (or more) containers, with the baiting process invisible for the primate, the experimenter points to the location where the food is hidden. Even if this gesture is combined with additional cues, such as eye gaze or gaze alternation and orienting the body towards the baited container, primates usually fail to use this information to find the food (Anderson et al., 1996; Call et al., 2000; Povinelli et al., 1999). Whether these negative findings can be explained by methodological issues (Barth et al., 2005; Mulcahy and Hedge, 2012), or whether this shows that primates do not understand others' pointing gestures as cooperative communicative acts helping them find the food (Hare and Tomasello, 2004), is currently unclear.

While this section focused on changes of research interests over time, the following part of this chapter specifically targets those aspects of primate gestural communication, which are currently investigated to find evidence for a gestural origin of human language.

6. What is known about primate gestural communication?

In the following section, the current knowledge of research into primate gestural communication is summarized and it is discussed if they provide support for a gestural scenario of language evolution.

6.1 Intentional use

To identify intentionality in primate gestures, several criteria have been adapted from research into gesture use of prelinguistic children (Leavens et al., 2005b). These criteria include the social use of gestures in the presence of an audience, the adjustment of gesture use depending on the recipient's behavior, as well as the persistent and elaborated gesture use in case the recipient does not respond. However, there is currently no agreement about which and how many of these markers of intentionality are necessary to define a gesture (Liebal et al., 2013b). In the following, each of these different markers of intentionality will be briefly discussed with regard to the existing findings.

Considering the social use of gestures in the presence of an audience, many experimental studies have demonstrated that in interaction with humans, great apes as well as several monkey species only gesture if an audience is present (Anderson et al., 2010; Blaschke and Ettlinger, 1987; Call and Tomasello, 1994; Hattori et al., 2010; Hostetter et al., 2001; Poss et al., 2006). Many of these and other studies showed that great apes and monkeys adjust their gesture use to the attentional state of the human, and only use visual gestures if the human is oriented towards them (Bourjade et al., 2014; Hattori et al., 2010; Liebal et al., 2004b; Maille et al., 2012; but see Povinelli and Eddy, 1996). If the human is present, but not attending to them, some studies reported that apes are more likely to use auditory signals to attract the human's attention (Cartmill and Byrne, 2007; Hostetter et al., 2001; Leavens et al., 2004; Poss et al., 2006), while other studies failed to find similar results (Liebal et al., 2004b; McCarthy et al., 2013; Theall and Povinelli, 1999). However, when great apes were given the opportunity to change their position in relation to that of the human experimenter, they preferred to move into the human's visual field instead of attracting the human's attention by using auditory signals (Lieba et al., 2004b).

Some researchers investigated if chimpanzees use their gaze – instead of gestural signals – to direct the human's attention to the food they want to obtain, by repeatedly looking back and forth between the human and the food item (Leavens and Hopkins, 1998; Tomasello et al., 1994). For example, Leavens and Hopkins (1998) found that at least 80% of the chimpanzees' indicative gestures (food beg and pointing gestures) were accompanied by such gaze alternation. Interestingly, this proportion even increased to 100% if chimpanzees vocalized simultaneously when producing a gesture. According to Leavens and Hopkins (1998), it remains an open question if the occurrence of gaze alternation depended on the number of different communicative means involved (gestures and/or vocalizations), or whether an increased proportion of gaze alternation and types of communication rather reflected their high motivation to get the food – and therefore the intensification of their communicative attempts. It is important to note that although some researchers use gaze alternation as markers of intentional communication (Leavens et al., 2005b; Schel et al., 2013b), an alternative explanation might be that primates simply gaze back and forth to check if the food is still there – rather than intentionally directing the human's attention towards the food (Liebal et al., 2013b).

In case the recipient does not respond to another's gesture, the gesturer can either repeat the very same gesture (persistence) or use a different, potentially more efficient gesture (elaboration) to elicit a response. Both persistence and elaboration are considered markers of intentional use, because they indicate that primates can use their gestures flexibly to achieve their goals (Liebal et al., 2013b). Regarding persistence, several studies have shown that both captive and wild chimpanzees continue to gesture in case there is no response to their initial gesture (Hobaiter and Byrne, 2011a; Liebal et al., 2004a; McCarthy et al., 2013; Roberts, Vick, and Buchanan-Smith, 2013), while this was not found for gorillas and orangutans (Genty and Byrne, 2010; Tempelmann and Liebal, 2012). Therefore, evidence for persistence in case of failed communicative attempts is mixed, and varies across great ape species. Similarly, research into elaborated gesture use produced inconsistent findings. While some studies failed to find evidence for the use of more efficient gestures in case of communicative failure (Genty and Byrne, 2010; Liebal, Call, et al., 2004a; Tempelmann and Liebal, 2012), an increasing number of studies with both wild and captive great apes shows that they are able to modify their gestures in case their goal is not or only partially met (Cartmill and Byrne, 2007; Leavens et al., 2005b; Roberts et al., 2013).

Taken together, gestures are, by definition, intentionally used signals, although the type and number of criteria used to access intentionality varies drastically across studies. Furthermore, while there is convincing evidence that both monkeys and great apes consider the presence of an audience and adjust their gestures to the attentional state of the recipient, findings for the use of attention-getting gestures to capture others' attention as well as for the persistent and elaborated gesture use if the recipient does not respond are rather inconsistent. Still, gesture researchers conclude that unlike vocalizations and facial expressions, gestures are intentionally used signals, and interpret this as evidence for a gestural origin of human language (Tomasello, 2008). This traditional dichotomy between intentional gestures in contrast to involuntarily produced vocalizations and facial expressions, however, is increasingly challenged by several studies (Crockford et al., 2012; Scheider et al., 2016; Schel et al., 2013b; Waller et al., 2015). The current lack of evidence for the intentional use of vocalizations and facial expressions as compared to the gestural modality might indicate "blind spots" in vocal and facial research, because of the different research interests across modalities (Slocombe et al., 2011) (see section 2). Therefore, although gesture researchers generally highlight that primates' ability to control the production of their gestures, but not vocal and facial signals, is important evidence for a gestural origin of human language (Tomasello, 2008), this conclusion might be too premature because of the current knowledge gaps regarding other communicative modalities (Slocombe et al., 2011).

6.2 Flexibility of use

Flexibility in gestural communication can be considered in different ways. While this section will specifically address the usage of gestures across different contexts, the combination of gestures into longer, potentially meaningful sequences is discussed in a separate section.

The term "means-end dissociation" (Bruner, 1981) has been adapted by primate gesture research to describe the use of one gesture in several contexts, or vice versa, the use of several gestures within one specific context (Tomasello et al., 1994). Therefore, this measure describes the extent of flexibility present in the use of a given repertoire for a variety of goals primates want to achieve. For example, great apes and siamangs in captive settings have been demonstrated to use between 50% and 75% of their gestures in two or more contexts (Liebal et al., 2004c; Liebal et al., 2006; Pika et al., 2003, 2005; Tomasello et al., 1997). For wild gorillas, it has been shown that they use 10 of their most common gestures in several contexts (Genty et al., 2009). Across species, the highest variety of gestures used within one specific context was observed in the play context. Orangutans also perform many of their gestures in the context of affiliation and in interactions about food (Liebal et al., 2006), whiles bonobos and chimpanzees use a high variety of gestures in the agonistic context (Pika et al., 2005; Tomasello et al., 1997). Tactile gestures are used more frequently for several purposes, while visual gestures are more likely to occur in a specific context, such as grooming, sexual behavior, or requesting food (Liebal et al., 2004; Liebal et al., 2006; Pika et al., 2003). Thus, visual gestures seem to function more often as "intention movements" (Tinbergen, 1952), which are used for a specific purpose, and which represent abbreviations of full-fledged behaviors (Tomasello et al., 1989). Attention-getting gestures, on the other hand, which are not necessarily used to capture others' attention, but to trigger them into action, are mostly tactile and auditory gestures (Liebal and Call, 2012).

A study with captive bonobos and chimpanzees found that the majority of their gestures is not associated with specific contexts, while facial and vocal signals are more likely to be context-specific (Pollick and de Waal, 2007). The authors concluded that their findings support the hypothesis of a gestural origin of language, since in contrast to facial and vocal signals, gestures are less context-specific and thus less tied to the (involuntary) expression of emotions, which "...makes gesture a serious candidate modality to have acquired symbolic meaning in early hominins" (Pollick and de Waal, 2007, p. 8185).

6.3 Gesture combinations

In searching for the roots of human language, gesture researchers have been interested in gesture combinations (or sequences) for two different reasons. First, "duality of patterning" is considered one of the design features of language (Hockett, 1960) and refers to the combination of a finite number of meaningless sounds (phonemes) into meaningful units (morphemes and words), which can be further combined into an (unlimited) number of utterances (phrases, sentences). "Productivity" is another design feature of human language and describes the ability to create new meanings by combining already existing utterances (Hockett, 1960). Therefore, much attention has been paid to the question if primates combine their signals into consecutive, meaningful sequences, which might have a different meaning than their single components, and whether such combinations are based on specific, potentially grammatical rules. The second reason for studying gesture combinations is that "...the extent to which animals have the flexibility to move beyond single signal production to combine signals into sequences is highly indicative of the potential for complexity in the system." (Liebal and colleagues (2013b, p. 317). Thus, although primates are able to create at least some new gestures and incorporate them into their individual repertoires, gesture combinations offer the possibility to further increase the flexibility of a rather limited gestural repertoire, by combining single gestures into larger sequences.

Gestures sequences are commonly defined as multiple gestures produced one after the other by one individual, towards the same recipient and the same goal (Liebal et al., 2004a; Tomasello et al., 1994). Studies vary, however, regarding the time interval in which one gesture must follow another one to be considered part of a consecutive sequence. For example, some studies used a 1 second time interval between two gestures (Genty and Byrne, 2010; Hobaiter and Byrne, 2011b), while other studies used a 5 second interval (Liebal et al., 2004a; Tempelmann and Liebal, 2012), or even up to 30 seconds (Roberts et al., 2013). This is important, since the observed occurrence and patterns of gesture sequences may vary drastically because of these different definitions.

The few existing studies all focused on great apes, in both captive and natural settings. For captive chimpanzees, it has been found that about one third of their gestures occur as part of a sequence, with two thirds of sequences consisting of two consecutive gestures (Liebal et al., 2004a). Almost 40% of these sequences are repetitions of the very same gesture. Considering the potential meaning of these combinations, there is no evidence that sequences are used for different functions than their single components, suggesting that gestures are not combined in meaningful ways to create new meanings. Rather than representing planned sequences used to increase the efficacy of certain gestures (e.g., by first using an attention-getting gesture to direct the recipient's attention towards the following gesture), it seems likely that sequences merely emerge because the recipient does not – or not appropriately – respond to the other's communicative attempts (Liebal et al., 2004a; McCarthy et al., 2013).

Similarly, for wild chimpanzees, Roberts and colleagues (2013) reported that they repeat gestures and even elaborate gesture use in case the recipient does not respond as expected way. Hobaiter and Byrne (2011b) used a slightly different approach and found that sequences of wild chimpanzees represent a way to increase communicative efficiency. Thus, gesture use changes over development, with chimpanzees shifting from using long, often redundant "rapid-fire gesture sequences" to more selective and efficient gesture use. While "rapid-fire gesture sequences", which consist of many different gestures and are produced without any response-waiting, are largely used by younger individuals, "bouts" often occur if the recipient does not react. In this case, chimpanzees are more likely to repeat the same gesture and wait for the recipient's response (Hobaiter and Byrne, 2011b).

For gorillas, a different picture emerged (Genty and Byrne, 2010). Like captive chimpanzees, they combine about one third of their gestures into sequences, mostly consisting of two gestures. There is also no evidence that sequences convey different meanings than the single gestures they are composed of. However, in contrast to chimpanzees, sequences in gorillas do not emerge because of an unresponsive recipient. Instead, Genty and Byrne (2010) suggest that gesture sequences in gorillas, which occur largely in the play context, provide the means to adjust their ongoing interactions. Tanner (2004), in her qualitative analysis of gesture sequences exchanged between a zoo-living pair of gorillas, came to a different conclusion, because she did not observe that gestures were repeated if there was no response to the initial gesture. Instead, she found that gorillas incorporate iconic gestures into their sequences, which depict the requested motion or the destination of the requested action, like the "armswing under", which ended touching the area between the legs, which according to Tanner (2004) indicates the location for sexual play. Gorillas seem to use such sequences and specifically the incorporated iconic gestures to communicate "...decisions on when, where and how to play...", which "...evolve gradually in the course of interaction." (Tanner, 2004, p. 18).

Finally, captive Sumatran orang-utans apparently seem to not consider the recipient's behavior, since they continue to gesture regardless of the recipient's response (Tempelmann and Liebal, 2012). It has therefore been suggested that sequences in orang-utans do not emerge because of the lack of a response, but they might be rather the result of high arousal, because they largely occur in the play context (Tempelmann and Liebal, 2012).

Taken together, although several great ape species have been reported to combine gestures into longer sequences, existing research provides little evidence that they represent meaningful combinations. Thus, at least in the gestural modality, there is no evidence for productivity and most likely also not for duality of patterning. However, note that up to date, very little is known about meaning (and meaningful components) in primates' gestural communication (see following section). Furthermore, although at least some studies show that sequences do not simply emerge because of the recipient's lacking response, there is currently no evidence that gesture combinations are governed by specific rules. However, at least some studies demonstrate that gesture sequences are used to adjust communicative means to the recipient's behavior, by either repeating a gesture or by producing a different gesture to elicit an appropriate response, providing further evidence for the flexible usage of great apes' gestural repertoires.

6.4 Meaning and reference

Meaning is broadly defined as the information an individual intends to convey to another individual. If a signal can be linked to specific meanings, this is referred to as semanticity, which is yet another design feature of language (Hockett, 1960). This relationship between a word and its meaning is thought to be of arbitrary nature (Hockett, 1960), although some would argue that there is more iconicity present in language than it is currently assumed (Kendon, 2016). The question if primates also use iconic gestures has therefore attracted much research (Douglas and Moscovice, 2015; Genty and Zuberbühler, 2014; Perlman and Cain, 2014; Tanner and Byrne, 1996), and is discussed later in this chapter.

The use of such linguistic terms to describe animal communication systems has been repeatedly criticized (Font and Carazo, 2010; Rendall et al., 2009; Scott-Phillips, 2008). Some argue that the use of these linguistic terms might be misleading, since there might be no language-like meaning in primate communication systems (Rendall et al., 2009). Others highlight that it is important to differentiate between message (which is the information the sender encodes in the signal and intends to convey) and meaning (which the recipient decodes from the signal and the contextual information) (Font and Carazo, 2010). In gesture research, however, this differentiation between the sender's message and the derived meaning is usually not made.

Currently, there seem to be two "camps" regarding the question if gestures do have specific meanings or not. As repeatedly described in this chapter, many researchers highlight the flexible nature of gesture use, with one gesture used in different contexts and several different gestures used to achieve the same goal (for an overview, see Call and Tomasello, 2007; Liebal et al., 2013b). The conclusion from this approach is that gestures might not have tight, specific meanings and that the corresponding information a gesturer intends to convey might be determined by the context in which an interaction takes place. Therefore, the meaning of a specific gesture might vary depending on the context. For example, the "extend arm" gesture used by orang-utans can function as a submissive signal when used by a subordinate towards a higher-ranking individual. Alternatively, when used in interactions between mothers and their infants, it may serve as an invitation to follow (Liebal et al., 2010; Liebal et al., 2006). For chimpanzees, de Waal (2003, p. 22) reports that "the begging gesture (...) has absolutely no meaning unless one can deduce its referent from the context", since they either direct this gesture to food possessors, while in conflicts they direct it at bystanders to request support.

Other researchers conclude that it is possible to assign specific meanings to specific gesture types (Cartmill and Byrne, 2010; Hobaiter and Byrne, 2014). In their approach, intended meaning is an equivalent to linguistic meaning in humans. For example, Hobaiter and Byrne, (2014) derived the intended meaning of a gesture from the "apparently satisfactory outcome". This is used as an approximation of the gesture's meaning, in a way that the gesturer's goal is inferred from the satisfying outcome of a gesture, which resulted in the cessation of the interaction. Based on this approach, they found that wild chimpanzees use 10 of the previously identified 66 gesture types (Hobaiter and Byrne, 2011a) for one specific outcome, while other gesture types were used for two or three of the total of 19 possible outcomes (Hobaiter and Byrne, 2014). Furthermore, 13 gestures were used in at least 70% for one specific outcome and were therefore classified as gestures with "tight meaning", while 11 gestures types were used in only 50% up to 70% for one specific outcome and were therefore gestures with "loose meanings". Twelve gestures were classified as ambiguous signals. Thus, gesture types varied with regard to the number of potential outcomes as well as the frequency of use for one of those potential outcomes. Based on these findings, Hobaiter and Byrne (2014) conclude that most chimpanzee gestures can be assigned to one or more intended meanings. A very similar approach was used by Cartmill and Byrne (2010) to study the meaning of gestures in captive Sumatran orang-utans. They used 40 of their total of 64 gestures to achieve one of six different outcomes, and thereof 29 gestures seemed to have only one specific meaning. Because of this frequent matching between the goal of an interaction and the corresponding outcome, the authors concluded that despite the considerable degree of flexibility, there is evidence for intended meaning in at least some of the orang-utans' gesture types (Cartmill and Byrne, 2010).

Roberts and colleagues (2013), however, seem to combine the approaches of these two "camps", since they conclude that gestures of wild chimpanzees have specific meanings, independent of the context, but recipients are able to flexibly interpret others' gestures by additionally using contextual information (Roberts et al., 2012a; Roberts et al., 2013).

Based on the current evidence, the conclusion whether primate gestures have specific meanings or whether they are flexibly used for different functions seems to depend on varying perspectives on the very same data set. Thus, great apes use most of their gestures in more than one context, while some of them represent context-specific gestures. However, while some authors focus on those gestures used for several functions and consequently highlight their flexible use (Call and Tomasello, 2007), others focus on the gestures used for one or few outcomes, and consequently emphasize the specific meaning(s) of these gesture types (Cartmill and Byrne, 2010; Hobaiter and Byrne, 2014).

The question if gestures have specific meanings is tightly linked to the question if a gesture can refer to a specific referent. *Reference* is a fundamental component of language; however, this term is very differently used across disciplines (Leavens et al., 2005b). While linguists usually use "reference" synonymously with symbolic reference, developmental psychologists focus on nonverbal reference in the form of pointing gestures (Bates et al., 1987; Camaioni, 2001). Furthermore, linguists usually emphasize that the relationship between a word and its referent is arbitrary (but see Kendon, 2016), while developmental psychologists will argue that for pointing gestures this relationship is not arbitrary, since it is determined by the triadic (spatial) relationship between signaler, recipient, and the external entity (Leavens et al., 2005b).

This different operationalization of reference in spoken language and gestural communication in humans is also evident in comparative research into primate communication, since reference is defined differently across modalities. Vocal researchers often focus on context-specific vocalizations "... to find the animal equivalent to referential words in human language"

(Liebal et al., 2013b, p. 399) in the form of functionally referential vocalizations. They need to be reliably produced in response to a specific stimulus (Evans, 1997) and perceived independently of the context, which means that listeners should respond (appropriately) to this vocalization even in the absence of the eliciting stimulus, in the same way they would respond to the actual eliciting stimulus (Macedonia and Evans, 1993). Although some scholars challenge this traditional strong focus on functionally referential vocalizations (Wheeler and Fischer, 2012), it remains an intensively researched topic (Kalan et al., 2015; Murphy et al., 2013; Schel et al., 2013a).

In the gestural domain, pointing gestures are considered referential signals. However, in contrast to functionally referential vocalizations, pointing does not necessarily refer to specific entities and thus has no one-to-one referential meaning. Instead, the meaning of a pointing (or other referential) gesture can only be interpreted if the gesturer and the recipient share a common ground (Liebal et al., 2013a; Tomasello et al., 2007). The nature of nonhuman primates' pointing gestures, however, is fiercely debated (Leavens et al., 2005a; Tomasello, 2006). With very few exceptions (Hobaiter et al., 2013; Vea and Sabater-Pi, 1998), the production of pointing gestures is limited to interactions with humans, where primates point to request food rewards which they cannot obtain otherwise. Although it has been shown that great apes may also point to the location of a hidden tool so that the human is able to find the food (Call and Tomasello, 1994; Zimmermann et al., 2009), their underlying motivation is not necessarily to *inform* the human. Thus, with the exception of language-trained apes (Lyn et al., 2011; Menzel, 1999), primates do not use their pointing gestures declaratively, but imperatively, since they want the human to get the food for them. Finally, there is little evidence that primates understand pointing gestures of humans (Miklósi and Soproni, 2006) or even conspecifics (Tempelmann et al., 2013) to find hidden food in both cooperative and competitive settings. Together these findings suggest that even though primates produce pointing gestures, they are mostly limited to interactions with humans, they are used in very specific contexts, and seem to serve a different function than pointing gestures in humans.

Iconic gestures are considered a special type of referential signals, and have received special attention in theories of language evolution (Arbib, 2005; Armstrong and Wilcox, 2007; Perlman and Cain, 2014). Unlike vocalizations, gestures "...can provide more obvious iconic links with objects and actions in the physical world" (Gentilucci and Corballis, 2006, p. 950), because hands (and arms) can be used to perform actions or to represent objects, which closely resemble real actions (Cartmill et al., 2012).

The exact nature and definition of an iconic gesture, however, varies across studies or might overlap with terms like representative, pantomimic, and mimetic (Douglas and Moscovice, 2015; Perlman et al., 2014). In general, iconic gestures either depict a motion or the shape of a referent (Tanner and Byrne, 1996). One of the first studies reporting the use of iconic gestures by a single gorilla male was published by Tanner and Byrne (1996), who found that this male performed gestures indicating the path of motion either in space or on the female's body to request further actions, such as moving into a specific direction. Later studies on other groups of gorillas did not find additional evidence for iconic gestures in this species (Genty et al., 2009; Pika et al., 2003). More recently, however, several studies have been published suggesting the use of iconic gestures not only in gorillas (Perlman, Tanner, and King, 2012), but also in bonobos (Douglas and Moscovice, 2015; Genty and Zuberbühler, 2014). For example, captive bonobos use a series of gestures incorporating iconic gestures, such as beckoning gestures, to indicate another individual to approach and then to jointly move to the indicated location to mate there (Genty and Zuberbühler, 2014). Some studies even claim that orang-utans and bonobos use pantomime (as a type of iconic gesturing) in a way that the corresponding action (referent) is re-enacted (Douglas and Moscovice, 2015; Russon and Andrews, 2011).

Taken together, different criteria are used to define reference across modalities. In gesture research, most studies have focused on primates' production and comprehension of pointing gestures, mostly in interactions with humans. There is currently little evidence that primates produce pointing gestures in interactions with conspecifics, and that they understand others' pointing gestures as referential means of communication. Furthermore, our current knowledge about iconic gesture use in nonhuman primates is still very limited, so that any conclusions regarding a gestural origin of human language seem to be too premature.

6.5 Acquisition of gestures

How primates acquire their gestures during ontogeny is an interesting topic to study, because this provides inside into the flexibility and "openness" of this communicative system. Different mechanisms have been suggested, each leading to different predictions regarding how gestures might be acquired (Genty et al., 2009; Schneider, Call, and Liebal, 2012b). For example, if gestures are genetically determined, we would expect that all individuals of a given species use rather similar repertoires, with little variability across individuals. This would also mean that individuals do not learn or create new gestures to incorporate them into their repertoire, resulting in relatively "closed", species-specific repertoires (Genty et al., 2009; Hobaiter and Byrne, 2011a). It is important to point out, however, that the genetic determination of gestures would not necessarily mean that structural properties of gestures or their use could not be modified over an individual's lifetime. Furthermore, genetic transmission does also not imply that all gestures have to be present right after birth (Rosati et al., 2014). On the other hand, if there is much variability of gestural repertoires across individuals, and if the number of gestures increases over lifetime, with new, mostly idiosyncratic elements being added to the existing repertoire, then gestures are most likely acquired by some other mechanism than genetic transmission (although a genetic component can still not be excluded) (Liebal et al., submitted).

Two scenarios have been proposed. First, gestures could be acquired by some form of social learning, with one individual acquiring parts of the behavioral repertoire of another individual (Whiten and Ham, 1992). In this context, imitation is discussed as one potential mechanism, which would result in very similar individual repertoires within a group, since individuals learn their gestures from each other, while across groups, repertoires are expected to vary (Tomasello et al., 1997). Second, if ontogenetic ritualization is the major mechanism underlying gesture acquisition, which involves the shaping of previously non-communicative behaviors into increasingly ritualized, communicative gestures (Call and Tomasello, 2007; Tomasello, 2008), we would expect a high degree of variability of individual repertoires within and across groups.

Only very few studies investigated how primates acquire their gestures, almost exclusively in great apes, in both natural (Fröhlich et al., 2016a, 2016b), and captive settings (Bard et al., 2014; Schneider et al., 2012a), resulting in rather mixed findings (Liebal et al., submitted). While some studies conclude that gestures of great apes are most likely genetically determined, because even across species, individuals shared very similar repertoires (Genty et al., 2009; Hobaiter and Byrne, 2011a; Schneider et al., 2012b), others conclude that gestures emerge in social interactions (Bard et al., 2014; Fröhlich et al., 2016b; Halina et al., 2013). Across existing studies, however, imitation does not seem to play any significant role in gesture acquisition.

There are a couple of reasons for these inconsistent findings. The logistics underlying data collection are often challenging, since it is difficult to observe several individuals, who are often not in the same group or zoo, over longer periods of time. As a result, sample sizes are very small. Observational periods are often too short to capture the individuals' complete repertoires, resulting in an overestimation of variability between individual repertoires. Furthermore, researchers use different criteria for defining a gesture as well as intentionality (see sections 2 and 3), and as a result, the number of identified gestures might vary drastically between studies, which complicates comparisons between studies of different research groups. Therefore, it is difficult to use the current evidence to either support or reject a gestural origin of human language.

7. What do primate gestures tell us about language evolution?

This chapter provided an overview of those aspects of primate gestural communication, which are frequently studied with the aim to find evidence for a gestural origin of human language. The current knowledge demonstrates that primates use their gestures intentionally and flexibly across a variety of different contexts, and that they adjust them to the recipient's behavior (Call and Tomasello, 2007). It is important to note, however, that the majority of gesture research focuses on great apes, often in captive settings (Slocombe et al., 2011). Furthermore, studies vary regarding the kind and number of criteria applied as markers for intentional use and they also operationalize flexibility in different ways. Still, it seems that primates' gestural communication is characterized by a high degree of flexibility in gesture production and usage, which seems to support a gestural origin of human language.

The evidence for rule-governed, meaningful combinations and referentiality – either in the form of pointing gestures or iconic gestures resembling actions or objects – is rather mixed. This also concerns mechanisms of gesture acquisition, which have been found to vary drastically across studies. Differences between species and/or studies are most likely explained by varying definitions and methodological approaches, with one of the most serious shortcomings representing the lack of shared definitions – such as for the term "gesture" – or the reliance on linguistic terms. Based on the current evidence, it seems rather difficult to draw any substantial conclusions regarding the potential origin(s) of human language.

What can we conclude from this current situation? Is it correct that "… studies of nonhuman animals provide virtually no relevant parallels to human linguistic communication … thus leaving any insights into language's origins unverifiable" (Hauser et al., 2014, p. 1)? This strong conclusion might not be justified – even if the current evidence is not sufficient to validate a gestural origin of human language, comparative research across different modalities convincingly showed that it is possible to identify building blocks to human language in the communication of other primates, rejecting the notion that language evolved from scratch in humans only. Furthermore, there are currently promising developments in the field of primate gesture research, which will – at the very least – broaden our knowledge about this modality, and in the best case, shed light on the origins of human language. The final section of this chapter will therefore summarize some recent developments as well as "blind spots" in primate gesture research.

First, there is an increase of studies investigating gestural communication of primates in their natural habitats (Douglas and Moscovice, 2015; Fröhlich et al., 2016a; Hobaiter and Byrne, 2011a; Pika and Mitani, 2006; Roberts et al., 2013). This addresses the challenge that most of our current knowledge about primate gestural communication is based on captive studies, therefore potentially lacking ecological validity. However, systematic studies are needed, which systematically compare gesture use in natural and captive settings, to estimate the impact of the environment on primates' repertoires and gesture usage.

Second, since flexibility and the use for different functions is an important aspect in gesture research, context-specific gestures, which are used to achieve one specific goal, have been largely neglected or might have been even excluded from further analyses. Consequently, it is possible that potentially referential gestures – which are linked to specific events, individuals, or objects, but are not used for more than one social function – have not been discovered yet.

Third, almost all existing gesture studies consider proximate aspects, such as the underlying cognitive mechanisms or developmental pathways of gestural communication. There are, however, virtually no studies which investigate ultimate aspects of gesture use, such as whether the use of specific gestures in "evolutionary urgent contexts" like consortships in the sexual context (Hobaiter and Byrne, 2012), has specific benefits for the gesturer, for example, in the form of increased numbers of mating partners. If we have a better understanding of the gestures' effect on the recipient, then we are better able to investigate the evolutionary function of gestures in terms of their costs and benefits. This, in turn, might enable us to develop more sophisticated gestural scenarios of language evolution, specifically with regard to the adaptive functions of gestural communication in the last common shared ancestor of humans and great apes.

Fourth, in comparative gesture research, gesture types are usually defined based on their function (e.g., 'food beg') rather than their structural properties (Roberts et al., 2012b). Therefore, it is currently unknown whether changes in these structural properties (e.g., arm extended with hand stretched versus arm extended with hand bent) result in a differential use – and thus potentially different meanings of these gestures.

Fifth, it is important to identify and differentiate the many different "components" of language and the abilities related to its use, to carefully select which aspects of primate gestural communication should be investigated and to decide which comparisons across species are meaningful at all (Fitch, 2005, 2010; Hauser et al., 2002). For example, although the criteria for the definition of intentionally used gestures are adapted from research into gestures of pre-linguistic children (Leavens et al., 2005b), there are very few studies which systematically compare gesture use in nonhuman primates and young children (Bohn et al., 2015; Liszkowski et al., 2012). In her review, Pika (2008) concludes that the majority of primate gestures are dyadic gestures, used to imperatively request immediate actions from the recipient, while children start to use triadic gestures from an early age on, which they also use declaratively to communicate about external entities outside the dyad, such as objects or events. Furthermore, many gesture types of primates closely resemble actions (Liebal and Call, 2012), while children from an early age on also use more abstract, ritualized gestures (e.g., 'wave good bye', head shakes for "no"). Therefore, it seems that the gestural communication of nonhuman primates is fundamentally different from gesture use in pre-linguistic children. However, because of the lack of systematic comparisons of the *same* gestural properties in nonhuman primates and humans, it seems difficult to conclude which aspects of their communication are informative with regard to hypotheses of language origins.

Finally, the review by Slocombe et al. (2011) showed that only few studies on primate communication investigate more than one modality at a time. However, given that language is multimodal (Kendon, 2004; McNeill, 1992), it seems unlikely that language evolved *either* from vocalizations *or* gestures *or* facial movements (Arbib et al., 2008). Therefore, recent publications suggest a multimodal language origin (Wacewicz and Zywiczynski, in press), with special attention paid to the role of orofacial movements. Thus, unlike facial expressions, orofacial movements seem to be voluntarily produced, because nonhuman primates have direct connections between the premotor cortex and the different nuclei controlling the jaws, lips and tongue (Jürgens, 2002). The importance of a multimodal approach to primate communication (Liebal et al., 2013b; Slocombe et al., 2011) is now increasingly acknowledged, since there is an increasing number of studies which investigate different modalities in a more integrated way (Ghazanfar and Logothetis, 2003; Habbershon et al., 2013; Leavens and Hopkins, 2005; Micheletta et al., 2013; Partan, 1999; Taglialatela et al., 2015; Taglialatela et al., 2011; Wilke et al., 2017). Although some questions and certainly methods might be modality-specific, a multimodal approach is essential to capture the complexity of primate communication and to use this knowledge to solve the "mystery" of language evolution.

Acknowledgements

Very special thanks go to Katie E. Slocombe, Bridget M. Waller and Anne M. Burrows for contributing their valuable expertise on "their" modality and for our fruitful, yet fierce discussions about different perspectives on primate communication. I also thank Manuela Lembeck for her help with the editing of the references and two anonymous reviewers for their helpful comments on earlier versions of this manuscript.

References

Anderson, J.R., Kuroshima, H., Hattori, Y., and Fujita, K. (2010). Flexibility in the use of requesting gestures in squirrel monkeys (*Saimiri sciureus*). *American Journal of Primatology*, 72, 707–714. doi: 10.1002/ajp.20827.

Anderson, J.R., Kuwahata, H., and Fujita, K. (2007). Gaze alternation during 'pointing' by squirrel monkeys (*Saimiri sciureus*)? *Animal Cognition, 10(2)*, 267–271. doi: 10.1007/s10071–006–0065-0.

Anderson, J.R., Montant, M., and Schmitt, D. (1996). Rhesus monkeys fail to use gaze direction as an experimenter-given cue in an object-choice task. *Behavioural Processes, 37(1)*, 47–55. doi: 10.1016/0376–6357(95)00074–7.

Arbib, M.A. (2005). From monkey-like action recognition to human language: An evolutionary framework for neurolinguistics. *Behavioral and Brain Sciences, 28(2)*, 105–124. doi: 10.1017/S0140525X05000038.

Arbib, M.A. (2016). Towards a Computational Comparative Neuroprimatology: Framing the language-ready brain. *Physics of Life Reviews, 16*, 1–54. doi: 10.1016/j.plrev.2015.09.003.

Arbib, M.A., Liebal, K., and Pika, S. (2008). Primate vocalization, gesture, and the evolution of human language. *Current Anthropology, 49(6)*, 1053–1076. doi: 10.1086/593015.

Armstrong, D.F., Stokoe, W.C., and Wilcox, S.E. (1995). *Gesture and the Nature of Language*. Cambridge: Cambridge University Press.

Armstrong, D.F., and Wilcox, S.E. (2007). *The Gestural Origin of Language*. New York & Oxford: Oxford University Press.

Bard, K.A., Dunbar, S., Maguire-Herring, V., Veira, Y., Hayes, K.G., and McDonald, K. (2014). Gestures and social-emotional communicative development in chimpanzee infants. *American Journal of Primatology, 76(1)*, 14–29. doi: 10.1002/ajp.22189.

Barth, J., Reaux, J.E., and Povinelli, D.J. (2005). Chimpanzees' (*Pan troglodytes*) use of gaze cues in object-choice tasks: Different methods yield different results. *Animal Cognition, 8(2)*, 84–92. doi: 10.1007/s10071-004–0235-x.

Bates, E., Benigni, L., Bretherton, I., Camaioni, L., and Volterra, V. (1979). *The Emergence of Symbols: Cognition and Communication in Infancy*. New York: Academic Press.

Bates, E., O'Connell, B., and Shore, C. (1987). Language and communication in infancy. In J. D. Osofsky (Ed.), *Handbook of Infant Development* (pp. 149–203). New York:: Wiley.

Benga, O. (2005). Intentional communication and the anterior cingulate cortex. *Interaction Studies, 6(2)*, 201–221. doi: 10.1075/is.6.2.04ben.

Bickerton, D. (1992). *Language and Species*. Chicago & London: The University of Chicago Press.

Blaschke, M., and Ettlinger, G. (1987). Pointing as an act of social communication by monkeys. *Animal Behaviour, 35(5)*, 1520–1523. doi: 10.1016/S0003-3472(87)80023-4.

Bohn, M., Call, J., and Tomasello, M. (2015). Communication about absent entities in great apes and human infants. *Cognition, 145*, 63–72. doi: 10.1016/j.cognition.2015.08.009.

Bourjade, M., Meguerditchian, A., Maille, A., Gaunet, F., and Vauclair, J. (2014). Olive baboons, *Papio anubis*, adjust their visual and auditory intentional gestures to the visual attention of others. *Animal Behaviour, 87*, 121–128. doi: 10.1016/j.anbehav.2013.10.019.

Bruner, J.S. (1981). Intention in the structure of action and interaction. In L.P. Lipsitt (Ed.), *Advances in Infancy Research* (Vol. 1, pp. 41–56). Norwood New Jersey: Ablex.

Burling, R. (1993). Primate calls, human language and nonverbal communication. *Current Anthropology, 34(1)*, 25–53. doi: 10.1086/204132.

Call, J., Agnetta, B., and Tomasello, M. (2000). Cues that chimpanzees do and do not use to find hidden objects. *Animal Cognition, 3(1)*, 23–34. doi: 10.1007/s100710050047.

Call, J., and Tomasello, M. (1994). The production and comprehension of referential pointing by orangutans (*Pongo pygmaeus*). *Journal of Comparative Psychology, 108(4)*, 307–317. doi: 10.1037/0735-7036.108.4.307.

Call, J., and Tomasello, M. (Eds.). (2007). *The gestural communication of apes and monkeys*. New York: Lawrence Erlbaum Associates.

Camaioni, L. (2001). Early language. In G. Bremner and A. Fogel (Eds.), *Blackwell Handbook of Infant Development* (pp. 404–426). Oxford, England: Blackwell.

Carpenter, C.R. (Ed.). (1940). *A field study in Siam of the behavior and social relations of the gibbon (Hylobates lar)* (Vol. 16). Baltimore, Maryland: The John Hopkins Press.

Cartmill, E.A., Beilock, S., and Goldin-Meadow, S. (2012). A word in the hand: Action, gesture and mental representation in humans and nonhuman primates. *Philosophical Transactions of the Royal Society B: Biological Sciences, 367(1585)*, 129–143. doi: 10.1098/rstb.2011.0162.

Cartmill, E.A., and Byrne, R.W. (2007). Orangutans modify their gestural signaling according to their audience's comprehension. *Current Biology, 17(15)*, 1345–1348. doi: 10.1016/j.cub.2007.06.069

Cartmill, E.A., and Byrne, R.W. (2010). Semantics of primate gestures: intentional meanings of orangutan gestures. *Animal Cognition, 13(6)*, 793–804. doi: 10.1007/s10071-010–0328-7.

Chomsky, N. (1966). *Cartesian Linguistics: A Chapter in the History of Rationalist Thought*. New York: Harper & Row.

Corballis, M.C. (2002). From hand to mouth: The origins of language. In M.H. Christiansen and S. Kirby (Eds.), *Language Evolution* (pp. 201–218). Princeton, New Jersey: Princeton University Press.

Crockford, C., Wittig, R.M., Mundry, R., and Zuberbühler, K. (2012). Wild chimpanzees inform ignorant group members of danger. *Current Biology, 22(2)*, 142–146. doi: 10.1016/j.cub.2011.11.053.

de Waal, F.B.M. (1988). The communicative repertoire of captive bonobos *(Pan paniscus)* compared to that of chimpanzees. *Behaviour, 106(3)*, 183–251. doi: 10.1163/156853988X00269.

de Waal, F.B.M. (2003). Darwin's legacy and the study of primate visual communication. *Annals of the New York Academy of Sciences, 1000*, 7–31. doi: 10.1196/annals.1280.003.

Douglas, P.H., and Moscovice, L.R. (2015). Pointing and pantomime in wild apes? Female bonobos use referential and iconic gestures to request genito-genital rubbing. *Scientific Reports, 5(13999)*. doi: 10.1038/srep13999.

Evans, C.S. (1997). Referential signals. In D.H. Owings, M.D. Beecher & N.S. Thompson (Eds.), *Perspectives in Ethology* (Vol. 12: Communication, pp. 99–143). New York & London: Plenum Press.

Fitch, W.T. (2005). The evolution of language: A comparative review. *Biology and Philosophy, 20(2)*, 193–203. doi: 10.1007/s10539-005–5597-1.

Fitch, W.T. (2010). *The Evolution of Language*. Cambridge: Cambridge University Press.

Fitch, W.T., de Boer, B., Mathur, N., and Ghazanfar, A.A. (2016). Monkey vocal tracts are speech-ready. *Science Advances, 2(12)*, e1600723. doi: 10.1126/sciadv.1600723.

Font, E., and Carazo, P. (2010). Animals in translation: Why there is meaning (but probably no message) in animal communication. *Animal Behaviour, 80(2)*, e1–e6. doi: 10.1016/j.anbehav.2010.05.015.

Fröhlich, M., Wittig, R.M., and Pika, S. (2016a). Play-solicitation gestures in chimpanzees in the wild: Flexible adjustment to social circumstances and individual matrices. *Royal Society Open Science, 3(8)*, 160278. doi: 10.1098/rsos.160278.

Fröhlich, M., Wittig, R.M., and Pika, S. (2016b). Should I stay or should I go? Initiation of joint travel in mother–infant dyads of two chimpanzee communities in the wild. *Animal Cognition, 19(3)*, 483–500. doi: 10.1007/s10071-015–0948-z.

Gentilucci, M., and Corballis, M.C. (2006). From manual gesture to speech: A gradual transition. *Neuroscience & Biobehavioral Reviews, 30(7)*, 949–960.

Genty, E., Breuer, T., Hobaiter, C., and Byrne, R.W. (2009). Gestural communication of the gorilla (*Gorilla gorilla*): Repertoire, intentionality and possible origins. *Animal Cognition, 12(3)*, 527–546. doi: 10.1007/s10071-009–0213-4.

Genty, E., and Byrne, R.W. (2010). Why do gorillas make sequences of gestures? *Animal Cognition, 13(2)*, 287–301. doi: 10.1007/s10071-009–0266-4.

Genty, E., and Zuberbühler, K. (2014). Spatial reference in a bonobo gesture. *Current Biology, 24(14)*, 1601–1605. doi: 10.1016/j.cub.2014.05.065.

Ghazanfar, A.A., and Logothetis, N.K. (2003). Facial expressions linked to monkey calls. *Nature, 423(6943)*, 937–938. doi: 10.1038/423937a.

Goodall, J. (1986). *The chimpanzees of Gombe: Patterns of behavior*. Cambridge, MA: Harvard University Press.

Habbershon, H.M., Ahmed, S.Z., and Cohen, Y.E. (2013). Rhesus macaques recognize unique multimodal face-voice relations of familiar individuals and not of unfamiliar ones. *Brain, behavior and evolution, 81(4)*, 219–225. doi: 10.1159/000351203.

Halina, M., Rossano, F., and Tomasello, M. (2013). The ontogenetic ritualization of bonobo gestures. *Animal Cognition, 16(4)*, 653–666. doi: 10.1007/s10071-013–0601-7.

Hare, B., and Tomasello, M. (2004). Chimpanzees are more skilful in competitive than in cooperative cognitive tasks. *Animal Behaviour, 68(3)*, 571–581. doi: 10.1016/j.anbehav.2003.11.011.

Hattori, Y., Kano, F., and Tomonaga, M. (2010). Differential sensitivity to conspecific and allospecific cues in chimpanzees and humans: A comparative eye-tracking study. *Biology Letters, 6(5)*, 610–613. doi: 10.1098/rsbl.2010.0120.

Hauser, M.D., Chomsky, N., and Fitch, T.W. (2002). The Language Faculty: What is it, who has it, and how did it evolve? *Science, 298*, 1568–1579.

Hauser, M.D., Yang, C., Berwick, R.C., Tattersall, I., Ryan, M.J., Watumull, J., Chomsky, N., and Lewontin, R.C. (2014). The mystery of language evolution. *Frontiers in Psychology, 5(401)*, 1–12. doi: 10.3389/fpsyg.2014.00401

Hewes, G.W. (1973). Primate communication and the gestural origin of language. *Current Anthropology, 12(1–2)*, 5–24. doi: 10.1086/204019.

Hobaiter, C., and Byrne, R.W. (2011a). The gestural repertoire of the wild chimpanzee. *Animal Cognition, 14(5)*, 745–767. doi: 10.1007/s10071-011–0409-2.

Hobaiter, C., and Byrne, R.W. (2011b). Serial gesturing by wild chimpanzees: its nature and function for communication. *Animal Cognition, 14(6)*, 827–838. doi: 10.1007/s10071-011–0416-3.

Hobaiter, C., and Byrne, R.W. (2012). Gesture use in consortship: Wild chimpanzees' use of gesture for an 'evolutionarily urgent' purpose. In S. Pika and K. Liebal (Eds.), *Developments in Primate Gesture Research* (pp. 129–146). Amsterdam: John Benjamins Publishing.

Hobaiter, C., and Byrne, R.W. (2014). The meanings of chimpanzee gestures. *Current Biology, 24(14)*, 1596–1600. doi: 10.1016/j.cub.2014.05.066.

Hobaiter, C., Leavens, D.A., and Byrne, R.W. (2013). Deictic gesturing in wild chimpanzees, (*Pan troglodytes*)? Some possible cases. *Journal of Comparative Psychology, 128(1)*, 82–87. doi: 10.1037/a0033757.

Hockett, C.F. (1960). The origin of speech. *Scientific American, 203(3)*, 89–96.

Hostetter, A.B., Cantero, M., and Hopkins, W.D. (2001). Differential use of vocal and gestural communication by chimpanzees (*Pan troglodytes*) in response to the attentional status of a human (*Homo sapiens*). *Journal of Comparative Psychology, 115(4)*, 337–343. doi: 10.1037/0735–7036.115.4.337.

Jürgens, U. (2002). Neural pathways underlying vocal control. *Neuroscience and Biobehavioral Reviews, 26(2)*, 235–258.

Kalan, A.K., Mundry, R., and Boesch, C. (2015). Wild chimpanzees modify food call structure with respect to tree size for a particular fruit species. *Animal Behaviour, 101*, 1–9. doi: 10.1016/j.anbehav.2014.12.011.

Kalan, A.K., and Rainey, H.J. (2009). Hand-clapping as a communicative gesture by wild female swamp gorillas. *Primates, 50(3)*, 273–275. doi: 10.1007/s10329-009–0130-9.

Kendon, A. (2004). *Gesture: Visible Action.* Cambridge: Cambridge University Press.

Kendon, A. (2016). Reflections on the "gesture-first" hypothesis of language origins. *Psychonomic Bulletin & Review, 24(1)*, 163–170. doi: 10.3758/s13423-016–1117-3.

King, B.J. (1999). *The Origins of Language: What Nonhuman Primates Can Tell us.* Santa Fe: School of American Research Press.

Kumashiro, M., Ishibashi, H., Itakura, S., and Iriki, A. (2002). Bidirectional communication between a Japanese monkey and a human through eye gaze and pointing. *Current Psychology of Cognition, 21(1)*, 3–32.

Kummer, H. (1968). *Social organization of hamadryas baboons. A field study.* Basel: Karger.

Kuroda, S.J. (1980). Social behavior of the pygmy chimpanzees. *Primates, 21(2)*, 181–197. doi: 10.1007/BF02374032.

Leavens, D.A., and Hopkins, W.D. (1998). Intentional communication by chimpanzees (*Pan troglodytes*): A cross-sectional study of the use of referential gestures. *Developmental Psychology, 34(5)*, 813–822. doi: 10.1037/0012–1649.34.5.813.

Leavens, D.A., and Hopkins, W.D. (1999). The whole hand-point: The structure and function of pointing from a comparative perspective. *Journal of Comparative Psychology, 113(4)*, 417–425. doi: 10.1037/0735–7036.113.4.417.

Leavens, D.A., and Hopkins, W.D. (2005). Multimodal concomitants of manual gesture by chimpanzees (Pan troglodytes): Influence of food size and distance. *Gesture, 5(1)*, 75–90. doi: 10.1075/gest.5.1.07lea.

Leavens, D.A., Hopkins, W.D., and Bard, K.A. (1996). Indexical and referential pointing in chimpanzees (*Pan troglodytes*). *Journal of Comparative Psychology, 110(4)*, 346–353.

Leavens, D.A., Hopkins, W.D., and Bard, K.A. (2005a). Understanding the point of chimpanzee pointing. *Current Directions in Psychological Science, 14(4)*, 185–189. doi: 10.1111/j.0963–7214.2005.00361.x.

Leavens, D.A., Hostetter, A.B., Weasley, M.J., and Hopkins, W.D. (2004). Tactical use of unimodal and bimodal communication by chimpanzees, *Pan troglodytes. Animal Behavior, 67(3)*, 467–476. doi: 10.1016/j.anbehav.2003.04.007.

Leavens, D.A., Russell, J.L., and Hopkins, W.D. (2005b). Intentionality as measured in the persistence and elaboration of communication by chimpanzees (*Pan troglodytes*). *Child Development, 76(1)*, 291–306. doi: 10.1111/j.1467–8624.2005.00845.x

Liebal, K., Bressem, J., and Müller, C. (2010). *Recurrent forms and contexts: Families of gestures in non-human primates.* Paper presented at the 4th International Society of Gesture Studies (ISGS), Frankfurt/ Oder, Germany.

Liebal, K., and Call, J. (2012). The origins of non-human primates' manual gestures. *Philosophical Transactions of the Royal Society of London B: Biological Sciences, 367(1585)*, 118–128. doi: 10.1098/rstb.2011.0044.

Liebal, K., Call, J., and Tomasello, M. (2004a). Use of gesture sequences in chimpanzees (*Pan troglodytes*). *American Journal of Primatology, 64(4)*, 377–396. doi: 10.1002/ajp.20087.

Liebal, K., Carpenter, M., and Tomasello, M. (2013a). Young children's understanding of cultural common ground. *British Journal of Developmental Psychology, 31(1)*, 88–96. doi: 10.1111/j.2044–835X.2012.02080.x.

Liebal, K., Pika, S., Call, J., and Tomasello, M. (2004b). To move or not to move: How apes adjust to the attentional state of others. *Interaction Studies,* 5(2), 199–219. doi: 10.1075/is.5.2.03lie.

Liebal, K., Pika, S., and Tomasello, M. (2004c). Social communication in siamangs (*Symphalangus syndactylus*): Use of gestures and facial expressions. *Primates, 45(1)*, 41–57. doi: 10.1007/s10329-003-0063-7.

Liebal, K., Pika, S., and Tomasello, M. (2006). Gestural communication of orangutans (*Pongo pygmaeus*). *Gesture, 6(1)*, 1–38. doi: 10.1075/gest.6.1.02lie.

Liebal, K., Schneider, C., and Lembeck, M. (submitted). How primates acquire their gestures: Current debates and the way forward. *Animal Cognition.*

Liebal, K., Waller, B.M., Burrows, A.M., and Slocombe, K.E. (2013b). *Primate Communication: A Multimodal Approach*. Cambridge: Cambridge University Press.

Liszkowski, U., Brown, P., Callaghan, T., Takada, A., and de Vos, C. (2012). A prelinguistic gestural universal of human communication. *Cognitive Science, 36(4)*, 698–713. doi: 10.1111/j.1551-6709.2011.01228.x.

Lyn, H., Greenfield, P.M., Savage-Rumbaugh, S., Gillespie-Lynch, K., and Hopkins, W.D. (2011). Nonhuman primates do declare! A comparison of declarative symbol and gesture use in two children, two bonobos, and a chimpanzee. *Language & Communication, 31(1)*, 63–74. doi: 10.1016/j.langcom.2010.11.001.

Macedonia, J.M., and Evans, C.S. (1993). Essay on contemporary issues in ethology: variation among mammalian alarm call systems and the problem of meaning in animal signals. *Ethology, 93(3)*, 177–197. doi: 10.1111/j.1439-0310.1993.tb00988.x.

MacKinnon. J. (1974). The behaviour and ecology of wild orang-utans (*Pongo pygmaeus*). *Animal Behaviour, 22(1)*, 3–74. doi: 10.1016/S0003-3472(74)80054-0.

Maestripieri, D. (1996a). Gestural communication and its cognitive implications in pigtail macaques (*Macaca nemestrina*). *Behaviour, 133(13–14)*, 997–1022. doi: 10.1163/156853996X00576.

Maestripieri, D. (1996b). Social communication among captive stump-tailed macaques (*Macaca arctoides*). *International Journal of Primatology, 17(5)*, 785–802. doi: 10.1007/BF02735264.

Maestripieri, D. (1997). Gestural communication in macaques: Usage and meaning of nonvocal signals. *Evolution of Communication, 1(2)*, 193–222. doi: 10.1075/eoc.1.2.03mae.

Maestripieri, D. (1999). The biology of human parenting: insights from nonhuman primates. *Neuroscience & Biobehavioral Reviews, 23(3)*, 411–422.

Maestripieri, D. (2005). Early experience affects the intergenerational transmission of infant abuse in rhesus monkeys. *Proceedings of the National Academy of Sciences of the United States of America, 102(27)*, 9726–9729.

Maestripieri, D., and Wallen, K. (1997). Affiliative and submissive communication in rhesus macaques. *Primates, 38(2)*, 127–138. doi: 10.1007/BF02382003.

Maille, A., Engelhart, L., Bourjade, M., and Blois-Heulin, C. (2012). To beg, or not to beg? That is the question: Mangabeys modify their production of requesting gestures in response to human's attentional states. *PLoS One, 7(7)*, e41197. doi: 10.1371/journal.pone.0041197.

Maynard-Smith, J., and Harper, D. (2003). *Animal Signals*. New York: Oxford University Press.

McCarthy, M.S., Jensvold, M.L.A., and Fouts, D.H. (2013). Use of gesture sequences in captive chimpanzee (*Pan troglodytes*) play. *Animal Cognition, 16(3)*, 471–481. doi: 10.1007/s10071-012–0587-6.

McNeill, D. (1992). *Hand and Mind: What Gestures Reveal About Thought*. Chicago: The University of Chicago Press.

Meguerditchian, A., and Vauclair, J. (2006). Baboons communicate with their right hand. *Behavioural Brain Research, 171*, 170–174. doi: 10.1016/j.bbr.2006.03.018.

Menzel, C.R. (1999). Unprompted recall and reporting of hidden objects by a chimpanzee (*Pan troglodytes*) after extended delays. *Journal of Comparative Psychology, 113(4)*, 426–434. doi: 10.1037/0735–7036.113.4.426.

Micheletta, J., Engelhardt, A., Matthews, L., Agil, M., and Waller, B.M. (2013). Multicomponent and multimodal lipsmacking in Crested macaques (*Macaca nigra*). *American Journal of Primatology, 75(7)*, 763–773. doi: 10.1002/ajp.22105.

Miklósi, Á., and Soproni, K. (2006). A comparative analysis of animals' understanding of the human pointing gesture. *Animal Cognition*, 9(2), 81–93. doi: 10.1007/s10071-005–0008-1.

Mitchell, R.W., and Anderson, J.R. (1997). Pointing, withholding information, and deception in capuchin monkeys (*Cebus apel-*

la). *Journal of Comparative Psychology, 111(4)*, 351–361. doi: 10.1037/0735-7036.111.4.351.

Mulcahy, N.J., and Hedge, V. (2012). Are great apes tested with an abject object-choice task? *Animal Behaviour, 83(2)*, 313–321. doi: 10.1016/j.anbehav.2011.11.019.

Murphy, D., Lea, S.E., and Zuberbühler, K. (2013). Male blue monkey alarm calls encode predator type and distance. *Animal Behaviour, 85(1)*, 119–125. doi: 10.1016/j.anbehav.2012.10.015.

Ouattara, K., Lemasson, A., and Zuberbühler, K. (2009). Campbell's monkeys concatenate vocalizations into context-specific call sequences. *Proceedings of the National Academy of Sciences, 106(51)*, 22026–22031. doi: 10.1073/pnas.0908118106.

Owren, M.J., Dieter, J.A., Seyfarth, R.M., and Cheney, D.L. (1992). Evidence of limited modification in the vocalizations of cross-fostered rhesus (*Macaca mulatta*) and Japanese (*M. fuscata*) macaques. *Developmental Psychobiology, 26(7)*, 257–270.

Partan, S.R. (1999). *Multimodal communication: The integration of visual and acoustic signals by macaques*. University of California, Davis, CA.

Perlman, M., and Cain, A.A. (2014). Iconicity in vocalization, comparisons with gesture, and implications for theories on the evolution of language. *Gesture, 14(3)*, 320–350. doi: 10.1075/gest.14.3.03per.

Perlman, M., Clark, N., and Tanner, J.E. (2014). Iconicity and ape gesture. In E. Cartmill, H. Lyn, & H. Cornish (Eds.), (Ed.), *The Evolution of Language. Proceedings of the 10th International Conference (EVOLANG 10)* (pp. 228–235). Singapore: World Scientific.

Perlman, M., Tanner, J.E., and King, B.J. (2012). A mother gorilla's variable use of touch to guide her infant: Insights into iconicity and the relationship between gesture and action. In S. Pika and K. Liebal (Eds.), *Developments in Primate Gesture Research* (pp. 55–72). Amsterdam/Philadelphia:: John Benjamins Publishing Company.

Pika, S. (2008). Gestures of apes and pre-linguistic human children: Similar or different? *First Language, 28(2)*, 116–140. doi: 10.1177/0142723707080966.

Pika, S., Liebal, K., and Tomasello, M. (2003). Gestural communication in young gorillas (*Gorilla gorilla*): Gestural repertoire, learning, and use. *American Journal of Primatology, 60(3)*, 95–111. doi: 10.1002/ajp.10097.

Pika, S., Liebal, K., and Tomasello, M. (2005). Gestural communication in subadult bonobos (*Pan paniscus*): Repertoire and use. *American Journal of Primatology, 65(1)*, 39–61. doi: 10.1002/ajp.20096.

Pika, S., and Mitani, J. (2006). Referential gestural communication in wild chimpanzees (*Pan troglodytes*). *Current Biology, 16(6)*, 191–192. doi: 10.1016/j.cub.2006.02.037.

Pinker, S., and Bloom, P. (1990). Natural selection and natural language. *Behavioral and Brain Sciences, 13(4)*, 707–784. doi: 10.1017/S0140525X00081061.

Pollick, A.S., and de Waal, F.B.M. (2007). Ape gestures and language evolution. *Proceedings of the National Academy of Sciences, 104(19)*, 8184–8189. doi: 10.1073/pnas.0702624104.

Poss, S.R., Kuhar, C., Stoinski, T.S., and Hopkins, W.D. (2006). Differential use of attentional and visual communicative signaling by orangutans (*Pongo pygmaeus*) and gorillas (*Gorilla gorilla*) in response to the attentional status of a human. *American Journal of Primatology, 68(10)*, 978–992. doi: 10.1002/ajp.20304.

Povinelli, D.J., Bierschwale, D.T., and Cech, C.G. (1999). Comprehension of seeing as a referential act in young children, but not juvenile chimpanzees. *British Journal of Developmental Psychology, 17(1)*, 37–60. doi: 10.1348/026151099165140.

Povinelli, D.J., and Eddy, T.J. (1996). Factors influencing young chimpanzees' (*Pan troglodytes*) recognition of attention. *Journal of Comparative Psychology, 110(4)*, 336–345. doi: 10.1037/0735–7036.110.4.336.

Rendall, D., Owren, M.J., and Ryan, M.J. (2009). What do animal signals mean? *Animal Behaviour, 78(2)*, 233–240. doi: 10.1016/j.anbehav.2009.06.007.

Riede, T., Bronson, E., Hatzikirou, H., and Zuberbühler, K. (2005). Vocal production mechanisms in a non-human primate: Morphological data and a model. *Journal of Human Evolution, 48(1)*, 85–96. doi: 10.1016/j.jhevol.2004.10.002.

Rijksen, H.D. (1978). *A Fieldstudy on Sumatran Orang Utans (Pongo pygmaeus abelii): Ecology, Behaviour and Conservation*. Wageningen: Veenman & Zonen.

Roberts, A.I., Vick, S.-J., and Buchanan-Smith, H.M. (2012a). Usage and comprehension of manual gestures in wild chimpanzees. *Animal Behaviour, 84*(2), 459–470. doi: 10.1016/j.anbehav.2012.05.022.

Roberts, A.I., Vick, S.-J., and Buchanan-Smith, H.M. (2013). Communicative intentions in wild chimpanzees: Persistence and elaboration in gestural signalling. *Animal Cognition, 16(2)*, 187–196. doi: 10.1007/s10071-012–0563-1.

Roberts, A.I., Vick, S.-J., Roberts, S.G.B., Buchanan-Smith, H.M., and Zuberbühler, K. (2012b). A structure-based repertoire of manual gestures in wild chimpanzees: Statistical analyses of a graded communication system. *Evolution and Human Behavior, 33(5)*, 578–589. doi: 10.1016/j.evolhumbehav.2012.05.006.

Rosati, A.G., Wobber, V., Hughes, K., and Santos, L.R. (2014). Comparative developmental psychology: How is human cognitive development unique? *Evolutionary Psychology, 12(2)*, 448–473. doi: 10.1177/147470491401200211.

Russon, A., and Andrews, K. (2011). Orangutan pantomime: Elaborating the message. *Biology Letters, 7(4)*, 627–630. doi: 10.1098/rsbl.2010.0564

Schaller, G.B. (1963). *The Mountain Gorilla: Ecology and Behavior*. Chicago: Chicago University Press.

Scheider, L., Waller, B.M., Oña, L., Burrows, A.M., and Liebal, K. (2016). Social use of facial expressions in hylobatids. *PLoS ONE, 11(3)*, e0151733. doi: 10.1371/journal.pone.0151733.

Schel, A.M., Machanda, Z., Townsend, S.W., Zuberbühler, K., and Slocombe, K.E. (2013a). Chimpanzee food calls are directed at specific individuals. *Animal Behaviour, 86(5), 955–965*. doi: 10.1016/j.anbehav.2013.08.013.

Schel, A.M., Townsend, S.W., Machanda, Z., Zuberbühler, K., and Slocombe, K.E. (2013b). Chimpanzee alarm call production meets key criteria for intentionality. *PLoS ONE, 8(10)*, e76674. doi: 10.1371/journal.pone.0076674.

Schneider, C., Call, J., and Liebal, K. (2012a). Onset and early use of gestural communication in nonhuman great apes. *American Journal of Primatology, 74(2)*, 102–113.

Schneider, C., Call, J., and Liebal, K. (2012b). What role do mothers play in the gestural acquisition of bonobos (*Pan paniscus*) and chimpanzees (*Pan troglodytes*)? *International Journal of Primatology, 33(1)*, 246–262. doi: 10.1007/s10764-011–9570-3

Scott-Phillips, T.C. (2008). On the correct application of animal signalling theory to human communication. In D.M. Smith, R.F. Smith and R.F. Cancho (Eds.), *The Evolution of Language. Proceedings of the 7th International Conference (EVOLANG7)* (pp. 275–282). Singapore: World Scientific.

Scott-Phillips, T.C. (2015). Nonhuman primate communication, pragmatics, and the origins of language. *Current Anthropology, 56(1)*, 56–80. doi: 10.1086/679674

Seyfarth, R.M., Cheney, D.L., and Bergman, T.J. (2005). Primate social cognition and the origins of language. *Trends in Cognitive Sciences, 9(6)*, 264–266. doi: 10.1016/j.tics.2005.04.001.

Seyfarth, R.M., Cheney, D.L., and Marler, P. (1980). Monkey responses to three different alarm calls: Evidence of predator classification and semantic communication. *Science, 210(4471)*, 801–803. doi: 10.1126/science.743399.

Slocombe, K.E., Waller, B.M., and Liebal, K. (2011). The language void: The need for multimodality in primate communication research. *Animal Behaviour, 81(5)*, 919–924. doi: 10.1016/j.anbehav.2011.02.002.

Slocombe, K.E., and Zuberbühler, K. (2005). Functionally referential communication in a chimpanzee. *Current Biology, 15(19)*, 1179–1784. doi: 10.1016/j.cub.2005.08.068.

Smith, W.J. (1977). *The Behavior of Communicating: An Ethological Approach*. Cambridge, Massachusetts and London: Harvard University Press.

Taglialatela, J.P., Russell, J.L., Pope, S.M., Morton, T., Bogart, S., Reamer, L.A., . . . Hopkins, W.D. (2015). Multimodal communication in chimpanzees. *American Journal of Primatology, 77(11)*, 1143–1148. doi: 10.1002/ajp.22449.

Taglialatela, J.P., Russell, J.L., Schaeffer, J.A., and Hopkins, W.D. (2011). Chimpanzee vocal signaling points to a multimodal origin of human language. *PLoS One, 6(4)*, e18852. doi: 10.1371/journal.pone.0018852.

Tanner, J.E. (2004). Gestural phrases and gestural exchanges by a pair of zoo-living lowland gorillas. *Gesture, 4(1)*, 1–24. doi: 10.1075/gest.4.1.02tan.

Tanner, J.E., and Byrne, R.W. (1996). Representation of action through iconic gesture in a captive lowland gorilla. *Current Anthropology, 37(1)*, 162–173. doi: 10.1086/204484.

Tempelmann, S., Kaminski, J., and Liebal, K. (2013). When apes point the finger: Three great ape species fail to use a conspecifics imperative pointing gesture. *Interaction Studies, 14(1)*, 7–23. doi: 10.1075/is.14.1.02tem.

Tempelmann, S., and Liebal, K. (2012). Spontaneous use of gesture sequences in orangutans: A case for strategy? In S. Pika and K. Liebal (Eds.), *Recent Developments in Primate Gesture Research* (pp. 73–91). Amsterdam/Philadelphia: John Benjamins Publishing Company.

Theall, L.A., and Povinelli, D.J. (1999). Do chimpanzees tailor their gestural signals to fit the attentional states of others? *Animal Cognition, 2*, 207–214.

Tinbergen, N. (1952). Derived activities; their causation, biological significance, origin, and emancipation during evolution. *Quarterly Review of Biology, 27(1)*, 1–32. doi: 10.1086/398642.

Tomasello, M. (2006). Why don't apes point? In N. J. Enfield and S. C. Levinson (Eds.), *Roots of Human Sociality: Culture, Cognition and Interaction* (pp. 506–524). Oxford & New York: Berg.

Tomasello, M. (2008). *Origins of Human Communication*. Cambridge, MA: The MIT Press.

Tomasello, M., Call, J., Nagell, K., Olguin, R., and Carpenter, M. (1994). The learning and use of gestural signals by young chimpanzees: A transgenerational study. *Primates, 35(2)*, 137–154. doi: 10.1007/BF02382050.

Tomasello, M., Call, J., Warren, J., Frost, G.T., Carpenter, M., and Nagell, K. (1997). The ontogeny of chimpanzee gestural signals: A comparison across groups and generations. *Evolution of Communication, 1(2)*, 223–259. doi: 10.1075/eoc.1.2.04tom.

Tomasello, M., Carpenter, M., and Liszkowski, U. (2007). A new look at infant pointing. *Child Development, 78(3)*, 705–722. doi: 10.1111/j.1467–8624.2007.01025.x.

Tomasello, M., George, B.L., Kruger, A.C., Farrar, M.J., and Evans, A. (1985). The development of gestural communication in young chimpanzees. *Journal of Human Evolution, 14(2)*, 175–186. doi: 10.1016/S0047-2484(85)80005–1.

Tomasello, M., Gust, D., and Frost, G.T. (1989). A longitudinal investigation of gestural communication in young chimpanzees. *Primates, 30(1)*, 35–50. doi: 10.1007/BF02381209.

Van Hooff, J.A.R.A.M. (1973). A structural analysis of the social behaviour of a semi-captive group of chimpanzees. In M. von Cranach and I. Vine (Eds.), *Social Communication and Movement: Studies of Interaction and Expression in Man and Chimpanzee* (pp. 75–162). London: Academic Press.

Vea, J.J., and Sabater-Pi, J. (1998). Spontaneous pointing behaviour in the wild pygmy chimpanzee (*Pan paniscus*). *Folia Primatologica, 69(5)*, 289–290. doi: 10.1159/000021640.

Wacewicz, S., and Zywiczynski, P. (2017). The multimodal origins of linguistic communication. *Language & Communication 54, 1–8*. doi: 10.1016/j.langcom.2016.10.001.

Waller, B.M., Caeiro, C.C., and Davila-Ross, M. (2015). Orangutans modify facial displays depending on recipient attention. *PeerJ, 3*, e827. doi: 10.7717/peerj.827.

Wheeler, B.C., and Fischer, J. (2012). Functionally referential signals: A promising paradigm whose time has passed. *Evolutionary Anthropology: Issues, News, and Reviews, 21(5)*, 195–205. doi: 10.1002/evan.21319.

Whiten, A., and Ham, R. (1992). On the nature and evolution of imitation in the animal kingdom: Reappraisal of a century of research. In P.J.B. Slater, R.J.S., C. Beer & M. Milinski (Eds.), *Advances in the Study of Behaviour* (Vol. 21, pp. 239–283). New York: Academic Press.

Wilke, C., Kavanagh, E., Donnellan, E., Waller, B.M., Machanda, Z.P., and Slocombe, K.E. (2017). Production of and responses to unimodal and multimodal signals in wild chimpanzees, *Pan troglodytes schweinfurthii*. *Animal Behaviour, 123*, 305–316. doi: 10.1016/j.anbehav.2016.10.024.

Zimmermann, F., Zemke, F., Call, J., and Gómez, J.C. (2009). Orangutans (*Pongo pygmaeus*) and bonobos (*Pan paniscus*) point to inform a human about the location of a tool. *Animal Cognition, 12(2)*, 347–358. doi: 10.1007/s10071-008–0194-8.

Zuberbühler, K. (2000). Referential labelling in Diana monkeys. *Animal Behaviour, 59*(5), 917–927. doi: 10.1006/anbe.1999.1317.

Zuberbühler, K. (2001). Predator-specific alarm calls in Campbell's monkeys, *Cercopithecus campbelli. Behavioural and Ecological Sociobiology, 50(5)*, 414–422.

Zuberbühler, K. (2002). A syntactic rule in forest monkey communication. *Animal Behaviour, 63(2)*, 293–299. doi: 10.1006/anbe.2001.1914.

Zuberbühler, K. (2005). The phylogenetic roots of language: Evidence from primate communication and cognition. *Current Directions in Psychological Science, 14(3)*, 126–130. doi: 10.1111/j.0963-7214.2005.00357.x.

W Tecumseh Fitch
University of Vienna, Austria

Dendrophilia and the Evolution of Syntax

"The human understanding is of its own nature prone to suppose the existence of more order and regularity in the world than it finds"

-- Francis Bacon (Bacon, 1620)

Abstract: This chapter reviews data showing that humans are unusual in their ability and propensity to attribute tree structures to data streams – a trait I call "dendrophilia". Unlike phonological rules, which can be captured by finite-state rules, syntactic rules require more powerful algorithms at the context-free level, which enable the flexible and extensible generation of tree structures. I review data supporting a key role for Broca's area, via its powerful connections between inferior frontal gyrus and the parietal and occipital lobes, as playing the computational role of an auxiliary memory (comparable to a "stack" in computer science) to support these more powerful algorithms in human "wetware".

Keywords: Animal communication, syntax acquisition, regular/supra-regular grammar, Broca Area

1. Introduction

A complete understanding of human language evolution will require a biologically grounded partitioning of the human ability to acquire, process and produce language into its many cognitive/biological components. These will include such general cognitive characteristics as perception, memory, categorization, generalization, rule learning, social cognition, motor planning and vocal or manual output, in addition to more specifically linguistic features as syntax or semantics. This broad set of abilities needed for language, as a whole, can be termed the "faculty of language in a broad sense" or FLB (Fitch et al., 2005; Hauser et al., 2002), and it is widely agreed that most of its sub-components are shared with other animals (Fitch, 2010; Hespos and Spelke, 2004; Pinker and Jackendoff, 2005; Seyfarth and Cheney, 2014). From an evolutionary viewpoint, this indicates that such shared features did not evolve specifically for language, but were either present among common ancestors, or evolved convergently in other clades for other purposes (e.g. vocal production learning in birds, or descended larynges in deer or lions).

We can further partition the FLB into those capabilities that are shared with our nearest living relatives, the great apes, and specifically with chimpanzees and bonobos. We can infer that this set of cognitive abilities was present in the last common ancestor (LCA) that we shared with chimpanzees, and thus had evolved prior to the separation of the *Pan* and hominin lineages roughly six million years ago. This set of abilities provides the starting point from which specifically human aspects of language must have evolved; we might dub this cognitive ability set (CAS) of the LCA the CAS_{LCA}.

By subtracting this CAS_{LCA} from the FLB, we arrive at a short list of cognitive skills that represent derived traits of humans – a quite brief list of novel traits whose evolution requires some specific explanation in any complete theory of language evolution. These are the *explananda* of a rich biological and evolutionary understanding of language. We can term this core set the FLD – the *derived* subset of the broad set of all capacities used in language. This subset is not the same as the "faculty of language in the narrow sense" (FLN) as previously defined by (Fitch et al., 2005; Hauser et al., 2002). The FLN refers to those traits that are not shared with any animal (including, say, birds, deers and lions) *and* are not used in other non-linguistic domains of human cognition (such as social, visual, or music cognition). This is a highly restrictive definition, that was hypothesized by Hauser, Chomsky and Fitch to be limited to a single general purpose recursive ability that can map to multiple input and output systems. Recent data indicating that recursive abilities are also available in two non-linguistic domains (visual pattern parsing and music) suggest that, by the strict definition above, even recursion is not part of the FLN (Martins et al., 2015; Martins et al., 2014). In contrast, FLD as defined here refers to those linguistically relevant traits that differentiate humans from other great apes, regardless of domain specificity. This is a more biologically meaningful category (because it does not lump convergently evolved traits in birds together with homologies among primates) and has a clear and unambiguous phylogenetic meaning: this is what theories of language evolution that take the LCA as their starting point need to explain.

Previous analyses of the comparative data have concluded that there are at least three broad areas in which we differ from chimpanzees and bonobos: signaling mechanisms, complex syntax, and semantics/pragmatics (Fitch, 2005, 2010). Signaling mechanisms are those involving speech production and (perhaps) perception (Fitch, 2000, 2009; Lieberman, 2007; MacNeilage and Davis, 2000). Semantic/pragmatic abilities have to do with meaning and inference, and include such factors as advanced theory of mind, pragmatic implicatures or ostensive signaling (Scott-Phillips, 2014; Stegmann, 2013). But the focus of the current chapter will be complex syntax.

2. Syntax – Shared and Derived Components

"Syntax" refers to our ability to arrange words into an unlimited variety of novel sentences that express specific meanings due to their structure. Note that although "syntax" in the English language is mostly accomplished via changes in word order ("dog bites man" vs. "man bites dog"), in many other languages including Latin or German word order plays a minor role, and such factors as case marking are the key determinants of syntactic structure. What is central to human syntax in general is the building up of complex hierarchical structures that combine a finite set of words and morphemes into an unlimited potential set of sentences, each of which has a different interpretation (Chomsky, 1957, 1965). Crucially, syntactic rules apply to these structures and *not* just the individual words they are made up of. Thus, we know that in the sentence "The boy who saw the dog chased the girl" it is the boy who chased the girl, despite the fact that the string "the dog chased the girl" is contained in that sentence, and that "boy" is further than "dog" from "girl". This is because we interpret "who saw the dog" as a phrase, the whole of which modifies "the boy", and which has nothing to do with the girl. This is just a simple example to show that the interpretation of syntax hinges on structure, not just word order or concatenations of words.

Linguists typically consider syntax separately from semantics, not because the two are not inter-related, but because it is easily possible to construct syntactically perfect sentences that are semantically deviant, or even uninterpretable (Chomsky's famous sentence "colorless green ideas sleep furiously" is a classic example). Thus, syntax is a very general rule-application system that can generate a huge number of sentences, only some of which are semantically useful or meaningful. It is also crucial that syntax applies across modalities, and works in precisely the same structure-determined way in signed language, speech, or written language. Thus, I will consider syntax here to be both modality-independent and separable from meaning. Thus, we can understand complex syntax as a type of learning, generalization and application of particular types of rules, and consider it in the broader cognitive context of rule learning, applying across sensory domains and equally applicable to perception and production.

A crucial next step is to attempt to characterize human syntax and to differentiate it from the rule-learning abilities that we share with other species. This attempt can profit greatly from a clear and unambiguous conceptual framework for characterizing different types of rules, in terms of their complexity, generative capabilities and their computational requirements. Such a framework is provided by **formal language theory**, a branch of mathematics founded by Alan Turing with his introduction of the Turing machine (Turing, 1937). Formal language theory is used today especially in computer science for clarifying the computational requirement and complexity of particular problems, algorithms or programming languages (Hopcroft et al., 2000; Minsky, 1967; Post, 1943). Formal language theory has long been applied to human language as well, and some refinements of the theory by Noam Chomsky and colleagues in the late 1950s provided the first attempts to precisely characterize the computational requirements for human linguistic syntax (Chomsky, 1956, 1957, 1959). These early attempts have been memorialized with the term "Chomsky hierarchy" which is still often used today to refer to a simple four-way classification of rule systems (Hopcroft et al., 2000; Sipser, 1997). Modern work in computational linguistics has progressed far beyond this classification (cf. Jäger and Rogers, 2012; Joshi et al., 1991), extending the Chomsky hierarchy beyond this original four-way categorization. This broader, modern framework is best termed simply the **formal language hierarchy** to distinguish it from the older Chomsky hierarchy (though the term "extended Chomsky hierarchy" is still sometimes used (Jäger and Rogers, 2012))

Today, computational linguistics has converged upon a clear and specific characterization of the abilities required to account for all known phenomena in human language syntax, known as the **mildly context-sensitive grammars** (Joshi et al., 1991; Stabler, 2004; Vijay-Shanker et al., 1987). These grammars, while vastly more restricted than a generalized Turing machine, have all of the capabilities needed to capture both the strings and the hierarchical structures of human syntax.

How do these well-defined abilities differ from the rule-learning capability of other animals? There is a rich comparative dataset now available concerning animal rule- and pattern learning in a wide range of species from pigeons and rats to chimpanzees, and abundant data make it clear that many species do have well-developed abilities both to learn rules, and to apply them to novel stimuli (generalization). However all of the rule-learning abilities demonstrated by nonhuman animals can be captured at a lower level of the formal language hierarchy, at the so-called "regular grammar" level (also called finite-state grammars). The type of regular rules that animals have been shown to learn include transition probabilities ("b follows a"), simple string concatenation (e.g. the $(AB)^n$ grammar discussed below), so-called "algebraic rules" like XXY (where XX denotes any syllable repeating itself), and simple long distance dependencies (such as AB*A, which means that there must be an A at both the beginning and end of the string with arbitrary Bs between). Thus, there can be little doubt that a variety of simple regular grammatical abilities represent part of the FLB which is shared very widely, including with apes, and are thus a component of the CAS_{LCA}. But grammars that go beyond this level – the "supra-regular grammars" – have not yet been conclusively demonstrated in any nonhuman animal species (see below).

We thus arrive at a clear characterization of the syntactic component of the FLD, the derived aspect of human syntactic abilities that differentiates our syntactic abilities from those of other species: this derived component is comprised of our **supra-regular capabilities.** In formal language theory, any type of grammar has, at its core, a finite-state automaton with a limited (but potentially large) set of states that could represent words or transitions between words and/or particular combinations of words – this is as true for a Turing machine as for a system implementing a simple regular grammar. What differentiates the supra-regular grammars is some sort of additional flexible memory, which can be used to store information relevant to past states and their "unfinished business". Thus, the key computational requirement differentiating human syntactic capabilities would be such an additional general purpose memory store. In formal language theory there are many different types of such additional memory, including stacks, queues and the "stack of stacks" needed to parse mildly context-sensitive languages. However, I would suggest that this is a place where formal language theory is most urgently in need of updating to a more biological and neurally grounded model of natural computation. Thus, rather than being concerned with whether this additional form of memory is a stack or a queue (Öttl et al., 2015; Udden et al., 2012) I think it is more profitable to simply think of it as a flexible type of working memory able to keep track of and recall past states over the span of a sentence.

From this more general viewpoint we can consider the animal syntactic capabilities demonstrated to date as being more akin to long-term memory for items and categories. These are learned and then retained indefinitely, and many studies show that animals can have very large long-term memory stores for such information. For example, pigeons and baboons can memorize thousands of individual images (Fagot and Cook, 2006) and dogs can memorize more than a thousand specific words and match them to objects (Griebel and Oller, 2012; Kaminski et al., 2004; Pilley and Reid, 2011). Despite this impressive storage capacity, what none of these species do is flexibly and temporarily build up novel and complex hierarchical combinations of these stored items – requiring a type of **structural working memory** able to generalize over many tokens or types and incorporate them into hierarchical structures (Caplan and Waters, 1999). I am obviously not claiming that animals lack working memory – many cognitive tasks (such as delayed match to sample tasks (Finch, 1942; Gazes et al., 2013), or simple object permanence (Pollok et al., 2000)) demonstrate that they do, but we are discussing here a particular class of working memory that is distinct from these more general abilities (cf. Caplan and Waters, 1999). Rather this is a specific form of working memory suited to assigning structures to input strings (for example, parsing a sentence) or converting some semantic structure into a syntactic structure and then outputting it as a string. Without such a structural working memory store, it is impossible to flexibly build up and manipulate tree-like structures – a crucial requirement of human syntax.

Because such terms as "structural working memory" are somewhat vague, and unambiguous terms like "supra-regularity" require a body of technical knowledge to be understandable, I have given this particular derived subset of human syntactic capabilities a more memorable name: **dendrophilia.** Dendrophilia (Fitch, 2014) is my term for the human proclivity to attribute tree-like structures to sensory patterns (composed from the Greek roots "dendro", meaning tree, and "philo-" meaning "loving, fond of, tending to") it is in short our "love of trees". By hypothesis, this capability applies across different sensory domains, and is as applicable to visual signs as spoken words, and indeed applies to music as well (Fitch, 2014). It is this component of complex, structural syntax that I hypothesize to be a relatively recent acquisition, post-dating the separation of the hominin lineage from that of other apes. Thus, dendrophilia is a core component of the human FLD.

3. Empirically Evaluating the Dendrophilia Hypothesis.

From a mathematical viewpoint it is relatively trivial to create rule systems that go beyond the regular level. For example, the simple grammar A^nB^n, where $n \geq 1$, generates string in which the number of As matches the number of Bs (examples include "AB", "AABB", "AAABBB", etc). This grammar is supra-regular because it requires n to be stored after the evaluation of the A string, and then recalled after the Bs have been counted to see if they are the same. In this case the only additional form of memory required is two simple integer counters, and a phrase comparison process by which the A and B components can be recognized and distinguished. Although this is in some sense the simplest possible supra-regular grammar, it seems to be beyond the capabilities of nonhuman animals. But there are many other more interesting (and more linguistically relevant) grammars that are supra-regular. These include the "mirror grammar" which can be represented $w\ w^R$, where w is any string and R indicates its mirror image, or the copy grammar $w\ w$. Both of these go beyond the capabilities of regular grammars, as do countless other possible rule sets.

Despite the relative ease of demonstrating mathematically that a grammar is supra-regular (often using some variant of the "pumping lemma", see Hopcroft et al., 2000; Jäger and Rogers, 2012) it is less trivial to demonstrate in a practical empirical experiment that some participant has in fact learned a supra-regular grammar. There are two essential empirical problems. The first is that mathematical proofs are based on allowing the length of strings (or substrings) to go to infinity: otherwise they can be captured by a large but finite set of states (amounting to a simple list of all strings). In reality of course, both our memories and our lifespans are limited (as is the patience of our human or animal participants) and so we cannot test them with infinite strings and attain this level of certainty in empirical experiments. However, what we can do is expose them to a training set of some limited-length strings, and then see if and how they generalize to longer (or shorter) strings. For example, if exposed to A^nB^n, with $n = 2$ or 3 (and strings of length 4 or 6), we can then see how our subjects react to A^nB^n, with $n = 4$ or 5. If they are able to generalize over n in this case, it cannot be because they simply memorized the previous patterns, or some 4 or 6 slot schema that generalizes over them. Rather, they must have inferred that there is some variable n which counts the As and Bs and must be the same for both sets. Thus, demonstrating generalization over n is crucial for demonstrating mastery of the A^nB^n grammar.

The second empirical problem is more subtle: we need to be sure that our subjects have not figured out some regular-level "trick" that provides correct answers most of the time, but does not actually encapsulate the supra-regular rule. Given that we don't expect our participants to be perfect – given enough trials, mistakes will be made due to distraction or boredom, even if participants have indeed inferred the correct grammar – we need some way to distinguish above-chance performance due to a regular-level grammar from below-perfect performance due to such errors.

For example, the first study to examine the A^nB^n grammar (Fitch and Hauser, 2004) contrasted it with a simple regular grammar $(AB)^n$, which means "repeat AB as many times as you want" (producing strings like "AB", "ABAB", "ABABAB", etc). Given these two grammars, and limited n, it is easy to come up with a regular rule that discriminates them. For example, checking whether there are any "BA" transitions within the string would suffice. Indeed, keas (a large-brained New Zealand parrot species) come up with precisely this regular rule to accomplish this discrimination. Similarly, after training with AABB and AAABBB strings, keas learned either the rule "starts with AA" or "ends with BB" (Ravignani et al., 2015). In order to exclude this or similar possibilities, it is necessary to test participants with various ungrammatical novel patterns, termed "foils", and show that they can exclude them. In the present case, for the A^nB^n grammar, it is enough to examine foils where the number of As and Bs differs (a "mismatched foil"). Indeed the keas in these experiments failed to reject such strings, showing that despite their above-chance performance during training, they failed to induce the intended supra-regular rule for these strings.

In summary, any clear demonstration of any grammar, and supra-regular grammars in particular, requires participants to both generalize to a novel *n* (novel string lengths) and to reject myriad foil stimuli that test for various simpler regular strategies.

4. Experimental Tests in Humans and Animals

Numerous published studies have demonstrated that humans successfully acquire supra-regular rules, including all of those mentioned above. Human mastery of the A^nB^n grammar has now been demonstrated in many labs, including in a few cases with generalization over *n* and appropriate rejection of mismatched foils (Bahlmann et al., 2006; Bahlmann et al., 2008; Hochmann et al., 2008; Ravignani, et al., 2013; Stobbe et al., 2012); other labs have found some mastery of this grammar but lacked these crucial test cases (cf. Fitch and Friederici, 2012; Perruchet and Rey, 2005). Human participants have also been shown to master mirror and copy grammars, which demand more sophisticated supra-regular abilities (de Vries et al., 2008; Öttl et al., 2015; Udden et al., 2012). These data combine with other studies in the musical domain (Rohrmeier et al., 2012) and a long history of computational linguistic studies (Huybregts, 1985; Shieber, 1985; Stabler, 2004) to confirm that humans have capacities above the regular level, and that these can be demonstrated in experimental studies in the laboratory.

Similarly, as already mentioned above, nonhuman animals of various species have been shown to master various finite-state grammars (reviewed in ten Cate and Okanoya, 2012). The type of regular rules that animals have been shown to learn include transition probabilities ("B follows A"), simple string concatenation (e.g. the $(AB)^n$ grammar), "algebraic rules" like XXY and simple long distance dependencies (Chen et al., 2015; Murphy et al., 2008; Ravignani et al., 2013; Sonnweber et al., 2015; Spierings and ten

Cate, 2016; Stobbe et al., 2012; Wilson et al., 2013). These data leave little doubt that the ability to learn simple regular grammars represent part of the FLB which is shared very widely, including with apes, and thus represents a component of the CAS_{LCA}.

In contrast, there are only three studies in the current literature claiming to show that animals can master supra-regular grammars, none of them credible. All of them are based on the A^nB^n grammar, which has often mistakenly been claimed to provide a test for "recursion" or "center embedding" (a misinterpretation of formal language theory, cf. Fitch, 2014; Fitch and Friederici, 2012). Of these a study in baboons by Rey et al., (2012) is the weakest, because it did not test for generalization over *n* (and indeed *n* was limited to 2 in these experiments). In this study 11 baboons were trained on a paired-associate task. Each baboon learned to associate a set of six arbitrary pairs of meaningless symbols, a1-a2, b1-b2, etc by pressing the symbols in order on a touch screen. After achieving 80% correct on each of these single pairs, the baboons were tested with two stimuli: they were first shown a1, and then b1. The key choice came next, when the baboons were simultaneously presented with both a2 and b2. The baboon had to select both of these stimuli to get a reward, with either order rewarded.

The central finding of this study is that baboons pushed b2 first roughly twice as often as a2 (in test2 the bias was weaker, about 3:2). In other words, the baboons showed a significant bias to match the most-recently presented stimulus (b), rather than the initially presented one, so that the pattern ab-ba was significantly (though see below) more frequent than ab-ab. This pattern held over individual animals, at least in the first test.

The authors claimed that this phenomenon demonstrates "recursion" and "centre-embedding" in baboons. But in fact this does not even demonstrate mastery of A^nB^n, a less-demanding goal, because it lacked the appropriate controls (extensions and mismatched foils). The most obvious alternative interpretation of the presented results is that the baboons show a simple recency effect, as is typical in memory experiments. The most recently presented pair remains more activated than the initially presented pair, and thus is completed first. This is predicted by many models of visual memory, and has no obvious relationship to embedding, recursion, or supra-regularity (for a more detailed and extensive critique, see Poletiek et al., 2016).

A second study with Bengalese finches was slightly better designed (Abe and Watanabe, 2011), in that the authors did test for generalization over *n*; more interestingly the neural basis of the rule-learning ability was investigated. Unfortunately this study did not test with mismatched foils and so cannot reject the possibility of a regular "trick" being used to achieve success on the task (again a more detailed critique, and various alternative explanations to those given by the authors, is given in (Beckers et al., 2012)). Clearly this study does not convincingly demonstrate supra-regularity either.

The best, and superficially most convincing, attempt to show mastery of the A^nB^n, grammar by an animal species was a study published in *Nature* by Gentner and colleagues (Gentner et al., 2006). This study is the only apparently positive result to have a convincing set of control stimuli, and showed both generalization and rejection of mismatched foils. This study investigated eleven starlings (*Sturnus vulgaris*) in an operant learning paradigm, where the A stimuli were starling "rattle" vocalizations and the Bs were "warbles". Starlings are songbirds with a quite complex song, where both males and females sing, and continue to learn new songs throughout their life. Thus, they seem to be a likely species in which complex rule learning might be expected. Training took a huge number of trials (between 9,400 and 56,000 trials, depending on the bird) and according to the authors "was slow by comparison to other operant song-recognition tasks". After training with $n = 2$ stimuli, nine birds showed successful generalization to $n = 3$ and 4. Furthermore, mismatched foils were tested (e.g. A^1B^3 or A^3B^2), and were successfully rejected by most of the individual birds (although successful here means simply that the birds accepted the correct strings (e.g., A^2B^2 or A^3B^3) significantly more often, not that they always rejected the mismatched strings. Nonetheless, this result seemed to demonstrate that starlings indeed had learned the supra-regular rule intended by the experimenters (although again, the authors erroneously claimed that this demonstrated recursion in these birds). This would have been a very exciting result, opening the door to investigations of the neural basis of supra-regularity in an animal species.

Unfortunately this positive conclusion was premature. The bad news came from a group of Dutch researchers in Carel ten Cate's laboratory in Leiden, who did an experiment very similar to that of Gentner et al (2006) but now using zebra finches (van Heijningen et al., 2009). Although the birds did seem to succeed on the task, using the same criteria and analyses as those used in the Gentner study, when birds were analyzed individually it was found that the "success" seen at the group level was not mirrored at the individual level, and that indeed each individual bird seemed to be using a different regular strategy to solve the problem. Only when birds were combined, resulting in a fictitious "super-bird" that correctly responded to all stimuli, did it appear that the group as a whole could correctly solve the task and learn the intended grammar. The same could be true in the starling study, since only averaged discrimination scores for all successful starlings combined were published for these crucial foils. After this critique appeared in a high-profile journal, it was followed immediately by a response letter to the editor by Gentner and colleagues, published in the same journal, but no reanalysis of the individual starling data has ever been published in the subsequent decade. This implies that the starling data too can be explained by individual birds adopting simpler regular strategies, that only when averaged together yield the appearance of success.

Multiple previous and subsequent studies have shown that the A^nB^n grammar is *not* learned by cotton-top tamarins, pigeons, zebra finches or keas (Fitch and Hauser, 2004; Ravignani et al., 2013; Stobbe et al., 2012; van Heijningen et al., 2009), based on the animals' failure to reject mismatched foil strings. As already mentioned, keas trained to discriminate between A^nB^n and $(AB)^n$ strings were able to successfully generalize, but failed to reject mismatched foils. A maximum likelihood analysis (Ravignani et al., 2015) showed that each individual bird came up with regular, bigram-based strategies to solve this task. For example when trained to respond positively to AABB and AAABBB strings, four out of five keas used a "BB recency" rule (choose stimuli with two Bs at the end), and the remaining bird used an "AA recency" rule (accept stimuli with two As at the beginning). While these strategies do allow birds to generalize to new stimulus lengths (accepting for example AAAABBBB), they do not allow the birds to correctly reject such mismatched stimuli as AABBB or AAABB. Furthermore, even with further training and feedback where each trial involved a matched and an unmatched stimulus, side-by-side, these highly intelligent birds were unable to master this task, which was mastered in less than ten minutes of exposure by humans (in comparison to months of training for keas or starlings). This task, apparently trivial to humans in either the auditory or visual domain, is extremely difficult for other species that have been tested.

In summary, all of the available human and animal data, as of late 2016, are consistent with the following proposals:

1) Both humans and animals are able, using diverse tasks and stimuli, to master various regular (finite-state) grammars;
2) Humans alone are able, again using multiple modalities and grammars, to master various supra-regular grammars.

These data are thus clearly consistent with the dendrophilia hypothesis, that humans have both an ability, and a propensity, to find higher-order tree-like structures in a set of input strings. Note that any individual human could have performed perfectly well, as the keas did, by using a regular strategy (like AA primacy or BB recency) to process the A^nB^n grammar: such strategies are perfectly compatible with the training data. But, although the occasional human subject appears to do so (as revealed by failing to reject mismatched stimuli, see (Hochmann et al., 2008; Stobbe et al., 2012)) the vast majority follow a supra-regular strategy nonetheless. The human data are thus consistent with the observation of George Miller, who long ago initiated the study of artificial grammar learning in his "Grammarama" project, that

> "*constituent structure languages {supra-regular languages} are more natural, easier to cope with, than regular languages... The hierarchical structure of strings generated by constituent-structure grammars ... would be easier for people than would the left-to-right organization characteristic of strings generated by regular grammars*" (p. 140, Miller, 1967)

5. The Neural Basis of Dendrophilia

I will end with a brief exploration of the neural basis of human dendrophilia. What neural changes occurred in our lineage to provide us the supra-regular computational machinery needed to process tree-like structures. Several initial remarks are in order. First, whatever circuitry is involved must apply to multiple modalities (both visual and auditory) and to both input and output (allowing both perception and generation of supra-regular, tree-formed stimuli). Second, if supra-regular abilities indeed are based on augmentation of "normal" finite-state capabilities by some accessory form of memory (counter, stacks, queues or the equivalent), we could expect to see a disjunction in the neural processing of regular and supra-regular grammars.

The most obvious place to look for neural circuitry involved in processing complex grammar is the inferior frontal gyrus (IFG), also termed Broca's area (though strictly speaking this term should be reserved for the left hemisphere). Paul Broca himself thought of this region as a speech output region (Broca, 1861). However, since the seminal work of Zurif and Caramazza in the 1970s, it has been known that patients with damage to this region also exhibit difficulties in processing sentences with complex syntax (Caramazza and Zurif, 1976; Zurif et al., 1972), and the involvement of the IFG in syntactic tasks is now a very robust result from brain imaging studies (Friederici, 2012; Grodzinsky, 2000; Musso, et al., 2003; Price, 2012; Pulvermüller, 2010; Udden et al., 2008). One recent result that is particularly convincing used a parametric design (which looks for correlations rather than using subtraction, and is thus immune to most of the critiques leveled at early brain imaging studies) where "chunk size" (the number of syntactically connected words in a string) was systematically varied from one to twelve (Pallier et al., 2011). Additionally, this study contrasted ordinary written French with "jabberwocky," in which nonsense nouns and verbs were used but syntactic structure was preserved. The findings were clear: although several regions showed increase activations with chunk size (presumably corresponding to semantic rather than syntactic binding), only the IFG was increasingly activated as chunk size increased even with the jabberwocky speech. Thus there can be little doubt that the IFG serves as an important "hub" in the neural network involved in processing complex syntax.

Comparative neuroanatomical research has corroborated and extended this hypothesis. In particular, Schenker et al. (2010) used painstaking microscopic examination of neuroanatomical slices to compare the size of IFG regions, defined by their cytoarchitecture, in humans and chimpanzees. They found that these regions, although present in chimpanzees, have been greatly expanded in the human brain, with Brodmann's area (BA) 44 being 6.6 times and BA 45 6 times larger than the same area in chimpanzees. For comparison, the primary visual area V1 of the human brain is just 1.4–1.8

times larger than in chimpanzees, depending on the hemisphere. Thus, these two IFG regions represent the most expanded regions of the human cortex that have been reported.

Several recent studies from Angela Friederici's lab have further explored these issues using the same grammars discussed above that have been tested with animals. When contrasting the A^nB^n and $(AB)^n$ grammars, these studies found (Friederici et al., 2006) that only the former, supra-regular grammar strongly activated the IFG, results confirmed by other work using complex grammars done in the same lab (Bahlmann et al., 2006; Bahlmann et al., 2008). This is consistent with the previous work discussed above. More interestingly, though, this study examined the *connectivity* of the areas maximally activated by these two grammars as well, using an imaging method known as diffusion-tensor imaging (DTI) to track white matter bundles emanating from particular regions. They found that the two parts of the frontal cortex activated by these two grammars had fundamentally different connectivity patterns, with Broca's area connecting, via dorsal pathways, to the parietal and temporal lobes, while the opercular regions associated with the regular grammar ran ventrally to the anterior temporal lobe alone.

Consistent with these human data, a comparative study by Rilling and colleagues (Rilling et al., 2008) used DTI to compare the connectivity of the IFG in macaques, chimpanzees and humans. They found that this dorsal white matter pathway (termed the arcuate fasiculus) has both expanded in size, and its connectivity greatly expanded in area, in humans relative to these other species. Indeed in the macaque there is very little arcuate fasiculus at all (and all frontal temporal connectivity seems to derive from the ventral pathway), while in the chimpanzee brain the dorsal pathway exists but ends in the parietal lobe and the posterior-most temporal lobe. In humans this pathway forms one of the largest cortico-cortical connections after the intra-hemispheric corpus callosum.

In summary, brain imaging results demonstrate that an important hub in the neural network that computes complex syntax in the human brain is formed by the IFG regions BA 44 and 45. Anatomically, these regions are the most expanded that are known, relative to chimpanzees, on both sides but particularly on the left (Schenker et al., 2010) and have also greatly expanded their connectivity via the dorsal arcuate fasiculus, relative to other primates (Rilling et al., 2008). This confluence of results strongly suggests that an expansion of the neural hub centered on the IFG was a crucial occurrence during hominin evolution, and represents a key prerequisite of human syntactic abilities, and supra-regular abilities in particular. It further suggests the tantalizing hypothesis that the IFG itself acts as the general-purpose memory system that is a computational requirement for supra-regular abilities, serving as the "stack" or "queue" that stores intermediate results from a more distributed finite-state system (Fitch 2014; Fitch and Friederici, 2012).

6. Implications and Conclusions

I have argued that a key component of the FLD is our derived capacity for complex syntax, specifically the supra-regular abilities that support human dendrophilia. Specifically, I suggest that we can partition human syntactic abilities into two components. The first is a finite state system, with a large long-term memory capacity that can store tokens and basic sequencing operations over tokens, or in some cases classes of tokens (e.g. combinations like bigrams or transitions from one token to the next). These are precisely the types of computational capacities that are required for human phonology (Heinz and Idsardi, 2013), itself a complex and rule-governed aspect of human language (Fitch, 2010; Heinz and Idsardi, 2013). The comparative data suggests that certain computations at this level of the formal language hierarchy are widely shared with other species. Although chimpanzee data remain scarce, they support the idea that regular-level processing is part of our CAS_{LCA} – part of our shared ape heritage (Sonnweber et al., 2015). How broadly precisely which operations are shared remains a topic for further research. Many regular-level phenomena are much more complex than the few regular grammars that have been tested so far in animals (Jäger and Rogers, 2012; Pullum and Rogers, 2006). Furthermore, most phenomena of human phonology have never been examined in animal models (Yip, 2006), so there could be uninvestigated aspects of phonology that turn out to be uniquely human (Pinker and Jackendoff, 2005).

The second component of human syntax is a supra-regular ability, and it is this alone that I hypothesize to be derived in the human lineage and thus part of FLD. These supra-regular computational abilities enable us to process multiple types of regularities and structures that would be inaccessible to purely finite-state systems, including such features as symmetry (mirror grammars) or arbitrary repetitions of novel stimuli (copy grammars), as well as long-distance phrasal correspondences and multiple levels of hierarchical embedding that are central to linguistic syntax. This supra-regular capability can be clearly characterized using formal language theory, and its availability to humans has been clearly demonstrated in multiple sensory domains (vision and audition) in many different laboratories.

Furthermore, I reviewed evidence supporting the neural hypothesis that these abilities are subserved by a network centered on the IFG. If this hypothesis is correct, one prediction is that the small and weakly connected IFG of chimpanzees can serve as a working memory for certain tasks, but not others, and that the key event in hominin evolution was an increase in the power and scope of this. This suggests that brain imaging experiments in chimpanzees might reveal what sorts of tasks activate the IFG in that species, and provide an indication of what types of computations might have served as precursors of complex syntax in early language evolution. Unfortunately, brain imaging in awake chimpanzees, especially performing a trained task, is anything but easy (Rilling et al., 2007), and we may be waiting for a long time to see results that clearly show IFG-specific activation (as opposed to large swathes of the brain, as in Taglialatela et al., 2008). Nonetheless, this prediction illustrates that we do not need to think of language evolution research as indulging in untestable "fairy tales" – using the comparative method and modern neuroscientific methods we can generate and test specific concrete hypotheses.

Why would dendrophilia have evolved? The capacity to infer and manipulate tree structures provides many potential benefits (and indeed, any computer science textbook offers dozens of useful algorithms that rely upon tree representations, e.g. Skiena, 2008). These include sorting, planning, strategizing and generalizing to unobserved data. In particular, the ability to manipulate trees gives its possessor a powerful tool to structure thought, to explore novel combinations of items, and thus to construct novel hypotheses about the world. Such a powerful and abstract ability could apply in many domains, from social cognition and reasoning about others, to tool construction and use, to complex motor planning. I think that the manifest diversity of applications of dendrophilia extend beyond language to include music and the visual arts (e.g. architecture and ornament) and thus represent a key component of the human "sense of order" (Gombrich, 1979) that enables us to structure input from the world, imagine unseen entities (whether ghosts or electrons) and is central not only to language but to all of human thought. Thus I think that it would not be particularly fruitful to ask "which use came first" or "which benefit is most important" – rather I suggest that dendrophilia in general is one of the core components of our mind that makes us human.

Acknowledgments

Research supported by European Research Council Advanced Grant "SOMACCA" (AdG #230604) to WTF. I thank Gesche Westphal-Fitch and an anonymous reviewer for comments on an earlier draft of the manuscript.

References

Abe, K., and Watanabe, D. (2011). Songbirds possess the spontaneous ability to discriminate syntactic rules. *Nature Neuroscience, 14*, 1067–1074.

Bacon, F. (1620). *Novum Organum.*

Bahlmann, J., Gunter, T. C., and Friederici, A. D. (2006). Hierarchical and linear sequence processing: an electrophysiological exploration of two different grammar types. *Journal of Cognitive Neuroscience, 18*, 1829–1842.

Bahlmann, J., Schubotz, R. I., and Friederici, A. D. (2008). Hierarchical artificial grammar processing engages Broca's area. *Neuroimage, 42*, 525–534.

Beckers, G. J. L., Bolhuis, J. J., Okanoya, K., and Berwick, R. C. (2012). Birdsong neurolinguistics: Songbird context-free grammar claim is premature. *NeuroReport, 23*, 139–145.

Broca, P. (1861). Nouvelle opservation d'aphémie produite par une lésion de la moitié postérieure des deuxième et troisième circonvolutions frontales. *Bulletin de la Societe d'Anatomie, Paris, 6*, 398–407.

Caplan, D., and Waters, G. S. (1999). Verbal working memory and sentence comprehension. *Behavioral and Brain Sciences, 22*, 77–126.

Caramazza, A., and Zurif, E. B. (1976). Dissociation of algorithmic and heuristic processes in language comprehension: Evidence from aphasia. *Brain and Language, 3*, 572–582.

Chen, J., van Rossum, D., and ten Cate, C. (2015). Artificial grammar learning in zebra finches and human adults: XYX versus XXY. *Animal Cognition, 18*, 151–164.

Chomsky, N. (1956). Three models for the description of language. *I.R.E. Transactions on Information Theory, IT-2*, 113–124.

Chomsky, N. (1957). *Syntactic structures*. The Hague: Mouton.

Chomsky, N. (1959). On certain formal properties of grammars. *Information and Control, 2*, 137–167.

Chomsky, N. (1965). *Aspects of the theory of syntax*. Cambridge, Massachusetts: MIT Press.

de Vries, M. H., Monaghan, P., Knecht, S., and Zwitserlood, P. (2008). Syntactic structure and artificial grammar learning: The learnability of embedded hierarchical structures. *Cognition, 107*, 763–774.

Fagot, J., and Cook, R. G. (2006). Evidence for large long-term memory capacities in baboons and pigeons and its implications for learning and the evolution of cognition. *Proceedings of the National Academy of Sciences, 103*, 17564–17567.

Finch, G. (1942). Delayed matching-from-sample and non-spatial delayed response in chimpanzees. *Journal of Comparative Psychology, 34*, 315–319.

Fitch, W. T. (2000). The evolution of speech: a comparative review. *Trends in Cognitive Sciences, 4*, 258–267.

Fitch, W. T. (2005). The evolution of language: A comparative review. *Biology and Philosophy, 20*, 193–230.

Fitch, W. T. (2009). Fossil Cues to the Evolution of Speech. In R. P. Botha and C. Knight (Eds.), *The Cradle of Language* (pp. 112–134). Oxford, UK: Oxford University Press.

Fitch, W. T. (2010). *The evolution of language*. Cambridge: Cambridge University Press.

Fitch, W. T. (2014). Toward a computational framework for cognitive biology: unifying approaches from cognitive neuroscience and comparative cognition. *Physics of Life Reviews, 11*, 329–364.

Fitch, W. T., and Friederici, A. D. (2012). Artificial grammar learning meets formal language theory: an overview. *Philosophical Transactions of The Royal Society B, 367*, 1933–1955.

Fitch, W. T., and Hauser, M. D. (2004). Computational constraints on syntactic processing in a nonhuman primate. *Science, 303*, 377–380.

Fitch, W. T., Hauser, M. D., and Chomsky, N. (2005). The Evolution of the Language Faculty: Clarifications and Implications. *Cognition, 97*, 179–210.

Friederici, A. D. (2012). The cortical language circuit: from auditory perception to sentence comprehension. *Trends in Cognitive Sciences, 16*, 262–268.

Friederici, A. D., Bahlmann, J., Heim, S., Schubotz, R. I., and Anwander, A. (2006). The brain differentiates human and non-human grammars: Functional localization and structural connectivity. *Proceedings of the National Academy of Sciences (USA), 103*, 2458–2463.

Gazes, R. P., Brown, E. K., Basile, B. M., and Hampton, R. R. (2013). Automated cognitive testing of monkeys in social groups yields results comparable to individual laboratory-based testing. *Animal Cognition, 16*, 445–458.

Gentner, T. Q., Fenn, K. M., Margoliash, D., and Nusbaum, H. C. (2006). Recursive syntactic pattern learning by songbirds. *Nature, 440*, 1204–1207.

Gombrich, E. H. (1979). *The Sense of Order: a Study in the Psychology of Decorative Art.* Oxford: Phaidon Press.

Griebel, U., and Oller, D. K. (2012). Vocabulary learning in a Yorkshire terrier: Slow mapping of spoken words. *PLoS ONE, 7*, e30182.

Grodzinsky, Y. (2000). The neurology of syntax: language use without Broca's area. *Behavioral and Brain Sciences, 23*, 1–71.

Hauser, M., Chomsky, N., and Fitch, W. T. (2002). The Language Faculty: What is it, who has it, and how did it evolve? *Science, 298*, 1569–1579.

Heinz, J., and Idsardi, W. (2013). What complexity differences reveal about domains in language. *Topics in Cognitive Science, 5*, 111–131.

Hespos, S. J., and Spelke, E. S. (2004). Conceptual precursors to language. *Nature, 430*, 453–456.

Hochmann, J.-R., Azadpour, M., and Mehler, J. (2008). Do humans really learn A^nB^n artificial grammars from exemplars? *Cognitive Science, 32*, 1021–1036.

Hopcroft, J. E., Motwani, R., and Ullman, J. D. (2000). *Introduction to automata theory, languages and computation* (2nd ed.). Reading, Massachusetts: Addison-Wesley.

Huybregts, R. (1985). The weak inadequacy of CFPSGs. In G. de Haan, M. Trommelen and W. Zonneveld (Eds.), *Van Periferie naar Kern* (pp. 81–99). Dordrecht, The Netherlands: Foris Publications.

Jäger, G., and Rogers, J. (2012). Formal language theory: refining the Chomsky hierarchy. *Philosophical Transactions of The Royal Society B, 267*, 1956–1970.

Joshi, A. K., Vijay-Shanker, K., and Weir, D. J. (1991). The convergence of mildly context-sensitive formalisms. In P. Sells, S. M. Shieber and T. Wasow (Eds.), *Processing of Linguistic Structure* (pp. 31–81). Cambridge, MA: The MIT Press.

Kaminski, J., Call, J., and Fischer, J. (2004). Word learning in a domestic dog: evidence for 'fast mapping'. *Science, 304*, 1682–1683.

Lieberman, P. (2007). The evolution of human speech: Its anatomical and neural bases. *Current Anthropology, 48*, 39–66.

MacNeilage, P. F., and Davis, B. L. (2000). On the origin of internal structure of word forms. *Science, 288*, 527–531.

Martins, M. D., Martins, I. P., and Fitch, W. T. (2015). A novel approach to investigate recursion and iteration in visual hierarchical processing. *Behavior Research Methods, online first*, 1–22.

Martins, M. J., Fischmeister, F. P., Puig-Waldmüller, E., Oh, J., Geissler, A., Robinson, S., et al. (2014). Fractal image perception provides novel insights into hierarchical cognition. *Neuroimage, 96*, 300–308.

Miller, G. A. (1967). Project Grammarama. In G. A. Miller (Ed.), *Psychology of communication*. New York: Basic Books.

Minsky, M. L. (1967). *Computation: Finite and infinite machines*. Englewood Cliffs, New Jersey: Prentice-Hall.

Murphy, R. A., Mondragón, E., and Murphy, V. A. (2008). Rule learning by rats. *Science, 319*, 1849–1851.

Musso, M., Moro, A., Glauche, V., Rijntjes, M., Reichenbach, J., Buchel, C., et al. (2003). Broca's area and the language instinct. *Nature Neuroscience, 6*, 774–781.

Öttl, B., Jäger, G., and Kaup, B. (2015). Does formal complexity reflect cognitive complexity? investigating aspects of the chomsky hierarchy in an artificial language learning study. *PLoS ONE, 10*, e0123059.

Pallier, C., Devauchelle, A.-D., and Dehaene, S. (2011). Cortical representation of the constituent structure of sentences. *Proceedings of the National Academy of Sciences, 108*, 2522–2527.

Perruchet, P., and Rey, A. (2005). Does the mastery of center-embedded linguistic structures distinguish humans from nonhuman primates? *Psychonomic Bulletin and Review, 12*, 307–313.

Pilley, J. W., and Reid, A. K. (2011). Border collie comprehends object names as verbal referents. *Behavioural Processes, 86*, 184–195.

Pinker, S., and Jackendoff, R. (2005). The faculty of language: what's special about it? *Cognition, 95*, 201–236.

Poletiek, F. H., Fitz, H., and Bocanegra, B. R. (2016). What baboons can (not) tell us about natural language grammars. *Cognition, 151*, 108–112.

Pollok, B., Prior, H., and Güntürkün, O. (2000). Development of object permanence in food-storing magpies (Pica pica). *Journal of Comparative Psychology, 114*, 148–157.

Post, E. L. (1943). Formal reductions of the combinatorial decision problem. *American Journal of Mathematics, 65*, 196–215.

Price, C. J. (2012). The anatomy of language: a review of 100 fMRI studies published in 2009. *Annals of the New York Academy of Sciences, 1191*, 62–88.

Pullum, G. K., and Rogers, J. (2006). Animal pattern-learning experiments: some mathematical background. *(unpublished PDF)*, 1–12.

Pulvermüller, F. (2010). Brain embodiment of syntax and grammar: Discrete combinatorial mechanisms spelt out in neuronal circuits. *Brain and Language, 112*, 167–179.

Ravignani, A., Sonnweber, R.-S., Stobbe, N., and Fitch, W. T. (2013). Action at a distance: dependency sensitivity in a New World primate. *Biology Letters, 9*, 20130852.

Ravignani, A., Westphal-Fitch, G., Aust, U., Schlumpp, M., and Fitch, W. T. (2015). More than one way to see it: Individual heuristics in avian visual cognition. *Cognition, 143*, 13–24.

Rey, A., Perruchet, P., and Fagot, J. (2012). Centre-embedded structures are a by-product of associative learning and working memory constraints: evidence from baboons (*Papio papio*). *Cognition, 123*, 180–184.

Rilling, J. K., Barks, S. K., Parr, L. A., Preuss, T. M., Faber, T. L., Pagnoni, G., et al. (2007). A comparison of resting-state brain activity in humans and chimpanzees. *Proceedings of the National Academy of Sciences, 104*, 17146–17151.

Rilling, J. K., Glasser, M. F., Preuss, T. M., Ma, X., Zhao, T., Hu, X., et al. (2008). The evolution of the arcuate fasciculus revealed with comparative DTI. *Nature Neuroscience, 11*, 426–428.

Rohrmeier, M., Fu, Q., and Dienes, Z. (2012). Implicit learning of recursive context-free grammars. *PLoS ONE, 7*, e45885.

Schenker, N. M., Hopkins, W. D., Spocter, M. A., Garrison, A. R., Stimpson, C. D., Erwin, J. M., et al. (2010). Broca's area homologue in chimpanzees (*Pan troglodytes*): probabilistic mapping, asymmetry and comparison to humans. *Cerebral Cortex, 20*, 730–742.

Scott-Phillips, T. C. (2014). *Speaking our minds*. New York: Palgrave Macmillan.

Seyfarth, R. M., and Cheney, D. L. (2014). The evolution of language from social cognition. *Current Opinion in Neurobiology, 28*, 5–9.

Shieber, S. M. (1985). Evidence against the context-freeness of natural language. *Linguistics and Philosophy, 8*, 333–344.

Sipser, M. (1997). *Introduction to the theory of computation*. Boston, MA: PWS Publishing.

Skiena, S. S. (2008). *The algorithm design manual* (Second ed.). New York: Springer Verlag.

Sonnweber, R. S., Ravignani, A., and Fitch, W. T. (2015). Non-adjacent visual dependency learning in chimpanzees. *Animal Cognition, 18*, 733–745.

Spierings, M. J., and ten Cate, C. (2016). Budgerigars and zebra finches differ in how they generalize in an artificial grammar learning experiment. *Proceedings of the National Academy of Sciences*, E3977–E3984.

Stabler, E. P. (2004). Varieties of crossing dependencies: structure dependence and mild context sensitivity. *Cognitive Science, 28*, 699–720.

Stegmann, U. E. (Ed.). (2013). *Animal communication theory: information and influence*. Cambridge, UK: Cambridge University Press.

Stobbe, N., Westphal-Fitch, G., Aust, U., and Fitch, W. T. (2012). Visual artificial grammar learning: Comparative research on humans, kea (*Nestor notabilis*) and pigeons (*Columba livia*). *Philosophical Transactions of The Royal Society B, 367*, 1995–2006.

Taglialatela, J. P., Russell, J. L., Schaeffer, J. A., and Hopkins, W. D. (2008). Communicative signaling activates "Broca's" homologue in chimpanzees. *Current Biology, 18*, 343–348.

ten Cate, C., and Okanoya, K. (2012). Revisiting the syntactic abilities of non-human animals: natural vocalizations and artificial grammar learning. *Philosophical Transactions of The Royal Society B, 367*, 1984–1994.

Turing, A. M. (1937). On computable numbers, with an application to the Entscheidungsproblem. *Proceedings of the London Mathematical Society, 42*, 230–265.

Udden, J., Folia, V., Forkstam, C., Ingvar, M., Fernandez, G., Overeem, S., et al. (2008). The inferior frontal cortex in artificial syntax processing: An rTMS study. *Brain Research, 1224*, 68–79.

Udden, J., Ingvar, M., Hagoort, P., and Petersson, K. M. (2012). Implicit acquisition of grammars with crossed and nested non-adjacent dependencies: Investigating the push-down stack model. *Cognitive Science, 2012*, 1–24.

van Heijningen, C. A. A., de Vissera, J., Zuidema, W., and ten Cate, C. (2009). Simple rules can explain discrimination of putative recursive syntactic structures by a songbird species. *Proceedings of the National Academy of Sciences, 106*, 20538–20543.

Vijay-Shanker, K., Weir, D. J., and Joshi, A. K. (1987). *Characterizing structural descriptions produced by grammatical formalisms*. Paper presented at the Proceedings of the 25th annual meeting on Association for Computational Linguistics.

Wilson, B., Slater, H., Kikuchi, Y., Milne, A. E., Marslen-Wilson, W. D., Smith, K., et al. (2013). Auditory artificial grammar learning in macaque and marmoset monkeys. *Journal of Neuroscience, 33*, 18825–18835.

Yip, M. J. (2006). The search for phonology in other species. *Trends in Cognitive Sciences, 10*, 442–446.

Zurif, E. R., Caramazza, A., and Myerson, R. (1972). Grammatical judgments of agrammatic aphasics. *Neuropsychologia, 10*, 405–417.

Joël Fagot, Raphaëlle Malassis,
Tiphaine Medam and Marie Montant
Laboratory of Cognitive Psychology, Institute of Language, Communication and the Brain, Aix-Marseille University, and CNRS

Comparing Human and Nonhuman Animal Performance on Domain-General Functions: Towards a Multiple Bottleneck Scenario of Language Evolution

Abstract: How come that *Homo sapiens* is the sole animal species communicating with a language (i.e., a human language)? Theorists of language evolution have mostly adopted a human-centered approach to address this question. This chapter discusses the limits of this approach and proposes an alternative that consists in studying the domain general functions that serve language comprehension and production from a comparative and evolutionary perspective. Special attention is given to domain general processes which allow humans and animals to integrate information in space and time, and thus develop perceptual and more conceptual abstract categories. This chapter presents illustrative studies that reveal the various aspects in which these integration processes differ in human and nonhuman animals. Finally, we discuss the source of these species differences and their potential implications for our understanding of language evolution.

Keywords: Animal communication, integration processes, categorization, language evolution, baboons

1. Introduction

1.1 Limits of strictly human-centered approaches

In the various scientific disciplines interested in language evolution – anatomy, physiology, paleoanthropology, linguistics, psychology, neurosciences, computer sciences – the precisely defined architecture of human language is classically used as the basic reference, the *canon,* to which all nonhuman communication systems are compared. Said differently, the communication system of *Homo sapiens* – an approximately 200,000 year-old isolated species (there are no other living species in the genus *Homo*) with almost 1.5 kg brain tissue, living in complex societies with a long history of cultural evolution – is used as the unique reference system in most comparative studies on language evolution (for a review, Hauser et al., 2002). Obviously, this makes sense given that the purpose of these comparative studies is not to build a descriptive catalogue of other animal communication systems but to better understand human language evolution. However, when pushed too far, such an anthropomorphic approach can be misleading because, by definition, nonhuman animals cannot equal human animals in performance when it comes to human language. A current caveat in human-centered comparisons consists in assuming non-explicitly that nonhuman cognitive architectures must *resemble* human cognitive architecture, in parts or as a whole. However, such an assumption can hold only if: 1) human and nonhuman cognitive architectures had followed similar evolutionary paths and were adapted to comparable environmental, social and biological constraints, and 2) the cognitive architecture of each species was a construction made of independent (non-interacting) cognitive components that are not sensitive to developmental and phylogenic factors. Given that every species has a unique cognitive architecture, it seems like a vain enterprise to search for strictly identical components in humans and nonhuman animals.

Consider one example: syntax. Different types of syntaxes have been formalized in Chomsky's hierarchy, from simple (finite state) grammars to complex (supra-regular) grammars. In an attempt to better understand human cognitive "uniqueness", major efforts have been put in the investigation of nonhuman species' ability to process supra-regular grammars (e.g., Fitch and Hauser, 2004; Gentner et al, 2006; Abe and Watanabe, 2011). Unsurprisingly, nonhuman animals do not equal *Homo sapiens* in that particular type of linguistic computation. The very attempt to search for strict human-like syntax in nonhuman animals implies that there is only one way to compute information in a complex communication system: the human way. This view does not consider the possibility that each species, even phylogenetically close to *Homo sapiens*, might have developed its own original and complex –possibly multimodal- cognitive architecture, which does not include human-like syntax. It also neglects the existence of interactions between the various components of a cognitive architecture whose effects increase over phylogenetical and developmental time scales (syntax might not exist as an independent computational subsystem, not even in humans, see Seidenberg and MacDonald, 1999). These interactions make it very unlikely that complex integrated levels of computation (syntax-like processes) in the cognitive architectures of two different species resemble each other.

We must acknowledge, however, that more and more studies are looking for simpler forms of syntax in nonhuman animals. This approach has recently gained interest in the field of ethology and to some extent in linguistics (for a review, Schlenker et al., 2016; see also Petkov and Wilson, 2012). It refers to basic learning principles that allow the extraction of combinatorial semantics, statistical regularities, or adjacent and nonadjacent dependencies according to which qualitatively comparable basic learning principles are supposed to hold for human as well as nonhuman communication sequence learning (e.g., Seidenberg et al., 2002).

The narrow anthropomorphic approach of language evolution that we question here is not only found in linguistics but has dominated other domains as well, such as comparative anatomy. Rapid progress of imaging techniques over the past 20 years has spurred a frantic search for the anatomical landmark of language in *Homo sapiens*. After a couple of unsuccessful attempts to teach human vocal language to nonhuman primates (e.g., Hayes, 1951), it has become clear that our closest cousins – chimpanzees – are not able to pronounce human phonemes. The first proposed explanation for the incapacity to produce speech was that their larynx was too high (e.g., Lieberman, 1968, 1975). The existence of a physical limitation for speech in chimps has been admitted in the research community for many years. However, recent studies suggest that vocal tract anatomy cannot suffice to explain the absence of speech in nonhuman mammals (for a review, Fitch 2010). Fitch et al. (2016) inferred from X-ray videos and a modeling approach that macaca's tract has the potential to produce a broad range of speech sounds. This was confirmed in one of our recent study in which we recorded the spontaneous vocalization of baboons (*Papio papio*, Boë et al., 2017). We found that the baboons produce sounds that share the acoustic F1/F2 formant properties of human [ɨ æ ɑ o u] vowels, and those baboon sounds were produced by movements of the tongue in a human-like articulatory space defined by two axes (anterior-posterior and superior-inferior, see Boë et al., 2017 for more details). Therefore, the inability of nonhuman primates to produce human phonemes is more likely due to differences in the neural circuits that command oro-facial muscles, or to other cognitive differences, than to differences in the anatomy of the vocal tracts.

Other anatomical factors have been proposed to explain the uniqueness of language in the human species. Most of them concern human-centered brain features. Neuroscientists have been looking for the homologue structures of Broca's area and Wernicke's area in the brain of various apes. Broca and Wernicke are important regions of the human brain whose injury provokes severe language disorders, the so-called Broca's and Wernicke's aphasias. Because of interspecific anatomical differences, direct comparisons of brain regions across species are difficult to make. Instead, researchers have compared brain asymmetries across species. In the human brain, the language function is mostly hosted in the left hemisphere. Broca's area, Wernicke's area and the *planum temporale* are bigger in the left than in the right hemisphere (Geschwind and Levitsky, 1968). Those asymmetries, that were first thought to be unique to humans, have been observed as well in nonhuman primates (e.g., Cantalupo and Hopkins, 2001; Cantalupo et al., 2003). In a recent paper, Gomez-Robles and collaborators (2013) propose that, at a whole brain scale, there is continuity in asymmetric variation between humans and chimpanzees: similar brain asymmetries exist in both species, even though the human brain tends to be a little more asymmetric and more sensitive to developmental constraints.

More recently, tremendous efforts have been put into the identification of precise brain regions or neural structures that might be unique to humans and explain the emergence of language in our species. For example, Leroy et al. (2014) have proposed that the superior temporal sulcus (STS) critically differs in human and nonhuman species: human STS is deeper in the right hemisphere compared to the left, and this depth asymmetry is not found in chimps. Given that STS is central in the perisylvian language region, this particular landmark could be a promising "human-only" candidate. However, as the authors acknowledge, the link between this anatomical feature and the language ability in humans remains loose: MRI anatomical measurements made on various groups of human subjects show that the STS depth asymmetry is bigger in men than in women, it persists in children with impaired language development, and is unchanged in adults with reversed language lateralization (*situs inversus*). To drive the point home, an even more recent study shows that the very same STS brain asymmetry exists in baboons (Meguerditchian et al., 2016). Further investigations on the relationship between variation in human brain anatomy (at a brain network scale) and variation in human language function are needed before one can make between-species comparisons and draw convincing conclusions about the role of precise human brain features in the evolution of human language.

The literature on human brain landmarks for language also raises a crucial question: what can gross anatomy tells us about fine-grained and complex interacting cognitive functions, in particular in a comparative perspective? Without a well-defined theoretical model of the anatomy-to-function relationship in each species that would make the comparison possible, the explanatory power of anatomy brain differences remains very limited. This would not be the case if we knew precisely the functional significance of those particular brain regions in their neural networks, in both species. For example, recent studies have shown that the homologue of Broca's area in chimps is involved in communicative behaviors (Taglialatela et al., 2008; 2011), however little is known about the type of computation it makes, or about the way it deals with a combination of communicative gestures and vocalizations, or about the way this region interacts with deeper structures involved in emotional vocalizations (Jürgens, 1979). Just like the linguistic/cognitive human-centered components, the anatomical landmarks of language -when strictly human-inspired (and functionally underspecified)- seem of very little explanatory value to understand language evolution in the human lineage, at least in the current state of our scientific knowledge.

The search for a unique (cognitive or anatomical) key factor at the origin of human language will inevitably lead to a stalemate. Brains, just like the cognitive function they host, are shaped by species-specific phylogenetical and developmental trajectories. They are complex systems and the differences between human and nonhuman brains and functions cannot be easily reduced to single "keys" components, especially when a strict human-centered definition of these components is applied.

1.2 An alternative ...

What could be the alternative to this narrow human-centered approach? In this chapter, we propose to step backward to enlarge our view of language evolution. Rather than focusing on the presence or absence of strictly human-inspired language features in nonhuman animals, we believe that an alternative is to examine the background of the language function, namely the inherited domain-general elements of "the machinery required to master human language" (Saffran and Thiessen, 2008), that might be used in nonhuman species for communicative and/or other purposes. Domain-general mechanisms correspond to learning devices that apply to a variety of different cognitive functions, as opposed to domain-specific mechanisms that are dedicated to specific cognitive functions. The opposition between domain-general *versus* domain-specific mechanisms was first proposed in a cognitive development framework where Skinner's view (1957) was opposed to that of Chomsky (1959). In the present chapter, we use the notion of domain-general mechanism in a more evolutionary perspective. This idea we defend here is that the human language-device is mainly made of domain-general elements, some of which are shared with other species either because they were present in a common ancestor or they result from convergence processes. These inherited cognitive components are very likely to take different forms in different species, as a function of the cognitive domain they are involved in, because different domains show different regularities and constraints that shape these components. However, we expect that close comparison between human and nonhuman performance will reveal what aspects of those domain-general cognitive components are shared across species. The underlying hypothesis we uphold here is that complex and phylogenetically recent cognitive functions, like language, are probably the product of intense re-use and re-combination of subsets of inherited anatomical, cognitive, behavioral components (Anderson 2010). Phylogenetically close species might share some (but not all) of these components, as a support of communication and/or other cognitive functions. For example, the serial organization and structuration of elements that we find in the processing of syntax might as well serve the planning of complex motor sequences in humans (Koechlin and Jubault, 2006), in other primates or in birds, including the sequences of bird's songs (Suzuki et al., 2016).

In the remaining part of this chapter, we will firstly describe some of the multilevel integrative processes that are critically involved in human language, which likely evolved from domain-general inherited functions. We will then consider and discuss the main findings of the literature on comparative cognition regarding these domain general mechanisms. In this context, we will pay special attention to the mechanisms by which animals integrate stimulus information at various levels, from low level perceptual grouping mechanisms that lead to global percepts, to more cognitively complex processing mechanisms that make it possible to associate meanings to objects or categories for example. Importantly, this chapter is not aimed at making an exhaustive review of the literature regarding these cognitive processes. Our goal is to document potential *species* differences and similarities concerning these processes, and to illustrate these differences and similarities by a selection of suggestive findings. In a final section of this paper, we will discuss the potential impact of these results on our understanding of language evolution.

2. Integration processes in nonhuman animals

2.1 Integration in time and space

Language production and comprehension involve the processing of a continuous flow of information, and a temporal integration of the different linguistic elements. There is an obvious connection between sequential learning and language, because these two cognitive processes require the extraction and further handling of elements occurring in temporal sequences. However, because most behaviors have a temporal structure, the capacity to relate events in the time dimension might derive from a domain-general function that is involved in many different contexts, beyond language. We will consider below four different aspects of sequence processing, which are the ability to (1) remember a sequence of events, and to process that sequence as temporally ordered, (2) to learn the transitional probabilities of items occurring in sequence, (3) to learn and process nonadjacent dependencies, and therefore to know that event A is followed by event B with intervening events between them, and (4) to gain information regarding the general structure of the sequence (i.e., syntax as concerns human language). For all these problems, we document below which aspects of sequence processing seem to be shared by animals and humans, and which aspects seem more restricted to humans.

2.1.1 *Serial list learning*

List learning corresponds to the capacity to remember items in a list, and to remember their temporal ordering. Nonhuman animals can learn and remember lists of items, (e.g., Terrace et al., 2003). Comparative investigations of serial list learning in animals, especially pigeons and monkeys, suggest that the memory of lists of items is not qualitatively different from what is found in humans. When taught that A<B, B<C, C<D, D<E, various nonhuman species have been able to properly order previously un-trained test pairs such as C<E. Pigeons and monkeys show serial position effects (Wright et al., 1985) sharing similarities with those of humans. Monkeys also show the symbolic distance effects found in humans: their latency to compare two items is shorter when the items are far apart in the list (e.g., AC vs CD; Colombo and Frost, 2001). All these discoveries show that animals are capable of developing representations of series of items based on their ordinal position. Although the succession of words in a sentence is not strictly equivalent to a succession of unrelated items, serial list learning capacity is probably necessary in both cases. These results encourage the view that the underlying features of the learned representations are shared by human and nonhuman animals, suggesting quite old evolutionary mechanisms for item list learning.

2.1.2 *Processing of adjacent dependencies and chunking*

Adjacent dependencies refer to a predictive relationship between one event and the event immediately following it in the sequence. Consider two three-item sequences, the sequences A-B-C and A-B-D, which are presented an equal number of times. In this very small corpus composed of two sequences, A is always followed by B and B is followed half time by C and half time by D. The transitional probability between A and B is thus equal to 1, while the transitional probability between either B and C or B and D is equal to .5. Consideration of the transitional probabilities is one of the mechanisms promoting the learning of auditory and visual sequences in humans (e.g., Hunt and Aslin, 2001). It is one of the mechanisms by which children learn word boundaries and segment speech streams into words: in a sentence, the transitional probability between the last syllable of a word and the first syllable of the following word is lower than the probability between two successive syllables within one word (Saffran et al., 1996).

Comparative studies have shown that a high transitional probability between two AB items facilitates the processing of the second B item in pigeons and monkeys. Froehlich et al. (2004), for instance, tested pigeons in a serial response time task requiring to peck a stimulus appearing sequentially at three possible locations and in a predefined order. The transitional probabilities between the stimulus locations were controlled in this task, and the authors report that response time to peck is a direct function of these probabilities. Thus, high transitional probabilities gave rise to short response times at the second location of the considered pair, while low transitional probabilities gave rise to longer response times. These authors directly compared their results on pigeons to Hunt and Aslin's study conducted on humans (2001), and report that although slower, pigeons processed information at roughly the same rate as humans, as reflected in similar overall regression slopes (Figure 1).

Figure 1: Average response times of pigeons and humans depending on transitional predictability of the items in a sequence. Results on pigeons are from Froehlich et al. (2004), those on humans are from Hunt and Aslin (2001). Also shown are linear regression lines for the two sets of data. Pigeons responded slower but processed information at roughly the same rate as reflected in similar overall slopes. Figure adapted from Froehlich et al. (2004).

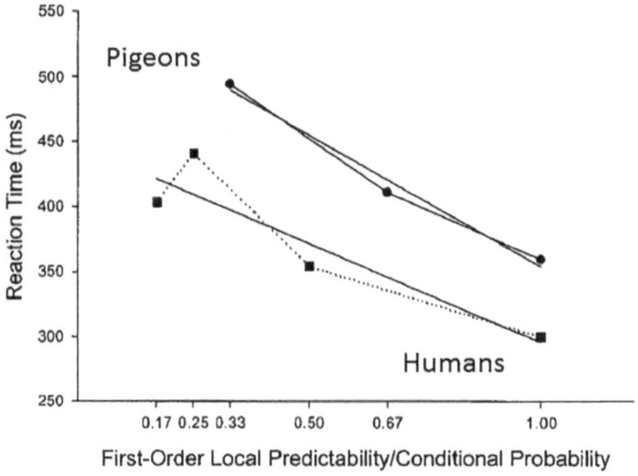

In humans, the processing of transitional probabilities between adjacent elements comes with another type of processing known as chunking that consists in clustering information within sequences (e.g., the syllables in a word). Chunking in humans occurs during the memorization of verbal items, but chunking is not restricted to verbal material, and can be found as well in the visual domain (e.g., Orbán et al., 2008). Experimental evidence suggests that the capacity to organize sequences in chunks is also in the scope of numerous animal species (rats: Fountain, 1990; pigeons: Terrace, 1991; tamarins: Hauser et al., 2001; baboons: Minier et al., 2016). In a recent unpublished experiment involving a serial response time task, we found for example that baboons organized 9-items sequences in three chunks of three items each, and that these chunks precisely included the items sharing the highest transitional probabilities (Minier et al., 2016). Therefore, is seems that human and nonhuman animals are prone to statistical learning making use of transitional probabilities to both segment streams of (visual or auditory) information and organize the elements composing these streams into chunks.

2.1.3 Nonadjacent dependencies

Equally important for language is the capacity to detect and learn nonadjacent dependencies. Consider the following sequence structure A-X-B: A is followed by a variable item X, and item X is systematically followed by B. Given this structure, there is a nonadjacent transitional probability of 1 between A and B. Learning nonadjacent dependencies is important for language. For instance, a listener has to detect the relation between the subject and the verb in a sentence despite the presence of intervening words such as adverbs. This capacity can also be useful in very different nonlinguistic contexts, for example when we have to detect a systematic relation between two events separated in time (e.g., the ring of the doorbell signaling that someone is coming), and irrespective of the intervening other events. Experiments have shown that humans and animals can both process nonadjacent dependencies in temporal sequences of events, although nonadjacent probabilities are more difficult to detect and learn that adjacent dependencies (humans: Newport and Aslin, 2004; tamarins: Newport et al., 2004; rats: Fountain and Benson, 2006). Moreover, there are similarities between species regarding the factors that affect the learning of nonadjacent dependencies (e.g., facilitatory effect of perceptual similarity of the nonadjacent elements in both humans and monkeys; humans: Creel et al., 2004; Gebhart et al., 2009; squirrel monkeys: Ravignani et al., 2013).

However, animals may have more difficulties when presented with complex sequences. Wilson et al. (2015) tested 2 monkeys and 33 humans using an auditory artificial grammar containing both adjacent and nonadjacent (long-distance) relationships. After an initial exposure to the sequences, the subjects from the two species were exposed to sequences containing violations of either the adjacent or both adjacent and nonadjacent relationships. Both species showed sensitivity to adjacent transitions, but only humans, and even roughly half of them, indicated significant sensitivity to nonadjacent dependencies. Wilson et al. (2015) concluded that in some conditions, nonadjacent probabilities are less salient in macaques than in humans. Although replications and extensions are required, this study suggests that, compared to monkeys, humans have a greater facility to deal with several dependencies of different types (i.e., both adjacent and nonadjacent) at the same time.

2.1.4 Learning of sequence structure

Learning the structure of a sequence requires the extraction of the relationships between the constitutive elements of that sequence. This kind of learning probably supports, among others, the encoding of grammatical and syntactic linguistic regularities (e.g., in most German sentences, the verb is at the end). However, such structural regularities also exist in many other (non-linguistic) domains, such as the motor domain. For instance, Byrne et al. (2001) reported that the preparation of food items requiring complex manipulations (thistle leaves) in wild gorillas follows a hierarchical sequential organization.

Marcus et al. (1999) have shown that 7-month-old infants can quickly learn that sequences of auditory stimuli follow an ABB or an ABA structure. The processing of such structures was also studied in nonhuman animals, but the results were quite inconclusive: zebra finches (Heijningen et al., 2012), rats (Toro and Trobalón, 2005) and even rhesus macaques (Procyk et al., 2000) do not seem to catch the difference between the ABA vs ABB structures. One study compared zebra finches and humans using the same experimental procedure and stimuli (Chen et al., 2015): zebra finches did not learn these two types of structures, while humans learned them readily. There is to our knowledge only one study in which an animal species could successfully learn sequence patterns of the ABA/ABB type. This study from Spierings and Ten Cate (2016) compared two avian species, budgerigars and zebra finches, and obtained positive results in the former. From this set of experiments, we can conclude that learning the structure of sequences might very well be in the scope of some nonhuman species, but this ability is clearly not as developed as it is in humans.

2.1.4 Integration in space

We documented above that human language function makes great use of temporal information, but spatial information is also crucially important for language. Infants at very young age learn that words and communicative gestures refer to entities in their immediate surrounding space. Words also make it possible to refer to objects that are spatially (or temporally) absent (the displacement feature proposed by the linguist Charles Hockett in 1960 to characterize human language). Finally, nonverbal forms of language, such as sign language or writing gestures strongly rely on an encoding of spatial information. No doubt that being capable to integrate and combine information in the spatial domain is also a domain general function important for a complex use of language.

In the late seventies, Navon (1977) has shown that human subjects tend to process the global shapes of visual object before they process their constitutive (local) details. This effect has been named the "global precedence effect" and is often considered to be an attentional phenomenon. Global precedence in humans was demonstrated in experimental research using large letters (global shape) made of smaller letters (local features). The degree to which animals perceive the global properties of the visual input in comparison to more featural ones has been an issue in animal cognition for some time. In our laboratory, we explored this effect in baboons (Deruelle and Fagot, 1998; Fagot and Deruelle 1997) using large shapes (square, circle, cross) made of smaller shapes (again square, circle, and cross) as stimuli. In our tests, the baboons were required to either match (Fagot and Deruelle, 1997) or identify (Deruelle and Fagot, 1998) these stimuli considering either their global of local structure. For comparative purposes, humans were also tested in the same experimental conditions.

Figure 2: Processing of the global / local stimulus structure in humans and baboons. Left: Illustration of the stimuli used with humans and baboons in Fagot and Deruelle (1997). This experiment required to match hierarchical stimuli considering their global or local structure (in this local trial, they have to match the circle made of circles, with the square made of circles, considering their common local features). Right: percentage correct obtained in humans and baboons in global and local trials. Humans showed an advantage to process the global structure of the stimuli, while baboons showed a local advantage. This local advantage in baboons is accounted for by a general difficulty to "group" the local elements of the hierarchical global/local stimuli into a global whole. Humans are much less sensitive than monkeys to the spatial distance separating the local elements (e.g., Deruelle and Fagot, 1998). Figure adapted from Fagot and Deruelle (1997).

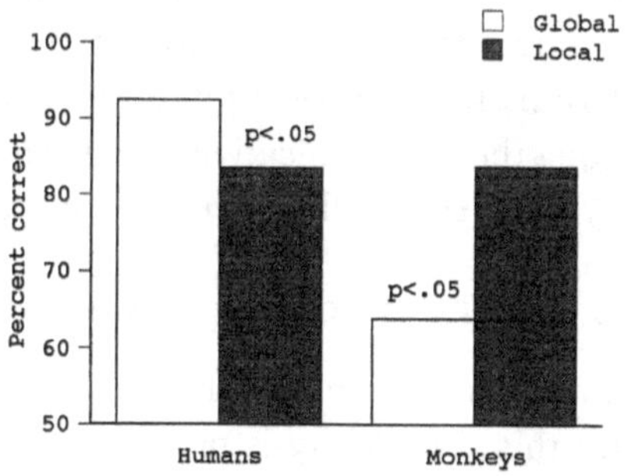

These experiments revealed a striking species differences in global–local processing (Figure 2). Human participants exhibited the global advantage already found by Navon (1977), whereas baboons demonstrated their best performance and fastest response times in the local condition. Several experiments were conducted to understand the cause of this human-baboon difference, which suggested that the performance of the baboons strongly depended on the distance separating the local elements (Deruelle and Fagot, 1998): when the distance was enlarged, the strength of the local bias increased and this effect was amplified in baboons compared to humans. This effect has been replicated many times in pigeons, capuchin monkeys and chimpanzees (for reviews, see Fagot and Barbet, 2006; Fagot and Parron, 2012).

At first glance, one might consider local precedence in animals as a purely perceptual/attentional phenomenon, but we propose that it is more than this. One of the main properties of language is that it can convey information about things that are not immediately present (spatially or temporally; Hockett, 1960). This displacement feature is crucial in the comparison between human language and other forms of primate communication. We have previously proposed (Fagot and Barbet, 2006) that a strong local bias limits the processing of the relation between and among objects. This effect is for instance demonstrated in Fagot and Parron (2010), showing that an increase in the separation of the distance between two bars of either identical or different colors limits the classification of this stimulus on a same/different relational basis. We suggest that a strong bias in favor of a local processing mode (in either the spatial or temporal domain) in nonhuman animals might place important constraints on their communicative systems. It reduces the possibility to make non (temporally or spatially) adjacent relations between or among the communicative signals, and between or among the communicative signals and the objects in the real word to which they refer, especially when they are far or absent.

2.2 Integration of stimulus dimensions and sensory modalities

Animals, including humans, live in a rich world of information, and dealing with this complexity is probably critical for the survival of every species. The processing of this complexity may be achieved by a variety of cognitive mechanisms, which includes, among others, the integration of the different perceptual dimensions into single entities (i.e., integration of stimulus dimensions), and the grouping of various exemplars of a given object into categories (e.g., object categorization; see a discussion of this issue below). Interestingly, these functions, which are of general adaptive value, are all critical for a multidimensional/multimodal system, such as language. For instance, speech comprehension requires that multiple prosodic (e.g., intonation, stress) and phonemic (e.g., voice onset time, place of articulation) dimensions present in the acoustic signal are processed and integrated.

Language comprehension is also achieved using a multimodal/multisensory mode of processing when various auditory and visual information are simultaneously integrated (e.g., for the mapping of the lip movements with the auditory signals).

2.2.1 Combining multiple stimulus dimensions

Discrimination tasks are often proposed to animals as a mean to assess their perceptual abilities. Typically, stimuli are presented to the subjects, and the subjects' behavioral responses to some perceptual dimensions or combinations of dimensions are reinforced, while responses to the non-relevant dimensions are not. Evidence suggests that animals can discriminate stimuli along various dimensions: color, shape, luminance, orientation of the visual objects, or the pitch of auditory stimuli. Animals can as well base their behavioral responses on combination of two or more stimulus dimensions. For instance, Cook (2001) showed that pigeons can learn to select the computer screen area where horizontal green lines are presented, while avoiding the screen areas showing non-green lines and green lines in a non-horizontal orientation.

In their review article, Lea and Wills (2008) comment on three main trends emerging from the literature on learned discrimination in nonhuman animals. The first one is that unidimensional discrimination is easier to learn than multidimensional discrimination based on combinations of features or conjunctions. Smith et al. (2012) have for instance trained monkeys to sort sine wave gratings depending on their orientations or both the orientation and spatial frequency considered in conjunction. Learning was much faster in the unidimensional than in the bi-dimensional test condition. Another example of this effect comes from research on conceptual discrimination by monkeys (e.g., D'Amato and van Sant, 1988) and the demonstration that discrimination performance relies strongly on an analysis of features, such as color, rather than on configurations of features. Lea and Wills' (2008) second conclusion is that when the stimuli are made of multiple relevant dimensions, nonhuman animals express a tendency to focus their attention on one dimension only, mostly when this dimension has sufficient discriminative values. There are also multiple examples of this trend in the literature. In our laboratory, we found that baboons discriminated computerized human faces considering exclusively the contour of the face (Martin-Malivel and Fagot, 2001) or pixel luminance information (Martin-Malivel et al., 2006), instead of the multiple levels of information (e.g., configural information, identity, etc.) that the facial stimuli may provide, and that humans process. Finally, Lea and Wills' (2008) report than even with ingenious experimental designs, attempts to force nonhuman animals to process multiple aspects of the stimuli mostly lead to failures. This can be nicely illustrated by Dépy et al. (1997). The baboons in this study were initially trained to discriminate between two categories of stimuli defined by the possession of any combination of two out of three possible binary features. Baboons could sort these two classes of stimuli to a good accuracy level, albeit after a long training process of several thousands of trials, but remained unable to take the three discriminative features into consideration to achieve this performance, two of the three features taking a leading role in the task.

Wang et al. (2015) recently recorded the brain activity (IRMf) of rhesus monkeys and humans in two test phases. In the first phase, subjects from the two species perceived passively sequences of four tones, the last one being either of a lower or higher pitch than the first three. After this habituation procedure, the same subjects perceived sequences violating the general structure used during the habituation phase. Thus, some test sequences contained a number of tones different from the habituation sequences (number deviant), some other sequences contained four tone units with identical pitch (sequence deviant), and a last set of sequences differed from the habituation sequences regarding both the number of items and pitch. In both species, homologous brain areas were particularly responsive to violations in number (intraparietal and dorso premotor areas), and sequence (ventral prefrontal and basal ganglia), but humans were the only primates showing a joint sensitivity to both factors in the perisylvian language region (bilateral inferior frontal and superior temporal gyri). One limitation of this study is that an absence of evidence monkeys is not the evidence of an absence. Although this study does not address directly the relationship between brain and behavior, its results suggest that the perisylvian region is involved in humans only in the integration of various stimulus dimensions contained in auditory sequences of stimuli.

2.2.2 *Combining multiple sensory modalities*

The question of multimodal integration warrants a discussion in this section, due to its relevance to the origin of language. Evidence suggests that the ability to integrate information across sensory modalities in not at all restricted to humans. Cross-modal integration was demonstrated for instance using task requiring the processing of multi-sensory stimuli (Lanz et al., 2013), cross-modal interference tasks in baboons (Martin-Malivel and Fagot, 2001), and cross-modal matching tasks in chimpanzees (e.g., Davenport and Rogers, 1970). Unfortunately, we are aware of only one study in which the performance of humans and nonhuman animals were directly compared in cross modal tasks using the same stimulus material. This experiment from Fagot et al. (2000) requested the subjects to categorize pictures of humans and baboons in one condition, and human and baboon vocalizations in another one. In this experiment, the subjects of the two species perceived a prime prior to the presentation of the stimulus to be categorized. Depending on the condition, the prime could be a picture or a vocalization of baboons or humans and three conditions were tested: intra-modal visual-visual priming, intermodal auditory-visual priming, and intramodal visual-auditory priming. Three subjects out of four in each species demonstrated intra-modal priming. Inter-modal priming was demonstrated in the three out of four human subjects in the auditory-visual condition, and all four in the visual-auditory condition, but it was only found in one baboon out of four in each intermodal condition, suggesting that inter-modal integration is more difficult in baboons than in humans. Given the small number of subjects involved in this study, a replication is warranted before drawing any firm conclusion on the evolution of intra-modal integration.

2.3 Categorization and conceptual integration

A central aspect of human cognition is our ability to form categories of various kinds of objects or mental entities. Categorization implies that the exemplars of each category are grouped into classes considering physical or more abstract properties. Categorization is a domain-general ability that is fundamental for a variety of more specialized functions (e.g., inference or decision making), including the language function considered in this book.

Historically, Herrnstein and Loveland (1964) were the first to demonstrate categorical abilities in nonhuman animals. They showed that pigeons could sort pictures of humans and pictures devoid of humans in two open-ended categories. Since this study, numerous papers have confirmed that various animal species can efficiently categorize stimuli considering low level perceptual stimulus dimensions, such as the pitch for auditory stimuli, and the color, shape, size, motion, orientation, luminance for visual stimuli (see for instance Berg and Grace, 2011, where pigeons were trained to categorize sine-wave disks considering their spatial frequency and orientation). However, the ability of nonhuman animals to apply categorical processes to more abstract – human-like – stimulus dimensions remains a matter of debate.

2.3.1 Equivalence classes: grouping arbitrary items within the same category

An important aspect of language is its arbitrariness. Arbitrariness corresponds to the fact that nothing in the physical form (acoustic properties) of most words refers to the objects they designate. For example, the word "car" does not "sound" like the vehicle it refers to. Therefore, words can refer to things in the real world, and things can refer to words, although there is no natural or necessary connection between them. In that case, words and objects are linked by a relation of equivalence, and the many exemplars of a given category of object (e.g., many different tables that vary in shape and color) can be categorized under a unique word label. The ability to form arbitrary connections between words and objects during ontogeny probably comes from the many co-occurrences of words-objects pairs that infants encounter during development. We presume that this ability relies on a more domain-general capacity to make associative (arbitrary) connections between items, or categories of items.

Experiments on "stimulus equivalence" have directly addressed the capacity of human and nonhuman species to group arbitrary items into categories, on the basis of their associative history (Sidman and Tailby, 1982). The prototypical design of stimulus equivalence experiments is shown in Figure 3. In stimulus equivalence experiments, the subject first learns a network of associations (shown in black in Figure 3) with repeated exposures to these associations. Then, probe trials (shown by the red arrows in Figure 3) test if the subject (1) can associate each stimulus to itself (e.g., associate A to A, reflexivity relation), (2) can revert the trained relations (e.g., associate B to A, symmetry), and can associate the stimuli that have a common associate (associate A to C, transitivity). According to Sidman and Tailby (1982), stimulus equivalence is fully shown if the subject demonstrates, without further training, the relations of reflexivity, symmetry and transitivity in post training trials. One may easily imagine the serious limits of a cognitive system that would fail in this task. The formation of equivalence classes has the power to permit to use stimuli interchangeably, and is probably the corner stone of complex symbolic thought.

Figure 3: Typical paradigm for experiments on stimulus equivalence. In this experimental design, the black arrows illustrate the trained associations. The dotted arrows illustrate the untrained associations that emerge in humans after an initial training phase (e.g., Sidman and Tailby, 1982).

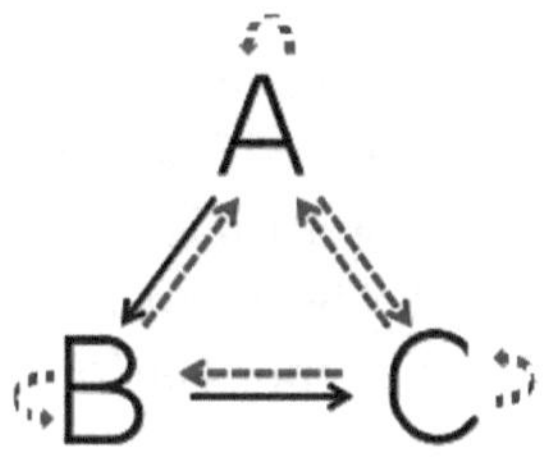

Equivalence relations emerge early in human infancy (23-month old, Lipkens et al., 1993). By contrast, the formation of equivalence classes seems especially difficult for nonhuman animals such as pigeons or monkeys, and among the three relations sustaining stimulus equivalence, the relation of symmetry seems the most difficult one to acquire. Lionello-DeNolf (2009) reviewed a total of 24 articles on symmetry testing, and found that the vast majority of these articles reported negative results. Moreover, the handful of articles reporting more positive findings all used a very small number of subjects who received special training procedures, such as long training combining symmetry and reflexivity trials (Frank and Wasserman, 2005) or forms of symbolic training (Pepperberg, 2006).

A recent paper from our research group nicely illustrates that difficulty (Medam et al., 2016). We trained baboons to associate pictures of bears and pictures of cars to two different shapes which served as category labels. The baboons demonstrated category learning in this task, but failed to respond correctly when asked to reverse the trained relation and to associate the category labels to the pictures of cars and bears. The immediate conclusion from this study and others (see Lionello-DeNolf, 2009) is that the preferred mode of processing in nonhuman animals is not favorable to the spontaneous emergence of symmetrical relations, which is fundamental for the emergence of equivalences classes. Arbitrariness, as defined by linguists, requires arbitrary, bidirectional associations between the words and their referent (e.g., the written "CAT" refers to the cat animal, as the cat animal refers to the "CAT"). Evidence suggests that nonhuman animals too can form arbitrary associations between various items with no logical connections between them, but they apparently have great difficulties to process these associations as bi-directional.

2.3.2 First order relations

Nonhuman animals can master a broad range of discrimination tasks, and some of them involve the processing of first-order relations. First-order relations refer to spatial or more abstract relations among objects, such as the fact that an object is above or below another one, or that two objects have the same functions. Particularly important for our linguistic system is the first-order relation of sameness/differentness. Our language makes great use of categories, and the abstract concept of sameness is essential for the development of verbal categories. The concept of sameness can provide the basis for the most complex cognitive operations, such as the conservation of volumes or areas, or analogical reasoning (see below). Without this concept, we would be unable to understand sentences such as "This is a cat!", and to get from this sentence the idea that the animal we see belongs to the cat category. We would as well be unable to understand sentences such as "It is warm again", suggesting a similarity between the current and past weather. Comparative psychologists have shown that a variety of nonhuman animal species, such as the chimpanzee (Premack, 1983), the baboon (Wasserman et al., 2001) and the rat (Wasserman et al., 2012), succeed in learning same-different relational tasks. However, the nature of the mechanisms supporting this competence in nonhuman animals, and their similarity with humans, remains unclear.

Wright and Katz (2006) asked the following question: how much training nonhuman animals need to form the same/different concept? To answer this question, they tested pigeons, capuchin monkeys and rhesus monkeys with the same test design. Animals from the three species saw two pictures in succession on a touch screen, and two kinds of trials were distinguished. In the "same" trials, the second picture was identical to the first one, while it was different from the first one in the "different" trials. In both kinds of trials, a white key was always displayed on the right of the second picture. When the first and second pictures were identical (identical trials), then the subject was asked to touch/peck the second picture to obtain a reward. If the second picture was different from the first one (different trials), then a touch/peck at the white response key was considered correct. The monkeys needed much less items (about 32) to develop the concept of sameness, than did the pigeons (256). The number of trials children would need in this task is not known, but studies have shown that children can categorize cats as different from dogs with only 12 training exemplars (Quinn et al., 1993), and by 10 months of age, they can form categories with only 7 or 8 training exemplars (Younger and Cohen, 1986). The data therefore suggest an evolutionary trend in this ability: humans would require exposure to a smaller number of items than the other animals, to form categories and develop same/different concepts.

The concept of sameness can be applied to a broad range of attributes, from the most perceptual to the most abstract ones, and another interesting issue in the comparative literature is to know if animals use the same kind of information as humans, when solving similar tasks requiring an abstract concept of sameness/differentness. This kind of questions has been addressed extensively by Wasserman and collaborators (see review in Wasserman et al., in press; Wasserman et al., 2004). These authors trained pigeons, baboons and humans to categorize displays resembling those of the top of Figure 4. They consisted in arrays of 16 icons which were either all same (same relation) or all different (different relation). After they received this category training, the subjects were tested with arrays containing mixtures of icons, in which some icons were duplicated a number of times in the array. The authors reasonned that if the subjects have formed the concept of sameness, then they should classify the arrays containing at least one item different from the others as "different arrays", irrespective of the fact that some icons are repeated. Figure 4 illustrates the most substantial findings of this set of experiments. This figure indicates the percentage of different responses with the mixtures, as a function of the entropy of the stimulus. Entropy in this experiment should be understood as a quantification of the perceptual variabity of the array: the all-same arrays have an entropy of 0, and the all different-arrays have the highest possible entropy value of 4 (mixtures have intermediate entropy values).

Figure 4: Use of perceptual (entropy) cues by pigeons, monkeys and humans, in a same/different discrimination task. The top panel shows the kind of displays employed in Wasserman and collaborators' experiments (Wasserman et al., 2004, left: same array, right: different array). The bottom figure shows that the same-different response of monkeys and pigeons is controlled in this task by the entropy of the arrays while 80% of the humans humans used more abstract cues in this task. Figure adapted from Wasserman et al. (2004).

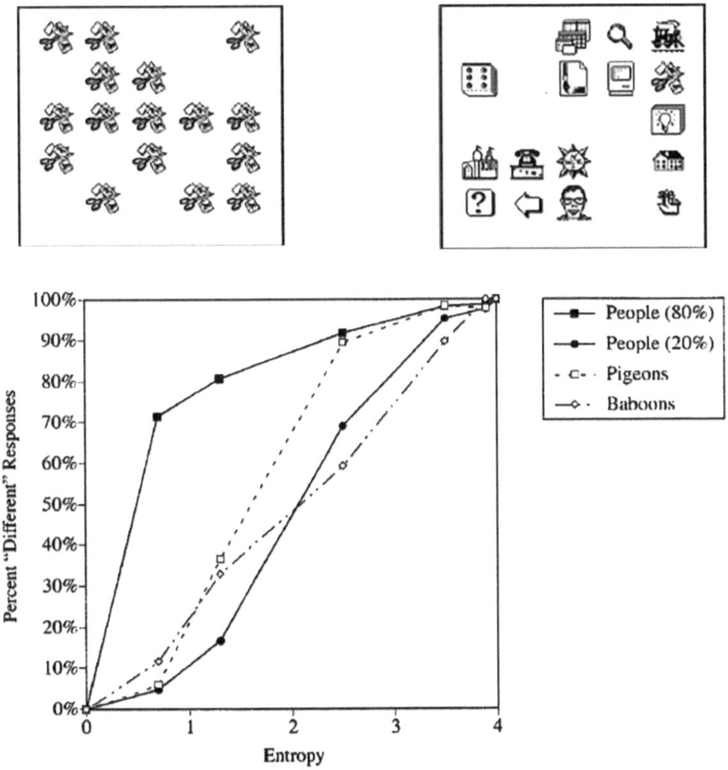

The bottom part of Figure 4 shows that the behaviour of the baboons and the pigeons was controlled by the entropy of the arrays, which is a perceptual cue. Response from a subset of humans (about 20% of the group) was also largely controlled by the entropy of the arrays, but this constraint was realeased in most of the subjects (80%) who treated the arrays containing at least one item different from the others as illustrations of the "different" concept. It can therefore be concluded that humans expressed more abstract judgments than pigeons and baboons in this task. We will not present in this chapter the full series of experiments using this kind of stimuli with animals (for a recent review, see Wasserman et al., in press), but the reader should be aware that clear demonstrations also exists that monkeys can also base their same/different responses on abstract cues independently of the entropy of the stimuli (Flemmig et al., 2013). Nevertheless, although several animal species seem capable of abstract same/different judgements, humans, more readily than other animals, apply qualitative, rule-based frameworks on the Same-Different discrimination task.

2.3.3 Analogical (second-order) relational processing

Our linguistic systems make great use of analogies, and our capacity to produce and understand analogies is considered by many as the "the Fuel and Fire of Thinking" (Hofstater and Sander, 2013). Developmental studies have shown that analogical reasoning is facilitated in children by the capacity to represent abstract relations in symbolic terms via linguistic labels (Christie and Gentner, 2014).

Most research on analogical reasoning in animals has used the Relational Matching-to-Sample task (RMTS: e.g., Fagot and Thompson, 2011) illustrated in Figure 5. In this task, the subject first perceives one pair of objects which are either identical or different. Two comparison pairs are then presented, and the subject must indicate the stimulus pair exemplifying the same (same or different) relation as the sample pair. In other words, the task can be conceptualized as "if AA then BB, and if AB then CD". Researchers in the domain of comparative cognition have tested several animal species, including pigeons, monkeys and apes using the RMTS task (for a review, see Wasserman, Castro and Fagot, in press). Most of these attempts failed (Thompson and Oden, 2000), but a handful of studies also provide more positive results, in particular in tests involving thousands of training trials (e.g., Fagot and Thompson, 2011).

Figure 5: Illustration of the relational matching task used, in baboons, with color (left part of the figure) and shape (right) stimuli.

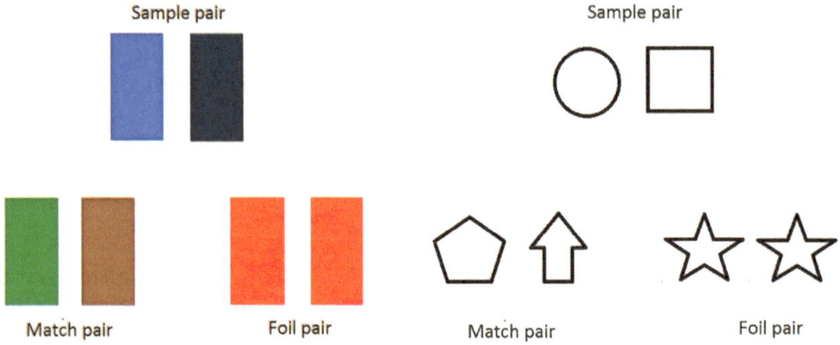

In our laboratory, we could demonstrate that the baboons can successfully solve the RMTS task with pairs of color patches as stimuli (Fagot and Parron, 2010, see Figure 5). In a different study (Fagot and Thompson, 2011), we could further demonstrate that baboons can also solve this RMTS task considering the shape of the items (Figure 5). Again, this cognitive feat also required an extensive training period (from 17 to 30 000 trials per subject). In both studies, the different generalization tests confirmed the real abstract nature of the processes at work in these two tasks. For instance, the baboons could continue to solve the task with a high level of performance when we used novel colors (Fagot and Parron, 2010), and novel shapes (Fagot and Thompson, 2011) as stimuli. However, although cognitive flexibility is suggested by these findings, the data also suggest limits in this processing. In Fagot and Parron's (2010) study, color cues were in fact proposed to 6 baboons, and 4 of these 6 subjects eventually learned the task. In Fagot and Thompson's (2011) study, the same task was given one year later to a larger group of subjects, including the 6 already tested in Fagot and Parron (2010). Six out of 29 baboons learned the RMTS task with shapes, but importantly, none of the baboons who had initially learned the task with colors could also learn it with shapes. In other words, learning the RMTS task with color cues did not help at all the subjects to learn. Generalization across domains is central to analogical reasoning in humans, and is probably what makes human reasoning so flexible. Generalization across domains allows us, for instance, to understand the meaning of a sentence like "atoms are like tiny solar systems" or "life is a gift, a chocolate box". Baboons – and probably other nonhuman primates as well – are quite flexible to process items and their relations within given domains, those they have been trained with (e.g., color), but are clearly not as skilled as humans to generalize across domains. This, we believe, is another factor that may greatly affect nonhuman animal's potential for developing elaborated forms of language.

3. Summary and Conclusions

The main goal of the chapter was to examine the origin of human language from the standpoint of comparative psychology. Language in its various forms (e.g., gesture, writing, speech) is a multi-level integrative process that requires, at the perceptual stage, the segmentation and grouping of perceptual information to extract the general meanings of the communicative signals. We have argued above that many of the integrative processes involved in the language function are in fact domain-general processes that can also be found in non-linguistic functions.

Considering that language uses a multitude of domain-general functions, examination of these functions in animals and especially in nonhuman primates should provide important information on the cognitive background that made it possible for language to emerge in our evolutionary history. Following this reasoning, we comparatively examined in this chapter a number of domain-general cognitive functions, which imply various forms of integration of perceptual/conceptual information. Among the considered functions, we examined the ability of animals to integrate information in time and space, to combine stimulus dimensions, to group objects into categories, and to develop conceptual/relational processes (first- and second-order concepts). Of course, this list is not exhaustive but we believe that it represents a significant selection of basic cognitive domain-general processes that serve language perception and comprehension.

The present overview of the literature allowed us to reach two main conclusions. First, we have identified clear-cut demonstrations that nonhuman animals are capable of grouping information in time and space, can combine stimulus dimensions, and can form categories at different levels of abstractness. Thus, we suggest that these functions are shared by humans and nonhumans, at least to some extent, and that they are not language specific (i.e., they have a long phylogenetic history). Second, the literature also reveals important differences in performance among nonhuman animals, and between nonhuman animals and humans. For example, evidence suggest that, compared to humans, in nonhuman animals the integration of information in time and space is more "local", and the capacity to integrate the information on a larger scale is more restricted. When we come to consider how animals integrate various stimulus dimensions, experimental evidence suggest that they tend to focus on some particular physical dimensions of the stimuli more than humans do, and they hardly combine information from different stimulus dimensions. When we come to consider more general categorization processes, it appears that nonhuman primates form categories but their categories seem to be more strongly tied to the perceptual input than those of humans, and abstract processes, when they emerge, need many more trials to develop, and/or do not generalize to untrained dimensions as readily as for humans. In other words, for most of the integrative functions we have considered in this chapter, nonhuman animals show behaviors that differ at least quantitatively, if not qualitatively from human behavior, and we propose that these differences might be the bottlenecks for the evolution of language. Obviously, evolution has no direction and mastering some human-like language is not and has never been an issue in nonhuman animals. However, it might be that a particular pattern of development of these domain-general functions was a prerequisite to the emergence of human language, and that favorable conjunctions occurred only once in the phylogeny of the primate group, giving rise to the human language. The idea that some domain-general nonlinguistic functions form the bottlenecks of language evolution will be further developed below.

To account for the evolution of language, many theorists have focused their attention on language specific functions, which were considered as key factors for the evolution of human language. For example, many traditional theories claimed that only humans have a low larynx (which was disconfirmed since), or that only humans have the ability to understand

and produce recursive structures allowing an infinite variability in language production. Here, we do not want to discount these explanations, but we think that they do not take the problem at its roots. At this point in the scientific endeavor, we think that it is now necessary to step back a bit to enlarge our view of the problem. Doing so, we can imagine two different scenarios on the origin of language.

The "language-first" scenario would be that our primate/prehominid ancestors had rather limited cognitive resources and it was the appearance of language that boosted their general cognitive capacities. Although there is no doubt that language is a booster for cognitive functions, we think that this scenario does not hold. To illustrate our rebuttal of this theory, imagine an animal cognitive system with a limited working memory span of N=1 item. How could a language system develop with such limited memory resources? A clear expansion of this memory system would be required before the animal would have a chance to develop a proto-language system. This very extreme example shows that language evolution requires a cognitive background to support it, and it is only when such a cognitive background has evolved – potentially as a response to ecological pressures – that a form of proto-language can have a chance to emerge.

The second possible scenario, called "prerequisites-first", is that language only emerged in our evolutionary history once critical domain-general functions had gained in cognitive power in our animal/pre-hominid ancestors. This scenario raises one major question: which domain general function(s), or combination of functions, must have evolved at first for the emergence of language? We have no clear answer to this question but can provide several lines of thinking on this issue. First, we note that studies focusing on the so called low level perceptual mechanisms showed important differences between the platyrrhines and catarrhines species, but very little differences are observed in perceptual functions in monkeys, apes and human species (see for instance Fobes and King, 1982 for a review of visual perception). This mere fact suggests that the evolution of these perceptual functions is probably not the factor that made the difference and triggered the evolution of speech in humans. Secondly, comparative experiments suggest more pervasive differences between humans and the other primates in two domains at least. The first domain of importance is the domain of working memory. Working memory in humans seems to depart from that of other animal species in several important aspects, for instance regarding the ability to process large amount of information in parallel (e.g., Fagot and De Lillo, 2011), or to process long-distance dependencies (e.g., Wilson et al., 2015). Although not discussed in the context of this chapter, working memory in humans may also qualitatively depart from that of the other animals in its use of a phonological loop facilitating memorization in the short term. The second domain for which strong differences emerge between humans and the other animals is the domain of attention. Nonhuman animals seem to focus on single stimulus dimensions more than humans do, and tend to have a more local mode of processing of the perceptual input than humans. There remains a debate on whether the increase in cognitive functions followed the phylogenetic order, from the remote prosimian species to the ape species phylogenetically closest to humans (e.g., Reader et al., 2011), or whether variations in cognitive power among the different primate species have occurred at multiple times in the course of evolution, in independent unrelated primates groups, for instance under the pressure of social factors such as the complexity of the social network (Dunbar, 1998). Discussing these hypotheses is out of the scope of the current chapter. However, whatever the source of this increment in cognitive power is, we propose that the language ability appeared in the evolution of primates at a point in time where domain-general cognitive capacities, especially those pertaining to attention and working memory converged and were sufficiently developed to permit its evolution. This idea is in line with recent usage-based theories suggesting that language could be acquired in humans by means of domain-general – evolutionary old – processes (Bybee, 2010; Tomasello, 2005).

From a more practical standpoint, we conclude from this chapter that the comparative investigation of non-linguistic, domain general functions should be considered of central importance in the debate on the evolution of language. Unfortunately, real comparative studies, in which humans and other species are tested on the same problems using the same tasks, are relatively rare in this literature, and most of them only concern a very limited number of species. Such studies will become mandatory to further test the hypothesis that the expansion of domain-general functions in nonhuman primates served as a basis for the evolution of language.

Acknowledgements

This project was supported by ANR-16-CONV-0002 (ILCB), ANR-11-LA-BX-0036 (BLRI) and ANR-11-IDEX-0001–02 (A*MIDEX).

References

Abe, K., and Watanabe, D. (2011). Songbirds possess the spontaneous ability to discriminate syntactic rules. *Nature Neuroscience, 14*, 1067–1074.

Anderson M. L. (2010). Neural reuse: a fundamental organizational principle of the brain. *Behavioural Brain Science, 33*, 245–266.

Berg, M.E., and Grace, R.C. (2011). Categorization of multidimensional stimuli by pigeons. *Journal of Experimental Analysis of Behavior. 95*, 305–326.

Boë, L.J., Berthommier, F., Legou, T., Captier, G., Kemp' C., Sawallis, T.R., Becker, Y., Rey, A., and Fagot, J. (2017). Evidence of a vocalic proto-system in the baboon (*Papio papio*) suggests pre-hominin speech precursors. *PLoS ONE, 12*, e0169321.

Bybee, J. (2010). *Language, usage and cognition.* Cambridge: Cambridge University Press.

Byrne, R.W., Corp, N., and Byrne, J.M. (2001). Estimating the complexity of animal behaviour: How mountain gorillas eat thistles. *Behaviour, 138*, 525–557.

Cantalupo, C., and Hopkins, W. D. (2001). Asymmetrical Broca's area in great apes. *Nature, 414*, 505.

Cantalupo, C., Pilcher, D., and Hopkins, W. D. (2003). Are planum temporale and sylvian fissure asymmetries directly related? A MRI study in great apes. *Neuropsychologia, 41*, 1975–1981.

Chen, J., van Rossum, D., and ten Cate, C. (2015). Artificial grammar learning in zebra finches and human adults: XYX versus XXY. *Animal Cognition, 18*, 151–164.

Chomsky, N. (1959). A review of B.F. Skinner's "Verbal Behavior." *Language, 35*, 26–58.

Christie, S., and Gentner, D. (2014). Language helps children succeed on a classic analogy task. *Cognitive Science, 38*, 383–397.

Colombo, M., and Frost, N. (2001). Representation of serial order in humans: A comparison to the findings with monkeys (Cebus apella). *Psychonomic Bulletin and Review, 8*, 262–269.

Cook R. G. (2001). Hierarchical stimulus processing in pigeons. In R. G. Cook (Ed.), *Avian visual cognition* [On-line]. Available: www.pigeon.psy.tufts.edu/avc/cook/

Creel, S. C., Newport, E. L., and Aslin, R. N. (2004). Distant melodies: statistical learning of nonadjacent dependencies in tone sequences. *Journal of Experimental Psychology. Learning, Memory, and Cognition, 30*, 1119–1130.

D'Amato, M. R., and Van Sant, P. (1988). The person concept in monkeys (*Cebus apella*). *Journal of Experimental Psychology: Animal Behavior Processes, 14*, 43–55.

Davenport, R.K., and Rogers, C.M. (1970). Intermodal equivalence of stimuli in apes. *Science, 1968*, 279–280.

Dépy, D., Fagot, J. and Vauclair, J. (1997). Categorization of three-dimensional stimuli by humans and baboons (*Papio papio*): Search for prototype effects. *Behavioral Processes, 37*, 299–306.

Deruelle, C., and Fagot, J. (1998). Visual search for global/local stimulus features in humans and baboons. *Psychonomic Bulletin and Review, 5*, 476–481.

Dunbar, R.I.M (1998). The social brain hypothesis. *Evolutionary Anthropology, 6*, 178–190.

Fagot, J., and Barbet, I. (2006). Grouping and segmentation of visual objects by baboons (*Papio papio*) and humans (*Homo sapiens*). In E. A. Wasserman and T. R. Zentall (Eds.), *Comparative cognition: experimental explorations of animal intelligence* (pp. 11–24). New York: Oxford University Press.

Fagot, J., and De Lillo, C. (2011). A comparative study of working memory: immediate serial spatial recall in baboons (*Papio papio*) and humans. *Neuropsychologia, 49*, 3870–3880.

Fagot, J., and Deruelle, C. (1997). Processing of global and local visual information and hemispheric specialization in humans (*Homo sapiens*) and baboons (*Papio papio*). *Journal of Experimental Psychology: Human Perception and Performance, 23*, 429–442.

Fagot, J., Martin-Malivel, J., and Dépy, D. (2000). What is the evidence for an equivalence between objects and pictures in birds and nonhuman primates? In J. Fagot (Ed.). *Picture Perception in Animals* (pp 295–320). East Sussex: Psychology Press.

Fagot, J. and Parron, C. (2010). Relational matching in baboons (*Papio papio*) with reduced grouping requirements. *Journal of Experimental Psychology: Animal Behavior Processes, 36,* 184–193.

Fagot, J., and Parron, C. (2012). Visual cognition in nonhuman primates. In O. Lazaverra, T. Shimizu and E.A. Wasserman (Eds.), *How animals see the world* (pp. 371–389). Oxford: Psychology Press.

Fagot, J., and Thompson, R. K. R. (2011). Generalized relational matching by Guinea baboons (*Papio papio*) in two by two-item analogy problems. *Psychological Science, 22,* 1304–1309.

Fitch, W.T. (2010) *The evolution of language.* Cambridge University Press.

Fitch, WT, de Boer, B., Mathur, N., and Ghazanfar, A.A. (2016). Monkey vocal tracts are speech-ready. *Science advances*, 2:e1600723.

Fitch, W. T., and Hauser, M. D. (2004). Computational constraints on syntactic processing in a nonhuman primate. *Science, 303* (5656), 377–380.

Flemming, T. M., Thompson, R. K. R., and Fagot, J. (2013). Baboons, like humans, solve analogy by categorical abstraction of relations. *Animal Cognition, 16,* 519–524.

Fobes, J.L., and King, J.E. (1982). Vision: The dominant primate modality. In J.L. Fobes and J.E. King (Eds.), *Primate behavior* (pp. 219–243). New York: Academic Press.

Fountain, S. B. (1990). Rule abstraction, item memory, and chunking in rat serial-pattern tracking. *Journal of Experimental Psychology. Animal Behavior Processes, 16,* 96–105.

Fountain, S. B., and Benson Jr., D. M. (2006). Chunking, rule learning, and multiple item memory in rat interleaved serial pattern learning. *Learning and Motivation, 37,* 95–112.

Frank, A.J. and Wasserman, E.A. (2005). Associative symmetry in the pigeon after successive matching-to-sample training. *Journal of Experimental Analysis of Behavior, 84,* 147–165.

Froehlich, A. L., Herbranson, W. T., Loper, J. D., Wood, D. M., and Shimp, C. P. (2004). Anticipating by pigeons depends on local statistical information in a serial response time task. *Journal of Experimental Psychology: General, 133*, 31–45.

Gebhart, A. L., Newport, E. L., and Aslin, R. N. (2009). Statistical learning of adjacent and nonadjacent dependencies among nonlinguistic sounds. *Psychonomic Bulletin and Review, 16*, 486–490.

Gentner, T. Q., Fenn, K. M., Margoliash, D., and Nusbaum, H. C. (2006). Recursive syntactic pattern learning by songbirds. *Nature, 440*, 1204–1207.

Geschwind, N., and Levitsky, W. (1968). Human brain: Left-right asymmetries in temporal speech region. *Science, 161*, 186–187.

Gomez-Robles, A., Hopkins, W.D., and Sherwood, C.C. (2013). Increased morphological asymmetry, evolvability and plasticity in human brain evolution. *Proceedings of the Royal Society: B, 280*, 20130575.

Hauser, M. D., Chomsky, N., and Fitch, W. T. (2002). The faculty of language: What is it, who has it, and how does it evolve? *Science, 298*, 1569–1579.

Hauser, M. D., Newport, E. L., and Aslin, R. N. (2001). Segmentation of the speech stream in a nonhuman primate: statistical learning in cotton-top tamarins. *Cognition, 78*, B53–B64.

Hayes, C. (1951), *The Ape in Our House*. New York: Harper.

Heijningen, C. A. A. van, Chen, J., Laatum, I. van, Hulst, B. van der, and Cate, C. ten. (2012). Rule learning by zebra finches in an artificial grammar learning task: which rule? *Animal Cognition, 16*, 165–175.

Herrnstein, R.J., and Loveland, D.H. (1964). Complex visual concept in the pigeon. *Science, 23, 549–551*.

Hockett, C. (1960). The origin of speech. *Scientific American, 203*, 88–111. Reprinted in Wang, William S-Y. (1982) Human Communication: language and its psychobiological bases, Scientific American pp. 4–12.

Hofstater, D., and Sanders, E. (2013). *Surfaces and essences: analogy as the fuel and fire of thinking*. New-York: Basic Books.

Hunt, R. H., and Aslin, R. N. (2001). Statistical learning in a serial reaction time task: Access to separable statistical cues by individual learners. *Journal of Experimental Psychology: General, 130*, 658–680.

Jürgens, U. (1979). Neural control of vocalization in nonhuman primates. In H.D. Steklis and M. J. Raleigh (Eds.), *Neurobiology of social communication in primates* (pp. 199–206). New-York: Academic Press.

Koechlin, E, and Jubault, T (2006). Broca's area and the hierarchical organization of human behavior. *Neuron, 50*, 963–74.

Lanz, F., Moret, V., Rouiller, E.M., and Loquet, G. (2013). multisensory integration in nonhuman primates during a sensory-motor task. *Frontiers in Human Neuroscience*, 7,799.

Lea, S. G. E., and Wills, A. J. (2008). Use of multiple dimensions in learned discriminations. *Comparative Cognition and Behavior Reviews, 3*, 115–133.

Leroy, F. et al. (2014). New human-specific brain landmark: The depth asymmetry of superior temporal sulcus. *PNAS, 112*, 1208–1213.

Lieberman, P. (1968). Primate vocalizations and human linguistic ability. *Journal of the Acoustic Society of America, 44*, 1574–84.

Lieberman, P. (1975). *On the Origins of Language: an Introduction to the Evolution of Human Speech*. New York: Macmillan.

Lionello-DeNolf, K.M. (2009). The search for symmetry: 25 years in review. *Learning and Behavior, 37*, 188–203.

Lipkens, G., Hayes, S. C., and Hayes, L. J. (1993). Longitudinal study of derived stimulus relations in an infant. *Journal of Experimental Child Psychology, 56*, 201–239.

Marcus, G. F., Vijayan, S., Rao, S. B., and Vishton, P. M. (1999). Rule learning by seven-month-old infants. *Science, 283*, 77–80.

Malivel, M. and Fagot, J. (2001). Cross-modal interference and conceptual categorization in baboons. *Behavioural Brain Research*, 122, 209–213.

Martin-Malivel, J. and Fagot, J. (2001). Perception of pictorial humans faces by baboons (Papio papio): effects of stimulus orientation on discrimination performance. *Animal Learning and Behaviour, 29*, 10–20.

Martin-Malivel, J. Mangini, M., Fagot, J. and Biederman, I (2006). Do humans and baboons use the same information when categorizing human and baboon faces ? *Psychological Science, 17, 599–607.*

Medam, T., Marzouki, Y., Montant, M., and Fagot, J. (2016). Categorization does not promote symmetry in Guinea baboons (Papio papio). *Animal Cognition, 19*, 987–99.

Meguerditchian, A., Marie, D., Love, S. A., Margiotoudi, K., Bertello, A., Lacoste, R., Roth, M., Nazarian, B, Anton, J.-L., and Coulon, O. (2016). Human-like brain specialization in baboons: an in vivo anatomical MRI study of language areas homologs in 96 subjects. In S.G. Roberts et al. (Eds.), *The evolution of language: Proceedings of the 11th International Conference* (EVOLANG11). New Orleans, LA, USA.

Minier, L., Fagot, J., and Rey, A. (2016). The temporal dynamics of regularity extraction in nonhuman primates. *Cognitive Science, 40*, 1019–1030.

Navon, D. (1977). Forest before the tree: The precedence of global feature in visual perception. *Cognitive Psychology, 9*, 353–383.

Newport, E. L., and Aslin, R. N. (2004). Learning at a distance I. Statistical learning of nonadjacent dependencies. *Cognitive Psychology, 48*, 127–162.

Newport, E. L., Hauser, M. D., Spaepen, G., and Aslin, R. N. (2004). Learning at a distance II. Statistical learning of nonadjacent dependencies in a nonhuman primate. *Cognitive Psychology, 49*, 85–117.

Orbán, G., Fiser, J., Aslin, R.N., and Lengyel, M. (2008). Bayesian learning of visual chunks by human observers. *Proceedings of the National Academy of Sciences USA, 105*, 2745–50.

Pepperberg, I.M. (2006). Ordinality and inferential abilities of a grey parrot (Psittacus erithacus). *Journal of Comparative Psychology, 120*, 205–216.

Premack, D. (1983). The codes of man and beast. *Behavioral and Brain Sciences, 6*, 125–137.

Petkov, C. I., and Wilson, B. (2012). On the pursuit of the brain network for proto-syntactic learning in nonhuman primates: conceptual issues and neurobiological hypotheses. *Philosophical Transactions of the Royal Society B: Biological Sciences, 367*, 2077–2088.

Procyk, E., Ford Dominey, P., Amiez, C., and Joseph, J. P. (2000). The effects of sequence structure and reward schedule on serial reaction time learning in the monkey. *Brain Research. Cognitive Brain Research, 9*, 239–248.

Quinn, P.C., Eimas P.D., and Rosenkrantz, S.L. (1993). Evidence for representations of perceptually similar natural categories by 3-month-old and 4-month-old infants. *Perception, 22*, 463–475.

Ravignani, A., Sonnweber, R.-S., Stobbe, N., and Fitch, W. T. (2013). Action at a distance: dependency sensitivity in a New World primate. *Biology Letters, 9*, 20130852.

Reader, S.M., Hager, Y. and Laland, K.N. (2011). The evolution of primate general and cultural intelligence. *Philosophical Transactions of the Royal Society, B, 366*, 1077–1027.

Saffran, J. R., Aslin, R. N., and Newport, E. L. (1996). Statistical learning by 8-month-old infants. *Science, 274*, 1926–1928.

Saffran, J.R., and Thiessen, E.D. (2008). Domain-general learning capacities. In E. Hoff and M. Shatz (Eds.), *Blackwell Handbook of Language Development* (pp. 68–86). New-York: Blackwell Publishing Ltd.

Schlenker, P., Chemla, E., and Zuberbühler, K. (2016). What do monkey calls mean? *Trends in Cognitive Sciences, 20*, 894–904.

Seidenberg M., and MacDonald M. (1999). A probabilistic constraints approach to language acquisition and processing. *Cognitive Science, 23*, 569–588.

Seidenberg, M., MacDonald, M., and Saffran, J. (2002) Does grammar start where statistics stop? *Science, 295*, 553–554.

Sidman, M., and Tailby, W. (1982). Conditional discrimination vs. matching to sample: an expansion of the testing paradigm. *Journal of Experimental Analysis of Behavior, 37*, 5–22.

Skinner, B. F. (1957). *Verbal Behavior*. New York: Appleton-Century-Crofts.

Smith, J. D., Berg, M.E., Cook, R.G., Murphy, M.S., Crossley, M.J., Boomer, J., Spiering, B., Beran, M.J., Church, B., Ashby, G., and Grace, R.C. (2012). Implicit and explicit categorization: A tale of four species. *Neuroscience and Biobehavioral Reviews, 36*, 2355–2369.

Spierings, M. J., and Ten Cate, C. (2016). Budgerigars and zebra finches differ in how they generalize in an artificial grammar learning experiment. *Proceedings of the National Academy of Sciences of the United States of America, 113*, E3977-3984.

Suzuki, T. N., Wheatcroft, D., and Griesser, M. (2016). Experimental evidence for compositional syntax in bird calls. *Nature Communication, 7*, 10986.

Taglialatela, J.P., Russell, J.L., Schaeffer, J.A., and Hopkins, W.D. (2008). Communicative signaling activates "Broca's" homologue in chimpanzees. *Current Biology, 18*, 343–348.

Taglialatela, J.P., Russell, J.L., Schaeffer, J.A., and Hopkins, W.D. (2011). Chimpanzee vocal signaling points to a multimodal origin of human language. *PLoS ONE, 6*, e18852.

Terrace, H. S. (1991). Chunking during serial learning by a pigeon: I. Basic evidence. *Journal of Experimental Psychology: Animal Behavior Processes, 17*, 81–93.

Terrace, H. S., Son, L. K., and Brannon, E. M. (2003). Serial expertise of rhesus macaques. *Psychological Science, 14*, 66–73.

Thompson, R. K. R., and Oden, D. L. (2000). Categorical perception and conceptual judgments by nonhuman primates: The paleological monkey and the analogical ape. *Cognitive Science, 24*, 363–396.

Tomasello, M. (2005). *Constructing a language*. Harvard: Harvard University Press.

Toro, J. M., and Trobalón, J. B. (2005). Statistical computations over a speech stream in a rodent. *Perception and Psychophysics, 67*, 867–875.

Wang, L., Uhrig, L., Jarraya, B., and Dehaene, S. (2015). Representation of numerical and sequential patterns in macaque and human brains. *Current Biology, 25*, 1966–1974.

Wasserman, E.A., Castro, L. and Fagot, J. (in press). Relational Thinking in Animals and Humans. In Call J.P. (Ed). *From Percepts to Concepts": APA Handbook of Comparative Psychology. New-York: American Psychological Association.*

Wasserman, E. A., Castro, L., and Freeman, J. H. (2012). Same-different categorization in rats. *Learning and Memory, 19*, 142–145.

Wasserman, E. A., Fagot, J., and Young, M. E. (2001). Same-different conceptualization by baboons (*Papio papio*): The role of entropy. *Journal of Comparative Psychology, 115*, 42–52.

Wasserman, E.A., Young, M.E., and Cook, R.G. (2004). Variability discrimination in humans and animals: implications for adaptive action. *American Psychologist, 59*, 879–890.

Wilson, B., Smith, K., and Petkov, C. I. (2015). Mixed-complexity artificial grammar learning in humans and macaque monkeys: evaluating learning strategies. The European *Journal of Neuroscience, 41*, 568–578.

Wright, A. A., and Katz, J. S. (2006). Mechanisms of same/different concept learning in primates and avians. *Behavioural Processes*, *72*, 234–254.

Wright, A. A., Santiago, H. C., Sands, S. F., Kendrick, D. F., and Cook, R. G. (1985). Memory processing of serial lists by pigeons, monkeys, and people. *Science, 229*, 287–289.

Younger, B.A., and Cohen, L.B. (1986). Developmental change in infants' perception of correlations among attributes. *Child Development*, *5*, 803–81.

Speech Production and Perception

Edited by Susanne Fuchs and Pascal Perrier

Vol. 1 Susanne Fuchs / Melanie Weirich / Daniel Pape / Pascal Perrier (eds.): Speech Planning and Dynamics. 2012.

Vol. 2 Anne Hermes: Articulatory Coordination and Syllable Structure in Italian. 2013.

Vol. 3 Susanne Fuchs / Daniel Pape / Caterina Petrone / Pascal Perrier (eds.): Individual Differences in Speech Production and Perception. 2015.

Vol. 4 Louis-Jean Boë / Joël Fagot / Pascal Perrier / Jean-Luc Schwartz (eds.): Origins of Human Language: Continuities and Discontinuities with Nonhuman Primates. 2017.

www.peterlang.com